Power Cable Technology

Power Cable Technology

Sushil Kumar Ganguli
Vivek Kohli

CRC Press is an imprint of the
Taylor & Francis Group, an informa business

CRC Press
Taylor & Francis Group
6000 Broken Sound Parkway NW, Suite 300
Boca Raton, FL 33487-2742

First issued in paperback 2023

Version Date: 20160119

ISBN 13: 978-1-138-32265-3 (pbk)
ISBN 13: 978-1-4987-0909-5 (hbk)
ISBN 13: 978-1-315-22253-0 (ebk)

DOI: 10.1201/b19574

Publisher's Note
The publisher has gone to great lengths to ensure the quality of this reprint but points out that some imperfections in the original copies may be apparent.

Visit the Taylor & Francis Web site at
http://www.taylorandfrancis.com

and the CRC Press Web site at
http://www.crcpress.com

Contents

Preface

History of Electric Cables

The first electric cable was used to detonate ores in a mine in Russia in 1812. In 1870, cables with rubber insulation were made. In 1880, DC cables using jute in pipes were invented. In 1890, Ferranti developed concentric construction of cables. In 1917, the first screened cables were used. The first pressurized paper cables were introduced in 1925. Polyethylene (PE) was invented in 1937, and by 1942, PE cables were introduced in the market. In the 1950s, PVC-insulated cables made their entry. XLPE (cross-linked polyethylene) was developed in 1963. In 1968, medium-voltage XLPE cables were employed by utilities in the United States. By late 1980, several improvements in the XLPE material were made with the introduction of extruded semiconducting materials. Further research focused on developing water-tree-retardant and super-smooth conductors and semiconducting materials to be used for HV and EHV cables.

In 1931, the first cable factory in India – Indian Cable Corporation (INCAB), a subsidiary to British Insulated Callender's Cables (BICC) in the United Kingdom – was established. At the time, the product range of the company was enamelled wires; rubber-insulated, tough-rubber-sheathed domestic wires; and cables and paper-covered wires. Raw materials were all imported. Between 1948 and 1950, INCAB started producing paper-insulated, low-tension 1100 V cables for Calcutta Electric Supply Corporation Ltd. (CESC) and Bombay Electric Supply and Transport Ltd. (BEST). Further requirements of LV (1100 V) and MV cables up to 33 kV (gas pressure type) were directly imported from the United Kingdom. After independence, an initiative was taken by the Government of India to augment power generation and distribution networks in urban and industrial areas. This primarily generated a higher demand for underground insulated low-voltage power cables. In the first and second five-year plans, the Government of India recommended to set up power cable factories within the country to reduce imported quantum substantially to save foreign exchange and to make the country self-sufficient in this sector. Between 1955 and 1965, a few major cable companies were established to manufacture paper-insulated 1100 V cables, along with a few other products, such as signalling and communication cables.

Apart from INCAB, a majority of these companies had technical collaboration with BICC, such as NICCO, Universal Cables and Asian Cables-Duncan's-PDIC. Oriental Power Cables Ltd. tied up with Sumitomo, Japan; Cable Corporation of India Ltd. came into collaboration with Siemens, Germany, and Industrial Cables with Phelps Dodge, USA. During the early 1960s and till late 1970s, collaborators closely monitored manufacturing and quality systems. Each one was trying to surpass the others on quality assessment. Between 1966 and 1970, medium-voltage (33 kV), paper-insulated cables were developed and produced in the country. All LV and MV cables were paper insulated, mass impregnated, lead sheathed and steel tape armoured and served, popularly known as DSTA & S cables. Galvanized steel wire armoured cables were manufactured to meet specific demands, such as for mining and other special applications. Most of the Indian power cable companies got orders from Russia for installation in Siberia under the India–Russia government cooperation. This kept cable companies busy for many years in manufacturing PILC cables.

At the initial stages of manufacturing and testing, the companies experienced many failures, which were analyzed critically to ascertain the root cause. After this, corrective actions were taken to improve the situation.

During the 1970s, PVC was introduced in India. Local insulating and sheathing grade PVC compound manufacturing was started by Calico Chemicals, NOCIL and Sriram Chemicals. With the coming of thermoplastic polymers like PVC and polythene, the processing technology underwent a complete change. Extrusion technology became popular. Jointing and installation became less hazardous. Ultimately, low-tension paper cables were replaced completely by PVC cables, followed by XLPE cables. During such transition periods, technologists were alert in maintaining the quality in line with specified requirements.

Unfortunately, over time, the quality of PVC compound started deteriorating, and small-scale entrepreneurs taking advantage of general extrusion process and easy testing procedures started using substandard grade polymer obtained from small manufacturers. Small-scale compounding firms started using scrap, low-grade fillers and plasticizers to offer cheaper versions of PVC compounds, though meeting specified parameters as stipulated in the Indian Standard specification. Naturally, larger Industries were not able to compete in the market. As a result they were also compelled to lower their standard to make cables at cheaper rates. Though, over time, some revision in specification was introduced, it was not sufficient to solve basic problems.

Polymer technology was taken to be an easy subject. Almost everyone started to become a compounding specialist. Cables made with such compounds generated higher transmission loss within the distributing systems. Failures were frequent. However, with the coming of Sioplas, systems to produce LT XLPE cable could partially salvage the misuse of market sentiment. Here, again, to produce quality material, extrusion technique needed to be properly understood. Skilled labour, quality equipment and proper handling of raw materials are essential to get satisfactory results.

Many varieties of cables have been produced in the country since the opening of the first cable factory in 1931. XLPE cables were introduced in India in 1978. Most of the top-level companies like M/S Asian Cables Ltd., Mumbai, Cable Corporation of India Ltd., Mumbai (CCI) and Indian Cable company Ltd., Jamshedpur, Bihar (ICC), and Universal Cables Ltd., Satna, MP(UCL) put up CCV lines up to 33 kV with steam curing and water cooling systems with the help of technology partners. The first commercial supply of 11 kV XLPE was made to a project of Kudremukh Iron Ore Company Ltd. While India was adopting steam cure technology, the development of XLPE process from steam cure to dry cure systems was happening in some of the bigger companies in the United States, Japan and Europe. This helped companies in taking greater leaps on commercialization with elevated voltages for XLPE cables. The technology also underwent a breakthrough with the development of a triple extrusion process which made high-quality XLPE cables free from possible dangers and failures, using closed circuit triple extrusion process at least in cables from 66 kV and upwards.

India started moving towards the era of manufacturing XLPE-insulated MV and EHV cables from the 1980s onwards. But a handful of the companies and their technologists are gearing up to raise the standard of their design and manufacturing processes. Not many of them can handle the process independently for EHV cables even 30 years after the introduction of XLPE cables. India is still depending on foreign technical support. With very little being spent by Indian companies in the development of EHV cables, we are still far behind the rest of the world. Globally, with only a few manufacturers for CCV/VCV lines, there is no Indian supplier who can design a CCV line and generate proper production facilities with infrastructures for EHV cables. In the early 1980s, a few Indian manufacturers, such as the Union Carbide of India and ICIC, tried to make XLPE compound locally but were soon shut down due to quality and cost considerations. XLPE and semiconducting compounds are currently imported. We have no local supplier for testing equipment like EHV resonance test set and PD testing for high voltages. A few attempts at high-power impulse tests were made with the help of the Indian Institute of Science, but on the whole, no one could supply world-class equipment for testing. Our industries work mostly like processing units and are not development oriented. Hence, the full implication of the problem that can generate within cables due even to a slightest deviation from the design and manufacturing technique cannot be fully understood by our technicians, thereby they are dependent on technical collaborators. Though India has come a long way producing electric cables, we are still behind in being self-sufficient.

There is a need to design and fabricate cable processing equipments as well on the basis of raw materials to be processed. An understanding of the finer points of a product's long-term performance is possible with locally available raw-materials and that would be advantageous to Indian cable manufacturers provided more focus is directed towards R&D rather than the companies performing quality check as a routine matter to pass products as satisfactory when test results are found to fulfil the criteria as per IS or IEC standards. More remains to be investigated on the inherent properties and the development side. It is expected that, over time, India will invest hugely in R&D rather than just manufacturing 'satisfactory' products. This is only possible when we become more conscious of our own responsibility and build confidence to declare that we can offer more than others.

Though the initial paragraphs dealt with the condition of Indian manufacturers of electric cables, the aim of this book is to make engineers and technicians understand more precisely the technology involved and the precautions they need to take during production to get a good-quality product at a reasonable cost. This can only be possible when the theory of designing a cable is in consonance with the practical application in every respect. Every detail of the product, its design criteria and even the construction of a particular cable processing machinery rests purely on a mathematically approved thought process. Our endeavour here is to establish this fact with as many cases as possible. Cable transmits electricity, is constructed with chemically processed materials and is shaped and finished by different mechanical processes. A successful cable technologist, therefore, should be prepared to handle complex admixture of all three branches of this technology – electrical, mechanical and chemical – simultaneously.

Apart from manufacturing the product, it is essential to extend the knowledge to how power cables are to be handled, installed and be joined and terminated at the given point. Generally, a lackadaisical attitude is adopted in dealing with the laying, installation and terminating a cable either with a transformer or within the kiosk of a consuming point. Cables are handled carelessly. Contracting firms do not understand the importance of an electric cable, and even supervising engineers are not properly trained. They are not made to understand that what they are dealing with can create havoc if not handled with utmost care and attention. One does not see the power flowing within, but can see its effect at the point where it is terminated with working equipment. At every joint and termination, power is lost in the form of heat. If proper care is not taken, thousands and millions of joints and terminations would dissipate enormous amounts of energy enough to light up a few cities.

Modern civilization depends on electricity, on uninterrupted power supply. The quality of life in any nation depends on it. Naturally, when selecting a domestic wire or cable, be it LT, MV, HV or EHV, or a cord for appliances, quality standards must be adhered to. Utmost attention needs to be paid during installation, jointing and termination. This calls for skilled technical manpower willing to stand on the field of work and take full responsibility and do an outstanding job to become an asset to the nation.

Computers can simulate a condition but the actual work has to be done at site by human power.

Authors

Sushil Kumar Ganguli, Bachelor of Science (Physics Honours) from St. Xavier's College, Calcutta University, 1952. Obtained Membership of Technical Institution, German Electro technical Association (VDE) 1960, (Vereinigte Deutsche Electrotechniker). He went to Germany in 1958. He was trained as a cable technologist in the designing, manufacturing and installation of telecommunication, power and high-frequency cable systems in reputed German cable industries. He was also entrusted with innovative activities of jointing systems to upgrade their techno-economic advantages. He returned to India with an overseas appointment and took charge as technical and production manager in Universal Cables Ltd., Satna, Madhya Pradesh. His immediate assignment was to redesign and develop railway traction cables with aluminium wire screening for the first time in India in collaboration with Research, Design and Standard Organisation (RDSO). This process replaced the expensive lead sheath, making the cable competitive in the market. Thereafter, he worked on redesigning airport lighting cables for runway shoulders. This led to the termination of import of such cables. Ganguli was also involved in the setup of rubber and elastomeric cable projects without any collaboration to produce all types of elastomeric cables to be used in different sectors, where mining and railways were given priority. He was part of the process that developed fibreglass composite cable jointing covers for the first time, which replaced heavy CI-Box. He took charge of the production and development work, particularly of 132 kV XLPE and rubber cables in Fort Gloster Industries, Calcutta. He interacted with Japanese technologists to adapt their systems to Indian conditions, which led to the modification of 132 kV jointing systems with CESC and Sumitomo.

From 1998, Ganguli worked as a consultant for the development of various cable projects and improvements in production techniques at Vidhya Kabelmats India Pvt. Ltd., Rewa. He remained as an advisor to the company in developing specific cable machinery and equipment as per customer requirements.

Ganguli joined as full-time technical advisor to Paramount Communication Cables, Bhiwadi, Rajasthan, to put up projects for aerial bunch cables and Sioplas-insulated cables up to 33 kV range (triple extrusion) and established MV cable lines (CCV) up to 66 kV. He also co-ordinated with AEI Cables Ltd., Birtley (Durham), UK (now taken over by Paramount Communications).

He advised Hindusthan Vidyut Products Ltd., Faridabad, to set up their 220 kV DCCV extrusion line and advised to set up working methods and progress. He was appointed by Uniflex Cables, Umbergaon, as technical

Vivek Kohli earned his BE in electrical engineering from the Birla Institute of Technology, MESRA, RANCHI India. He has more than 39 years of extensive experience as CEO/Executive Director/President, Business head and Chief technology officer in all functions of business, including P&L, business and product, market development. He has an all-round proficiency in plant operations, project management of power conductors/cables and related fields. His career started in August 1978 with one of the largest business houses, RPG Cables Ltd., near Mumbai, as Chief operations manager. He later joined HVPL near Delhi as Vice president (Technical). Kohli gained experience in all types of manufacturing of light-duty, HT, LT and telephone cables, and was designated as General manager (manufacturing) with Polycab, one of the largest electrical cable manufacturing companies in India. He also gained extensive technological experience on XLPE-insulated EHV cables by implementing a project involving cables up to 220 KV range at Fujairah (UAE) for Ravin Cables Ltd., a subsidiary of Prysmian Group, Italy. From August 2008 to October 2012, he was Business head and Chief Technology officer at a leading multinational business group, Sterlite Technologies Ltd., with facilities at multiple locations and a group turnover of US$10 billion. Kohli was instrumental in expanding power conductors business and set up 220 KV cables business. At present, Kohli is CEO/Executive Director/President (Strategic New Business and Projects) in a very large diversified business incorporation known as RP Mody Group. He has made several technology agreements to bring in new technologies in Power conductor and led group IT and HR function. He is sitting board member in Hindusthan group for several companies. During his professional career, Kohli was involved in various strategic planning projects, spearheading value analysis/value engineering, and has undertaken various improvement projects by initiating new process developments. He conceptualized to implement measures in operating procedures by adopting single-point monitoring systems to optimize resource/capacity utilization. He was also involved in the evaluation of capital expenditure by adopting proposals to secure the best interest of the plant by increasing productivity. Significant achievements were made in the field of manufacturing and supplying FIRST XLPE cables in India in 1978 at RPG Cables. An initiative was taken to develop and commercialize first HTLS conductors in India, and developed India's first low-loss, high-voltage MV power cables. Development was initiated to produce products like EHV cables from 66 to 132 kV at RPG Cables. To develop his technical knowledge, Kohli underwent overseas industrial training to initiate more innovative ideas, for which he visited companies like Maschinenbau Scholz GmbH & Co., and Weniger, Germany, where process control software and computerization of the extrusion process for CCV line were undertaken, both in West Germany. He had been to Phelps Dodge International Corporation to study their system of operation in Thailand for EHV cables. He also took the opportunity to be at British Insulated Callenderia Cables plants in the United Kingdom to get familiarised with super tension cables technology for power cables. It was also imperative to understand the processing ability of machinery and equipment. To that effect, some study was undertaken at the premises of Trőster in their plant in Germany. During the course of technical activities, the following publications were initiated to impart up-to-date knowledge to the young technical professionals of India: a paper at the *International Meet on Power and Telephone Cables* at IPCL Baroda on 9 and 10 February 1989 on 'Influence of processing parameters on performance of XLPE cables'; a paper at a seminar on EHV cables manufacturing, 'The cable industry – Perspective 2000', held at the NSE Complex, Mumbai, in January 1997. The following professional affiliations are attributed to Kohli for outstanding contribution to the field of cable technology:

- Corporate member of the Indian Institute of Plant Engineers.
- Member of the Society of Power Engineers.
- Member of Cigré, France.
- Member of the Belgian-Luxembourg Business Association.
- For initiating innovative ideas in cable processing techniques and for extensive knowledge in production/process planning/engineering and project planning and managerial skills, Kohli was certified as assessor for RPG Quality Award and was appointed as certified assessor for project examination at the University of Mumbai.

1

Introduction

An electrical power generating station is the heart of a power distribution system, and transmission lines its arteries. Without a transmission system, the generated power cannot be brought to the points of consumption and hence would be no value. It is interesting to note that the commercial sale of electricity began more than 100 years ago, attributable to Thomas Edison.

Copper was the first metal used to transmit electricity in the early 1880s. Copper conductors, however, had to face the challenge of weight. Later it has to face competition against Aluminium. Further with the price of metal being relatively high, the overall cost of the overhead transmission lines could not be made economically viable, particularly for long-distance ones. Naturally, Aluminium conductors became popular because of their low weight and cost. It was in California, in 1894, that aluminium was first allowed to be used as a conducting material. However for underground cables Copper remains as conducting material for western countries and middle-east supply systems.

The transmission systems can be divided into two categories and as follows:

1. Overhead transmission lines made of all-aluminium wire stranded conductors (AAC), aluminum wire stranded conductor steel reinforced (ACSR), all-aluminium alloy conductors (AAAC), and aluminum conductors alloy reinforced (ACAR). These are long-distance, high-voltage transmission systems installed across countries.
2. Insulated underground power cables of different voltage grades used for supplying power within urban/rural areas.

Discussion in this book will be confined to power cables and their technical features during the design, manufacturing, and performance stages. The machinery employed is also of importance. It is interesting to note that processing units are designed and manufactured to meet physical and electrical requirements. Naturally, these units may need to be customized in almost all the cases. The subject is a vast one. Efforts will be made to incorporate as much variations with the maximum possible technical discussion.

'Cables' in the early days were designated for flexible ropes made of jute or steel wires and used mainly for tugging and anchoring ships. The same term is now used for long and flexible insulated electric power and communication lines. The term 'cable', however, is attached to the prefix 'electric' to distinguish it from normal ropes. Even now, though the manufacturing of cables involves an intricate technology, people still see it as a simple rope-like product. One expects that it should supply power as per one's requirement. However, there are several technical points and conditions which need to be considered while designing, manufacturing, and installing cables. Many users who do not understand its full significance tend to belittle the technological importance of the product. At times, engineers too ignore these facts and do not highlight the technical features in their paperwork. Therefore, purchasing/commercial authorities are unable to put stringent conditions when buying cables. Electrical systems should not be taken lightly, and it is in the interest of the state and utility, and the users, to ensure trouble-free power at all times.

A lot of technical input is needed for the manufacturing technology to build a cable. Due to a lack of understanding, several manufacturers push low-quality products into the market at a price that does not even meet the material cost of a quality manufacturer. Naturally, low-cost offers are considered under tight budgets and economic conditions and authorities are only concerned about low-cost products. It needs to be understood that the ultimate sufferers are the nation, the utility, and the user. All this leads to a 20%–25% power loss from just transmission lines, be them overhead or underground.

The early development of underground power cables resulted from a long and experimental research work, which was started simultaneously in different countries.

Initially, underground cable works were undertaken to develop communication systems to connect distant islands in Europe, particularly in the United Kingdom, for administrative purposes. This was later extended to develop power cable systems to supply electricity to distant lands as well. Underground electric power cable made way for a more aesthetic look to human establishments, like townships and cities, and kept people away from coming in contact with bare electric wires and cables, avoiding untoward accidents. At hazardous working sites, such as in a mine or within a shaft, where excavation, drilling and blasting are to be undertaken, the intermittent shifting of power points becomes necessary. Power lines should be moved within the work site frequently without endangering the life of workmen and ensuring that no fire hazard breaks out while work is in progress.

This is an impossible task if bare electrical conductors are installed on poles. Naturally, such lines should be insulated and made flexible to withstand all sorts of critical conditions. A major development was also called for when power was to be supplied to nearby islands of England, France, Italy, Sweden, Norway and the like from the power stations situated on the main land. The lines were to cross the sea and the ocean. At times, the system had to withstand severe tidal force and stormy weather and even attacks from wild fish.

While designing a power cable, the following properties are required to be attributed:

1. The cable has to be sufficiently flexible in order to be laid underground in a safe way, or at any place in constricted areas and bends, without being damaged.
2. It should withstand sufficient pulling strength.
3. The longer the length, the less costly the installation would be. The more the number of joints, the higher the chances of failure and higher costs.
4. The cable should withstand jerks, shocks and impacts of falling debris or rocks in case of a natural catastrophe.
5. The cable should have the following electrical features:
    a. Low conductor resistance
    b. High dielectric strength of an insulating material
    c. Low thermal resistivity
    d. Low dielectric loss factor
    e. Capacity to withstand intermittent overloading and predetermined short-circuit conditions
    f. Lower dimensions to bring down the cost
    g. Materials that shall last for more than 40 years
6. Cables should have the following chemical features:
    a. Should be impervious to moisture, water and chemicals
    b. Should be able to withstand hostile environments
    c. Should have a high thermal stability against ageing and not interact with the material used in their manufacturing
    d. Should contain materials that are fire resistant and low smoke in areas prone to fire accidents

Apart from these qualities, cables may be customized with additional features to meet specific purposes.

The history of the development of insulated cables started with the need for a suitable underground communication system. This required a proper conducting material, such as copper, and an insulating material that would not allow leakage of power, not absorb moisture and remain mechanically stable yet flexible. Initially, watertight gutta-percha was the insulating material used. The cable was stiff with a single core and considerable diameter. With time, the demand for multichannel system rose, resulting in the design of a multicore cable with a reduced insulated conductor diameter, twisted and laid together. This was achieved by insulating a conductor with a wax-impregnated cloth. Bunched or laid-up cores were protected by drawing them in a steel or lead pipe. Wax, however, cracked when bent sharply and melted as the temperature rose within the conductor and the surroundings. Gutta-percha when used in land had two major defects. First, it was susceptible to oxidation, and, second, it would soften and deform eve at a slightly higher temperature. Next, jute and hessian cloths impregnated with bitumen were used. The limitations here were the inability to withstand high temperatures and the problem of bending. The material was hard and stiff which limited the length of production. Jointing was also difficult. Transmission lines for telephone and telegraph need a more sophisticated insulating system. During the intermediate period, compounds of natural rubber were developed and were brought in as an insulating material. But the limitation was that a large power supply could not be undertaken, as the technology available for processing rubber cables was not adequate. For communication cables, the performance of natural rubber was unacceptable.

By this time, different types of industrial complexes came into existence, and the demand for the supply of power increased. Along with the communication system, power cable development was also initiated. Ferranti laid 10 km of gutta-percha-insulated cables in England. The cables' disadvantage was that it was stiff. Large power transmission lines could not be constructed with this material. Accelerated ageing was another problem.

These problems were solved in 1875 with the finding that paper could be used as an insulating material provided it did not come in contact with moisture. Wheatstone and Cooks proposed lead as an alternative sheathing material. Further investigations revealed that if an insulated conductor or a bundle of conductors were protected by drawing a lead pipe over the assembly, not only the insulated conductor would remain free from ingress of moisture but the cable would also remain flexible. But the weight of the cable increased considerably. In spite of this, the suitability of lead as a protecting material became an established fact. In 1890, a lead sheath was introduced. Initially, a lead pipe was drawn over the cable core by drawing dies. Later in 1895, a regular lead sheath was introduced by employing an extrusion process. Simultaneous efforts were made to produce insulating paper with an improved quality. However, paper as such retained moisture that needed to be removed. This led to the development of drying and impregnating processes. Paper drying under vacuum was initiated for communication cables, and the processes of drying and impregnation were developed for power cables.

As for a lead sheathing, pure lead is soft and is often attacked by microorganisms. Normally, lead alloy is used to protect the cable core. A sheath could also get damaged under mechanical forces. Naturally, a metal sheath needs to be protected by applying a moisture proof layer, called bedding, which consists of a hessian cloth soaked in bitumen. Mechanical protection can be provided by applying an armour of steel wire or steel tape. Steel wire or tapes in turn were protected against corrosion by applying a bitumen-coated hessian cloth or a layer of jute. In order to prevent cable layers from sticking together, a coating of lime was applied. This kind of cable was popularly known as the PILC DSTA or PILC SWA cable.

In the course of time, during the Second World War, drastic changes occurred in the insulating material technology. Plastic in the form of PVC was introduced. Polythene came into the market as an insulating material, which could compete with paper. Its limitation was its temperature-withstanding property. Simultaneously, synthetic elastomer (synthetic rubber) was developed to manufacture flexible cables for special applications. This development continued step by step, bringing in more new materials and cables to cater to individual and specific requirements.

Some of the important developments are shown chronologically:

1880	Electric cable with gutta-percha, rubber cables, paper cables and 10 kV cables were introduced by Ferranti.
1892	British Insulated Callender's Cables introduced shaped conductors.
1914	Höchstadter Screen Cable made three-core cables up to 33 kV.
1926	Emanuelli's pressurization with an oil field cable, up to 66 kV.
1930	Siemens introduced first PVC cable in collaboration with Bayer, Germany.
1943	First 132 kV oil-filled cable installed in service.
1949	Introduction of mass impregnated non-draining compound (MIND).
1950	a. Commercial PVC/PE cables manufactured b. Aluminium sheath cable and aluminium as a conductor material
1956	Cross-linked polyethylene introduced with low coefficient of expansion.
1960	EPR – ethylene propylene rubber entered the arena of high-voltage cables.
1970	Cables containing thermoplastic and thermoset compounds were commercialized, which took over the market rapidly.

In India, the first cable company ICC (Indian Cable Company Ltd.) was established in Jamshedpur by BICC (British Insulated Callender's Cables, UK) in 1931.

An electric cable, at a glance, looks very simple, with a round rope-like contour, but it takes various complex procedures, in designing and processing, to produce a cable fit for a particular purpose. It needs to fulfil given electrical parameters that are different for different applications. To achieve the desired quality, an appropriate material for insulation, sheathing and armouring needs to be selected. While manufacturing, the mechanical characteristics of all the materials used, along with those of the finished cable, are to be taken into consideration. This means that to make oneself conversant with the technology of making electric cables, one is required to know all the three technological aspects – that is electrical, chemical and mechanical – and the installation methods involved in designing and manufacturing. One important aspect that remains intrinsically associated with cables is electrical and mechanical 'tension'. Cables transmit electrical power under tension (voltage), and while manufacturing, they must be processed under controlled tension to achieve best results. During laying and commissioning, cables are to be handled under proper tension to avoid installation problems.

Today, the application of electric cables has spread over a wide range of products, with each type of application calling for a particular design feature and the same type of cable requiring different treatments in different environments.

A mining cable can have a different construction given its application. A normal power cable can be constructed with an aluminium conductor; a PVC insulation; with two or three, three and half or four cores being laid up together to form a cable core; covered with a PVC bedding; and armoured and finished with a PVC covering. This cable is allowed to have an operating temperature of 70°C. Short circuit rating is 130°C. Such cables are suitable for operating at 660/1100 V. PVC cables of 6.6 kV were attempted, but because of the development of new materials and the thermal limitation of PVC, these cables are no longer in use. Normally, for a general distribution system, the cable can be armoured with galvanized steel strips. When a cable is to be laid in trenches or ducts, or in areas where corrosion and pulling tension are high, the cable is armoured with steel wires. In the case of a mining cable, double wire armouring is provided. To give an adequate short circuit protection armour resistance of 75% of that of the main conductor, galvanized iron wires need to be replaced with tinned copper wires. The choice of the number of copper wires depends on the size of the cable and short circuit requirements.

The cross-sectional properties of a conductor should be chosen based on its current-carrying capacity. At times, a voltage drop or a short circuit may call for a higher cross-sectional area. As PVC is soft and has a high dielectric constant, a cable made with PVC insulation has a higher dielectric loss factor and hence cannot be used for higher voltage ranges. Naturally, the alternative is polythene, which has a lower dielectric constant. Polythene is a soft material and cannot be operated beyond 70°C. However, cables up to 132/220/400 kV were manufactured in France using polythene as an insulating material. But due to thermal limitations, the current-carrying capacity of such cables is limited. Thereafter, XLPE with a higher rating factor came into the market. Here too, different constructions should be adopted for different applications. Cables with different cross-sectional areas, different voltage ranges, different screening factors and different armour designs need to be considered and made as per customers' requirements.

In order to control process parameters, different types of instrumentation cables are constructed and produced. For taking measurements, temperature thermocouple cables are developed. For moving vehicles like drag lines and excavators in mines, flexible high-power and high-voltage cables with elastomeric insulation, pliable armour and elastomeric sheath are constructed. Within this type, many varieties are produced depending on environmental conditions and users' demands. There are flexible coal cutter cables, drill cables, shuttle car cables, miners' lamp cables and so on, with each version and group having many variations.

There are various types and sizes of control cables: cables for ship wiring, subsea cables, defence cables, locomotive cables, oil rig cables and aircraft wiring cables. Cables are used in numerous fields with copper or aluminium as conductors; round or shaped; stranded or flexible; armoured or unarmoured; insulated with PVC, XLPE, polythene, PTFE, elastomer; and with different voltage ranges. Every day, some constructional features should be considered in the manufacturing of cables to satisfy customers' requirements. There are so many varieties that one will get lost in counting the types and numbers.

Today, every type of cables has become a specialized domain. A single person will not be able to handle all the areas and assimilate them.

With progressive development of various polymers and their uses in the modern world, emphasis will be on polymeric insulated cables such as XLPE, while considering the current trend of cable development – design and manufacturing and their applications and limitations. In this context, the details presented in the following chapters will provide an insight into the subject matter.

2

Basic Materials for Manufacturing Various Types of Electric Power Cables

The designing and manufacturing of cables are complex multidimensional and multifunctional processes. Different materials are to be selected and used to satisfy diverse requirements while considering the nature of use, surrounding environments, climatic conditions and power supply systems. Naturally, the designer and processing engineer should have a thorough understanding of the materials, their characteristics, behaviour under different conditions, process ability and long-term stable performance. Hence, there should be proper selection of materials for a given type of product.

2.1 Conductor Material

The transmission line is made of conducting media. Metals like iron, zinc, nickel, lead, aluminium, copper, silver and gold can be used as conductors. If the same amount of power is allowed to be transmitted through an individual wire made out of each of the previous elements, having the same cross-sectional area, it can be found that during transmission, the heat generated will be different in different metals. Silver will have the lowest temperature rise with the highest conductivity. This is due to the different amount of resistance offered by different metals during transmission of power.

Thus, while the choice of a material for creating a conductor primarily depends on its conductivity, the following factors should also be considered to produce and procure material at a reasonable cost:

1. Conductivity of metal
2. Quantity of material available on the Earth's crust
3. Cost of production of primary metal from the ore stage
4. Easy processing ability of refined electrical grade metal during processing (ductility)
5. Malleability in order to impart flexibility during processing and installation

6. Sufficient strength to withstand hazardous working conditions
7. Resistant to affluent and adverse environmental conditions

It is difficult to obtain a metal with all of its qualities at their peak values. Some compromises are to be made considering economic viability.

It is of prime importance that the conductivity of a metal should be high (purity) so that the maximum given amount of power can be transmitted through a minimum possible area with minimum loss of power (as heat energy). This will allow the metal to be used in a rational way. In this respect, silver is found to be ideal. But the availability of metal just as an electrical conductor is not enough to meet the world's demand. Further, the cost of production of primary metal is higher. In contrast, iron is cheap and could be made easily available. To transmit the same amount of power as that by silver, by restricting heat loss to the same degree, the size of the steel wire will have to be very large, stiffening the conductor. The processing cost and the weight of the conductor will be prohibitive.

In comparison, copper and aluminium have qualities that meet all the parameters. Though copper stands as the number one metal, aluminium, on the other hand, has the advantages of lesser weight, lower cost and unrestricted availability.

Given in Table 2.1 are comparative values of a few metals with their electrical and mechanical parameters.

The aforementioned comparison shows that copper and aluminium are the metals that retain economic viability.

While selecting materials, a study on their properties was undertaken to understand their chemical, mechanical and electrical characteristics under different conditions.

2.1.1 Copper

It is known that copper has been used since the early days of civilization. The earliest specimen of cast copper was excavated in Egypt and Babylonia dating back to 4000 BC and those at Mohenjo-Daro and

TABLE 2.1

Electrical and Physical Values of Few Selected Metals Used for Manufacturing Electric Cables

Sl. No.	Material Description	Density (cm^3)	Resistivity σ at 20°C (Ω mm^2/m (10^{-8}))	Temp. Coeff. α of Res. per°C	Tensile (MN/m^2)
1	Silver	10.470	1.626	0.0041	
2	Copper	8.890	1.724	0.0039	>225
3	Aluminium	2.703	2.803	0.0040	>70–150
4	Lead	11.370	21.400	0.0040	
5	Steel	7.860	13.800	0.0045	>500

TABLE 2.2

Material	Atomic Number	Density at 15°C	Melting Point (°C)	Vapour Colour
Copper	29	8.96	1083	Green
Silver	47	10.47	960	Blue
Gold	79	19.3	1063	Yellowish green

Harappa to about 5000 BC. The Inca and Mayan civilisations in South America were rich in the use of copper.

The chemical name of copper is 'cuprum' and is derived from the name of the place Cyprus. The reddish-yellow metal obtained from the place was called Cyprium, which then transformed to 'cuprum' – Cu – copper.

In the periodic table, copper is found in group '1 B' within the odd series placed with silver and gold (Table 2.2). General properties are:

The varying valency of copper, silver and gold is due to the fact that these are transitional elements (Figure 2.1). The inner group of 18 electrons of copper can be drawn upon for one or two valence electrons. Due to this, copper reacts to produce both monovalent and bivalent compounds. The electron on the outer shell is available for free movement under slight electromagnetic pressure. This is the reason these three metals show almost identical properties as a conducting material, silver being the most conductive.

The availability of copper as different oxides and sulphides is higher than that of silver and gold. There are several methods available for reducing ores to metallic copper. Copper is refined by an electrolytic process. A purity of 99.99%

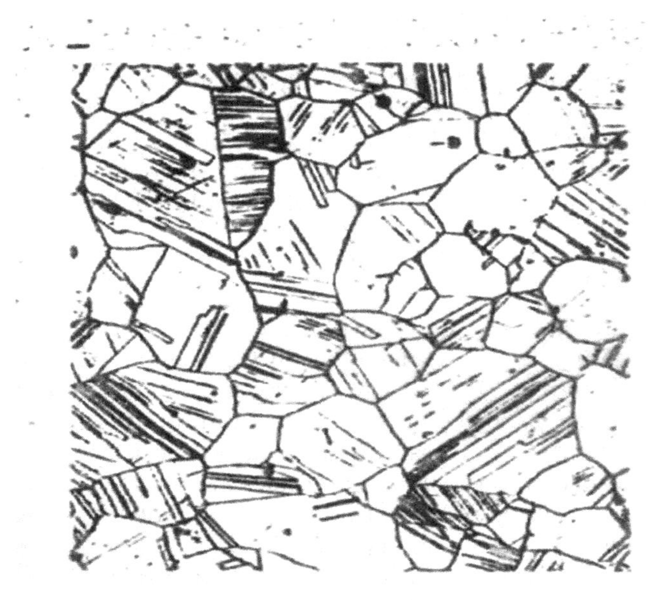

FIGURE 2.1
Soft working copper crystal.

can be achieved at times. Generally, 99.95%–99.98% of copper is available in the market and is used in electrical transmission and distribution and in appliances. The metal in its pure state is very ductile and malleable. It can be shaped as a very thin foil or wire in continuous length, without breaking, by annealing intermittently. The density of high-conductivity copper varies from 8.87 to 8.91 and occasionally from 8.83 to 8.94. The presence of 0.03% of oxygen will cause a reduction of density by about 0.01. Hard-drawn copper has about 0.02% less density than annealed copper.

Nowadays, the analysis of copper density and its behaviour at different stages has gained importance because of the demand and the rising price in the market. The International Annealed Copper Standard (IACS) provides the accepted value for copper with 100% conductivity. This standard is expressed in terms of mass resistivity as 0.15328 Ω g/m^2 or the resistance of a uniform round wire 1 m long having 1 g weight at 20°C. On this basis, volume resistivity becomes 0.017241 Ω mm^2/m and conductivity 58 S mm^2/m at 20°C. Based on this value for cable design, the resistance and input of copper are considered during manufacturing. Apart from these values, the coefficient of resistance and the thermal resistivity of metal are also taken into account while designing a conductor for manufacturing cable.

It is of utmost importance to note that if the metal remains exposed to a moist atmosphere for a long time, it may tarnish and become greenish-black upon interacting with oxygen and sulphur. In this case, it reacts with hydrogen sulphide in the presence of oxygen. Modern methods of continuous casting and up-cast casting to produce a wire rod eliminate the formation of any oxide film on the surface. These bright wire rods can directly be used for the drawing of wires. Nowadays, oxygen-free copper rods are manufactured in up-cast mills, regularly producing high-purity wire rods.

Wire rods thus produced have a diameter of 8 or 6 mm and tensile strength of a minimum 210 MPa, with elongation being min 32%. The surface of rolled wire shall be fairly smooth, free from the inclusion of foreign contamination, depression, scales and twists, etc. and shall generally conform to IS 9731.

2.1.2 Aluminium

Between 1909 and 1912, aluminium was used to manufacture conductors by British Insulated Callender Cables Ltd. (BICC), England. Aluminium was not used much until 1930. After copper prices rose very high, the use of aluminium became extensive in the 1950s. Aluminium is present all over the Earth's surface as clay and is abundantly available as bauxite. Bauxite is digested with caustic soda solution to produce alumina, Al_2O_3. This is then dissolved in fused cryolite. Pure aluminium is obtained by an electrolysis process. It is a trivalent element. The electrons of outer orbit of this element are available for movements. Unlike copper, the mechanical strength aluminium is limited. It can be cold-formed during working without changing its conductivity. Aluminium has the advantage of being lighter, with a specific weight of 2.703 against that of copper (8.89). This advantage, however, gets offset by aluminium's lower conductivity (61). Further, the current-carrying capacity and thermal resistivity do affect the design criteria. However, all these get compensated by its price advantage over that of copper. But, for the same current rating, a higher size conductor has to be selected. This will naturally require a cable with a higher diameter, and the consumption of raw materials required for subsequent processes becomes higher. Even then, at present, considering the price of copper and its overall weight, aluminium cables are found to be somewhat more competitive both weight and price wise.

The aluminium used for making cables is procured in the form of a wire rod, in coils weighing approximately 2–3 tons. Wire rods are formed by rolling in a Properzi mill from a primary ingot. The material has a purity of not less than 99.5%. Resistivity is less than 0.02803 for electrolytic grade. Generally, ½H or ¾H grade aluminium is accepted for making cables. Normally, the resistance value of aluminium does not change appreciably during its working.

The wire rod procured is uniform, with a smooth surface, without any dents and scale formation, cracks, laps, pipes, etc.

The one factor to be considered is that aluminium, during working, gets coated by a very thin layer of oxide film instantaneously, which acts as insulating media. This oxide film protects the metal from atmospheric influence and corrosion. During jointing, this film must be removed thoroughly to achieve good electrical contact. For long time, aluminium was not considered as a power cable conductor because of this problem. Due to aluminium's inherent oxide layer, jointing of conductor to conductor became difficult. The rising price of copper and the steady depletion of its reserves forced manufacturers to use aluminium. In the meantime, Alcan Ltd came up with a solution for jointing the metal by removing the oxide layer on the surface. This was an organic flux (Eyre Smelting Flux no. 7) and a solder made of an alloy of zinc–tin and lead (Eyre Solder no. 375). Eyre Smelting Co. was a subsidiary of Alcan Ltd.

India has come a long way in using aluminium as a power cable conductor and has a large deposit of bauxite; the production capacity of aluminium is sufficient to meet the nation's requirements in all respects.

The material is made tough and corrosion proof by alloying it with a small percentage of magnesium (0.5%) and silicon (0.5%–0.6%). This alloy is used extensively to make an all-aluminium alloy conductor (AAAC) overhead transmission line, to be installed along the sea coast where a corrosive saline atmosphere prevails year-round. This particular alloy does not get affected under such conditions. The tensile strength of alloy is very high, and the overhead conductor does not require to be reinforced with a stranded steel core made of galvanised steel wires. Nowadays, aerial bunch cables are manufactured extensively using the AAAC conductor as the suspension wire to support the cable between poles, and to be used as neutral, and earth as required.

TABLE 2.3

Grade of Alloy	Heat Treatment Temp. (°C)	Approx. Time of Heating (min)[a]	Quench[b]	Ageing Temp. (°C)	Time of Ageing
2014[c]	500–510	15–60	Water	170	10 h
2024	485–500	15–60	Cold water	Room	4 days[d]
6061	515–525	15–60	Cold water	155–160	18 h
7075	460–470	15–60	Cold water	120	24 h

[a] This depends on the size and amount of materials. In India, a 9.5 mm rod for 2 tons of coil weight is heated for 20–30 min max.
[b] This is done within a minimum time, say, within 15–20 s, and the coil is dropped in water from furnace.
[c] This is the type used in India as per IS 14255.
[d] More than 90% ageing occurs on the first day of ageing.

For the AAAC conductor, a primary alloy rod, as supplied by the manufacturer, is heat-treated and tempered to get the required strength. The primary rod is first drawn to a size of 5.5 mm in a bull block wire drawing machine and then made into a coil. These coils are then electrically heated in a furnace to 500°C–550°C for a given period of time. Thereafter, hot coils are dropped instantaneously in cold water for quenching. This process is called solution treatment, by which crystalline structures of magnesium and silicon are distributed uniformly within the metal. During this process, the metal becomes extremely hard. To achieve its required strength, 5.5 mm coils are drawn to the required size of the wire in a wire drawing machine. The wire thus produced is tempered in an ageing oven where it is heated at a predetermined temperature and for a given period of time to achieve the desired tensile property. Some of the heat treatment parameters for a few popular grades of alloy are given in Table 2.3.

2.2 Insulating Materials

Finding a suitable insulating material was a long-drawn promotional activity carried out by repeated trial and testing. Suitability had to be ascertained through actual application. Choosing the insulating material for development happened after innumerable results were analysed. On the basis of the numerous data obtained, the accepted quality parameters of each material were standardised.

It is shown in Chapter 1 how the development of cables took place step by step with the introduction of new material. Problems were experienced during development, but solutions were also worked out simultaneously. To make a product with the desired quality, machinery and equipment were developed hand in hand to cater changing demands. The skills of personnel were also updated with the technological progress.

Insulating materials should have the following qualities:

1. It should withstand high AC impulse voltage and should remain stable for long-time performance.
2. Its dielectric power loss should be as low as possible.
3. It should have excellent treeing resistance and corona resistance.
4. It should be flexible and withstand mechanical abuses and adverse environmental conditions.

2.2.1 Paper

Though paper as an insulating material for electric cables is now almost obsolete, it has retained its importance in the understanding of any significant development of insulating media. Some aspects of the material are described for the purpose of knowledge.

Paper was known to be an insulating material from early days. It is a product obtained from natural sources such as pine wood or manila wood that has a long chain of cellulose structure. Wood logs are first cut to small pieces and fed into a digester, mixing with sodium hydroxide, calcium and other chemicals to eliminate unwanted mineral compounds and impurities, forming into a pure pulp. The pulp thus formed is then led over a very fine screen, eliminating excess water and then calendared to form a smooth, plain sheet with the required thickness. During calendaring, density and porosity are controlled with care. Paper density is varied from 0.85 to 1.20 g. Tensile strength is also maintained lengthwise at 77 g/cm^2 and the transverse side at 36 g/cm^2. The thickness of the paper ranges from 0.04 to 1.45 mm according to the voltage and application system. The water content during lapping over the conductor should not exceed more than 4%. Absolute dry paper also cannot be lapped as paper would break intermittently. The permittivity of paper in this condition is around 5–6 width porosity and cannot be used as such. During initial operation, paper is lapped over the conductor leaving a butt gap between the edges. While lapping in layers, butt gaps of the lower layer are successively covered by those on top. Covering is registered on a 70/30 basis. The formed butt gap must be uniform and is essential for sliding paper layers one above the other on the conductor during bending, winding and unwinding while adjacent edges do not not press onto each other damaging the insulating layer. The edge of the tapes must be very smooth and free from burrs. This can be ensured during slitting by using sharp circular cutters. Lapping must be smooth and free of crease and folds, as well as tight. For high-voltage or extrahigh-voltage cable, an inner layer of paper of very less thickness, say 0.04 mm, is used to make insulation compact and void free. The thickness is gradually increased in a graded manner to allow a more porous form and also to restrict the number of layers. Otherwise, the paper lapping process would be enormously long, and the process of impregnation would also take a very long time, increasing the production cost substantially.

2.2.2 Impregnation and Impregnating Oil

Paper, being porous, absorbs moisture and gets ionised easily. Naturally, the breakdown strength becomes less. In order to fill in the voids, moisture- and contamination-free mineral oil is used to impregnate paper. Contamination-free mineral oil has a low permittivity and is used in conjunction with paper to make a composite impregnated paper insulation. The dielectric constant of the combination is around 3–4. After the application of paper as the insulation over the conductor, the core is wound on a perforated tray. Two or three such trays are placed in a hollow, cylindrical, vertical chamber. After closing the lid of the chamber, vacuum is applied to extract moisture and trapped air from insulation. The degree of vacuum is determined as per the voltage grade of the conductor. The chamber is heated to 120°C which is kept constant throughout. Temperature is not raised so that the papers do not get brittle. After attaining the required level of vacuum, pre-heated contamination- and air-free filtered oil is introduced inside the chamber under vacuum. The cable is kept in oil until it is ensured that the oil has penetrated and impregnated the paper thoroughly, filling all voids. The tank is then allowed to cool down by circulating water from outside and sometimes under pressure of an inert gas like nitrogen. Temperature is brought down to 45°C–50°C. Immediately after bringing out the trays, the cable core is covered with a metallic sheath like lead, corrugated aluminium or corrugated steel tape, thus protecting the core from coming in contact with outside moisture and air. The dielectric power factor of such cables is good, but during installation and operation of a cable in an inclined gradient, the oil drains down to the lower part of the cable exerting an unwanted pressure. At the same time, the upper part of the gradient can get dried up because the oil migrates to the lower part. This phenomenon can lead to the formation of localised void causing a failure spot within the cable. To obviate such a situation, a mass-impregnated non-draining (MIND) compound was developed by BICC, England, in collaboration with Dussex Campbell. Though the dielectric properties of this compound were somewhat lower, it retained a longer service life as it does not migrate while the cable is installed in an inclined position. At a later stage, a poly-isobutylene (popularly known as PIB) compound was developed, which had the same non-draining quality but had a much lower dielectric constant and better insulating properties. The tan δ value of the compound was generally found to be in the order of 0.0003 against 0.005 for the rosin oil MIND compound. This compound was used to manufacture cables up to and including 33,000 V. Above this voltage, oil pressure or gas pressure cables were considered, as MIND and PIB compounds are not suitable for use at such higher voltage ranges. High- and extrahigh-voltage gas and oil pressure cables are not discussed as these cables are not manufactured in India.

2.2.3 Synthetic Polymers as Insulating Material

2.2.3.1 History of Development

The geographical position of Europe forced it to develop various synthetic materials like marine blue, insulating paper, synthetic drugs and chemicals. Like in the case of the Germany in Central Europe which was surrounded by hostile neighbours, countries without much natural resources had to import raw materials for industrial uses and finished products for home consumption, making the state dependent on others. Naturally, efforts were directed toward finding alternative sources of basic raw materials that can compete and substitute natural products. Developments thus were initiated to find synthetic products, compelling these countries to develop scientific and technological research centres and establish industries that would make them self-sufficient as far as possible. To cite an example, the first synthetic blue was developed by Bayer Farben Fabrik AG in Germany, at a time when England dominated the market of blue pigment, extracted and made from natural sources. During the First World War, Germany realised the necessity of accelerating scientific development to give a boost to its economy. From that time onwards, efforts were made to find alternative insulating materials for electrical equipments, not only in Germany but in other European countries and the United States, and to replace products obtained from natural sources, as they were becoming costly in a market monopolised by a few countries. One of these materials was insulating paper for manufacturing power and telephone cables. The shortage of paper became acute before the Second World War, and during that time polyvinyl chloride (popularly known as PVC), discovered by Eugen Baumann in 1872, was modified to make it a useful insulating material. By 1938, polythene (PE) resins were discovered by Gibson and Fawcett at ICI, UK. Their invented process required very high pressure to produce the material in a reactor. In time, Professor Ziegler in Germany developed a method to manufacture PE at a much lower pressure, which is now popularly known as the Ziegler process. Thereafter, many synthetic thermoplastic and thermosetting materials came into the market. Apart from Europe, the United States too fell in line in search of synthetic polymers. From then onwards, the march continued relentlessly. Many thermoplastic and thermosetting materials were developed, the trend of development continuing even today. Thermoplastic materials deform when heat is applied, but regain original form when heat is removed. Thermosetting materials do not deform when heated after they are set by the cross-linking process.

2.2.3.2 Polythene

Polythene, a thermoplastic material with a lower density, can be processed in an extruder. It is obtained by the polymerisation of ethylene in the presence of oxygen at high pressure. PE was first prepared by Bamberger and Tschirner before 1900 by the decomposition of diazomethane. Subsequently, PE was produced by ICI at high pressures, as low-density PE. Later, more than one process was developed, out of which the Ziegler process became popular. By controlling pressure and using a different catalyst, PE of various molecular weight and density could be formed; the base material used was ethylene. Ethylene is obtained directly from petroleum refineries. The Ziegler process uses an organometal catalyst such as titanium tetrachloride ($TiCl_4$) and an aluminium alkyl such as tri-ethyl aluminium [$(C_2H_5)_3Al$] for the polymerisation of an ethylene monomer.

$$CH_2{=}CH_2 \rightarrow -CH_2-CH_2-CH_2-CH_2-CH_2-$$

The Ziegler process gives a somewhat lower density, such as 0.94–0.95, than the Phillips process. Later, a material with a density of 0.96 was produced, which had a straight-chain molecule crystalline structure.

This material was a very good insulating material; it's general characteristics are listed in Table 2.4.

The first polythene-insulated 63 kV cable was produced in France. Subsequently, they manufactured PE insulated 132 kV cable for a short period. This PE had a low density. From 1930 onwards, this material found considerable use in the production of power cable in the continent. Later, PE was extensively used in the United States in medium-tension cables as an insulating material.

PE becomes soft with increasing temperature; as such, the normal operating temperature of a conductor in a cable is settled at 70°C. It can withstand low temperatures up to −30°C in normal conditions. A specially prepared compound can withstand up to −50°C. The material is manufactured with a density range of 0.92–0.93, such as a low-density polymer; the medium density ranges from 0.94 to 0.95 and high density from 0.96 to 0.98. High-density PE is difficult to process under normal conditions. It is modified by blending with a small amount of antioxidant, ethylene vinyl acetate (EVA), or ethylene propylene rubber (EPR). The material is highly sensitive to light and ultraviolet rays. By blending with a small amount of butyl rubber or EPR, PE is prevented from getting oxidised rapidly. If it is blended with >2% carbon black, the material can be protected from degradation under the influence of ultraviolet rays and sunlight. An antioxidant and a modifier are incorporated when used with a colourant. At high temperatures, the material is

TABLE 2.4

Insulation resistance	10^{16} (Ω cm)
Surface resistance	10^{14} (Ω)
Dielectric constant ε	2.3–2.5
Loss angle tan δ at 50 Hz	>0.0001
Breakdown voltage (kV/mm)	25–30
Tensile strength	120–150 kg/cm²
Elongation approx.	400%–600%

attacked by saltpeter and concentrated sulphuric acid. At 70°C, the material can get dissolved in carbon tetrachloride. At 50°C, the material gets slightly softer and starts absorbing sulphur, oil and other impurities, degrading its electrical properties. By blending with polyisobutylene, the material can be modified to gain excellent antioxidant and ozone-resistant properties. Nowadays, PE is modified in different ways by combining it with a small amount of elastomeric polymer to make it suitable for specific applications. In France, an extrahigh-voltage cable of 132 kV was produced, but the operating temperature had to be restricted to 70°C, which was too low and was not advantageous. Impulse voltage also had to be lowered along with short-circuit rating. However, this material was found to react with peroxide at high temperatures to form a cross-linked product which became popular as cross-linked polythene (XLPE), which had all-round excellent characteristics.

2.2.3.3 Cross-Linked Polythene

XLPE is a thermosetting material. The base material of this solid dielectric is PE. PE can be made a thermosetting material by curing under temperature and pressure and blending with selective organic peroxide and antioxidants. Al Gilbert and Frank Precopio invented XLPE in March 1963 at the GE Laboratories. It has been found that by cross-linking PE molecules, the properties of the compound can be improved to a great extent. The operating temperature could be raised from 70°C to 90°C–95°C. The short-circuit temperature is improved from 150°C to 250°C. The electrical stress value is enhanced considerably. The AC breakdown voltage is raised to 30 kV/mm, and the basic impulse level can be increased up to 55/60 kV when the compound is made super clean – that is completely contamination free. This compound can be cross-linked with the help of a peroxide having high melting temperature, such as dicumyl peroxide or benzoyl peroxide, under high temperature and pressure. The voltage range can be from 1100 V to 500 kV depending on the nature of the compound. Various modifications in the basic polymer structure are made to increase the productivity of the cable manufacturing process. The compound has also been modified to limit the exudation of by-products during cross-linking. This is particularly significant for high voltage (HV) and extra high voltage (EHV) cables, where degassing needs to be done to make insulation free of methane and other unwanted gases generated during the curing cycle. To make the insulated core free of such inflammable gases, it is to be heated in a closed chamber at 70°C–80°C under controlled conditions so that these gases do not explode during curing, installation and under service conditions. The manufacturing of HV and EHV cables became successful with the use of XLPE, replacing oil- or gas-filled paper insulation, because of its relative straight line technological handling procedure and the economy in its use as an underground power cable. The design of extruded solid dielectric cables requires less material; offers easy installation, jointing and terminations; and allows one-time continuous processing.

During 60's and middle of 70's, however, insulation was found to be weak, with low resistance to partial discharge. Internal void and gas pockets were relatively large and numerous. Consequently, while designing a cable for HV and EHV installations, relatively high stress value per millimetre basis could not be considered. Void formation is an intrinsic phenomenon for a plastic or elastomer. During the extrusion process, volatile gas exudes from compounds at higher temperatures, making the substance spongy. This can be contained by applying pressure while the temperature is raised to a certain level during extrusion.

With more experience and greater working ability, it became possible to increase the working voltage by modifying compound structures and processing technology and observing their breakdown voltages on a large number of cable samples of different grades. By applying Weibull's statistical relations, these values were established for safe performance. Further, it was found that to increase stress values and to design cables to achieve comparable economical parameters by bringing down insulation thickness to a reasonable level, clean and super clean insulating compound must be produced making it free from contaminants and foreign particles. It was also noticed that during processing in a continuous catenary vulcanization (CCV) line, if a neutral gas is applied at a high pressure, quality could improve considerably. Though people were apprehensive that by steam curing the compound is likely to absorb too much unwanted moisture/water, it was not fully the case. The cause of the relatively higher discharge rate was due to the large size and numerous voids present within the insulation. It was necessary to restrain the formation of large-size void structures and their numbers. This could not be achieved as steam pressure could not be raised beyond the temperature limit of 250°C which is <18 bar. The actual pressure applied was 10–12 bar max, which lies within the temperature limit of 183°C–190°C. To exude air/gas trappings relatively at a higher rate and minimise void formation and their sizes, higher temperature and pressure were needed. This was to accelerate curing time so that the internal expansion of the insulating material could be restricted to a minimum. It also led to an increased line speed simultaneously. However, this would require superheated steam and would be very costly, needing higher safety requirements. During subsequent development, it was found that by using nitrogen gas, temperature can be varied independent of the applied pressure. Utilising this advantage, temperature was raised to the required degree (400°C–450°C), and pressure was adjusted to improve the quality of insulation, containing the dimensions and number of voids per unit area to a minimum level, thereby raising the stress-withstanding capacity of the insulating material and allowing proper reduction in insulation thickness. Cooling under water or gas pressure for a reasonable length of time also improved performance criteria. The quality of performance was observed by analysing the failure due to the formation of the 'treeing'-type electrical breakdown. The greater part of the exact mechanism of the formation of the tree remained unexplained.

As explained, in early stages, manufactured XLPE cable failure was frequent and did not meet the calculated expected life span of 20–25 years. Rigorous analysis and R&D work revealed that most of the failures were initiated by treeing-type electrical breakdown. 'Treeing' is a general term for fault progressing under electrical stress through a section of solid dielectric. It is associated with AC or impulse voltage, but has also been found where direct voltage at high stress is applied during extreme wet conditions. There are three main types of treeing phenomena:

1. Water treeing
2. Electrical treeing
3. Electrochemical treeing

Basically, within the polymer, some amount of void formation is present due to the entrapment of evolved gas and water during extrusion and curing. While passing through the curing zone, a certain amount of water is given out, as the polymer reacts with the catalyst and curing agent. Further within the water cooling zone, some amount of moisture is also absorbed by insulating media due to the retention of excess water for a certain period of time. When voids are somewhat larger in size, retaining moisture, water particles get ionised as the cable is charged. Ionised molecules striking against the wall of the cavity penetrate within the insulation body radial parallel to the electric field and migrate towards the region of the highest electrical stress by a mechanism of 'dielectrophoresis' (movement of polarisable water molecule in an electric field), forming a bush-like pattern (called water tree) and travelling towards outer and inner conducting media, such as the outer or inner semiconductor, creating a fault condition.

Electrical and electrochemical trees differ significantly from the water tree in appearance. If a contaminated particle is present within insulating media, high electrical stress will develop within the solid dielectric (Figure 2.2).

Any contamination during insulation will form an ionised particle and any water particle surrounding it will ionise to initiate a discharge condition. These discharges come out in the form of a fine thread-like structure on both sides trying to reach conductive layers such as the conductor or semiconducting areas. As the field intensity increases, the structure forms a bush-like structure and in time initiates failure (Figure 2.3).

Electrochemical trees are formed by water in the dielectric containing solutions. Electrochemical trees differ in appearance, having a bush- or broccoli-type growth. After cutting normally and in the direction of growth, no channels will be found. Their structure consists of very fine paths along which moisture penetrates under the action of the voltage gradient.

Water tree and electrochemical trees are formed at a much lower voltage due to the presence of moisture, whereas electrical trees are found to form at a higher level of voltage gradient. Naturally, where there is no water, the failure level comes down considerably, and to ensure a longer operational

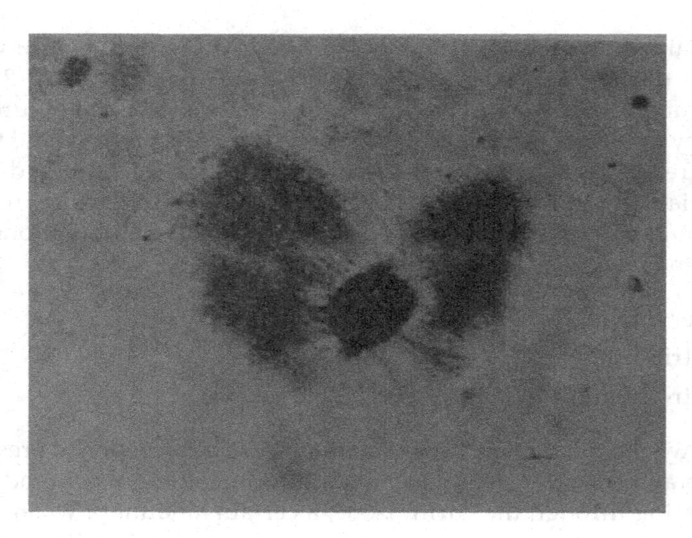

FIGURE 2.2
Formation of a bush-like structure.

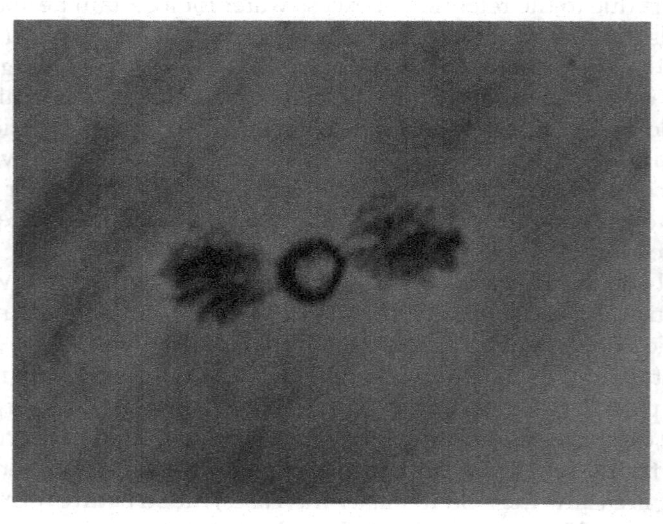

FIGURE 2.3
Formation of a bow-like structure.

life, the number and size of voids are restricted to a minimum level , and so are moisture and water. Further, the contamination level must be reduced to its lowest possible count.

In order to reduce the formation, retention and influence of moisture and water within insulating media, suppliers have developed a water tree–retardant compound having special additives which inhibits water and moisture during curing and cooling operations.

2.2.3.4 Peroxide Cure XLPE Compound

In the beginning, PE, along with a measured quantity of antioxidant and dicumyl peroxide or benzoyl peroxide, was introduced in the extruder separately. Other materials were mixed within the extruder before covering the conductor. The process at times created problems, and materials were found to be getting scorched within the machine. Subsequently, PE was grafted with antioxidant and curing agents, and the material became safer to handle.

The initial process of producing grafted XLPE starts with passing the polymer through a very long stainless steel pipe reactor, under very high pressure and temperature. This steel tube is termed as 'high-pressure tubular reactor'. The reactor is designed to give out polymers with constant properties, free of any contaminants and chemical elements that can degrade the dielectric strength of the material. Naturally, the selection of the polymer along with the reacting agents must be free of any impurities and should be fed in sequence, as required for proper controlled reaction. The polymer thus formed is to be conveyed to the compounding equipment in such a manner that it shall not get contaminated on the passage length. It should also be ensured that, within the passage of movement, it does not get entangled and form dust particles due to the rubbing effect. This leads to adopting a gravity or dense-phase conveying system. Before the incorporation of cross-linking and stabilising chemicals, the material has to be passed through extremely fine filters.

At every intermediate step, and also at the final stage, samples are taken and analysed for quality consistency and cleanliness. Physical and chemical properties, along with extrudability and cross-linking properties, are determined before final approval and shipment. Packaging and warehousing are done under extremely clean atmosphere to ensure quality production.

Naturally, it is also imperative that the users take extreme care while unpacking and feeding materials into the extruder. This leads to the installation of a dust-free pressurised enclosure and a closed-circuit material feeding system.

2.2.3.5 Moisture-Cured XLPE

In case of the peroxide system, XLPE is processed in a CCV line, where, to achieve economical productivity, a longer length must be processed at one go (machine to run for a few days continuously), as during starting and stopping, some amount of length remains in an uncured state and would be rejected as scrap. Naturally, the ratio of scrap to good length will be large when shorter lengths are produced. In case of wet or silane process, these constraints are eliminated. This being a batch process where extrusion and curing are done separately, the production of shorter length poses no problem. At present, compounds up to 11 kV ranges are available. Some 33 kV compounds are in the process of development but are yet to be commercialised.

There are three types of system by which moisture curing can be activated by introducing a suitable silane as the main component.

1. *Sioplas process*: A selected silane is compounded with peroxide and polymer. In this process, silane gets grafted within a polymer chain. A catalyst to initiate the cross-linking process, along with an antioxidant, a processing aid and a colouring agent, is added during extrusion as a master batch. The shelf life of this product is very limited. Further, during processing, it is not possible to stop the machine for a while and change extrusion tools to start a new size of cable as the compound will get scorched within a very short period. The mixed compound also cannot be left on the floor in an exposed condition as the material will start absorbing moisture from the atmosphere, degrading its process ability. This becomes more acute during humid conditions in a tropical atmosphere.
2. *Monosil process*: In this case, all the ingredients such as silane, peroxide, antioxidant, accelerator and colour are added together along with the polymer and fed into the extruder at the same time. The total reaction takes place within the extruder along with the extrusion.
3. *Ethylene vinyl silane copolymer process*: In this case, during polymerisation, silane is introduced into the polymer chain and the compound comes out in a ready-to-use form. The catalyst, colour and accelerator as a master batch, is added during extrusion. The shelf life of this material is very stable. During processing, the material can be retained in the extruder for a long time controlling the temperature. Cable size can be changed without any degradation to the material. Insertion of silane in the polymer chain makes it a more homogeneous and uniform product.
4. *Post curing process*: The conductor insulated with any of the previous types of compounds is kept on the shop floor at least for 8–12 h, to stabilise the condition of the material at room temperature, while insulation sets to its maximum dense form bringing homogeneity, preventing the formation of normal water pockets and allowing moisture to diffuse uniformly. At the end of the moisture diffusion process, curing is allowed to continue under steam or water bath at about 80°C–90°C for a certain period of time depending on the thickness of coating. After this, cores are kept again on the floor for conditioning at least for 12 h in order to bring down the moisture level to a minimum. If the process is followed properly, the moisture content in the XLPE (sioplas) can be contained within 300 ppm. The material can then be almost compared to CV/CCV cured cores.

2.2.3.6 Solar Power for Curing Sioplas

In tropical countries, insulated cores can be kept in a double-walled specially constructed glass chamber, under intense sun radiation. Water too can be kept within the chamber in a metal tray. The water in the tray evaporates slowly and creates a humid condition within to accelerate the cross-linking process. This process saves electrical power/steam considerably (Table 2.5).

The properties of the XLPE compound on the extruded conductor after curing shall comply with the values specified in IS 7098-part 2 for cables from 3.5 to 33 kV. The material above 33 kV shall comply with IS 7098-part 3 for cables up to 220 kV.

The compound used needs to be clean and contamination free, which is ideal. Factually, the compound produced and delivered by the supplier has to be tested for all physical and electrical properties no doubt, but to ensure a good quality production, the supplied polymer also needs to be checked for the level of contamination. This is checked optically on a very long thin film made out of virgin material, or from a representative core sample. The computerised test apparatus is used to ascertain the contamination level and can be used to examine different thicknesses while also computing the number of faults and voids of all the samples with the help of a comparative data chart for evaluation. Normally for HV and EHV cable compounds, suppliers specify the contamination level.

At the users' end, the method of handling and conveying a compound to the hopper must be closed circuit and contamination free. To fill in the hopper vacuum, devices are used to suck the material from the container in a closed-circuit method to avoid manual handling. For feeding a measured quantity of semiconducting compound, the dosing units used are heated with hot dry filtered air under a controlled temperature. Apart from the previous measures, extrusion room must be kept dustproof and partially pressurised with dry filtered air.

TABLE 2.5

Characteristics of XLPE Compound

Insulation resistance	10^{17} (Ω cm)
Surface resistance	10^{14} (Ω)
Dielectric constant ε	2.3–2.5
Loss angle tan δ at 50 Hz	<0.008
Super clean compound	<0.0001
Breakdown voltage (kV/mm)	45–50
Tensile strength	120–140 kg/cm^2
Elongation approx.	300%–500%

2.2.3.7 Recent Development on XLPE Compound

2.2.3.7.1 RT-XLPE Compound

During extrusion, void formation within a polymer cannot be avoided but can be minimised in number and size by

1. Using a super clean compound
2. Controlling the melt pressure within the extruder having a slightly high compression ratio than normal polymers
3. Curing and cooling under pressure

Even then, a certain amount of void comes into existence due to absorption of moisture and formation of gas pockets during reaction with curing agents. Further, a polymer chain becomes softer under the application of heat while it takes time to reassociate within the chain, by which time minute gaps may form within the chain links. Within all these pockets, a certain amount of water molecules can get entrapped. The electrical degradation process of 'treeing' in polymers occurs in the presence of electrical stress and moisture. This is the basic cause of nonmechanical degradation and cable failure. Water treeing processes are considered to occur in the following phases:

1. Initiation due to uneven stress at interface and contamination
2. Moisture initiating growth even in the ppm level under stress
3. Any surge leading to bridging of electrical trees

Various experiments have been conducted to analyse the nature of failure, and after a long statistical evaluation, considering the breakdown voltage and their nature, it was concluded that the life of XLPE-insulated MV and HV cable is to be fixed within 25–30 years in service.

In order to contain a water tree formation, insulation development is being carried out to inhibit water molecules within the insulation during the initial stages of processing. A water tree–retardant compound has been developed to be used to effectively prevent the formation of water tree for a long time (called TR-XLPE compound). The effectiveness of this compound has been evaluated through the Accelerated Water Treeing Test as per ANSI/ICEA S-649-2004. In this case, the cable sample was kept immersed in water for a 1 year under elevated temperature applying a rated voltage, after which the condition of the sample was examined, and the results obtained were satisfactory. It is now considered on testing various samples produced with TR-XLPE that the cables made with the TR-XLPE compound can have a life span >40 years, reducing failure rate and maintenance cost while providing an uninterrupted performance.

2.2.3.7.2 *Semiconducting Compounds*

The wavy surface of a stranded conductor increases electrical stress levels at the crest point. The stress values increase with the increase in operating voltage. Nowadays, compacted conductors with a smoother surface are used. In spite of this, some imperfections at the edges of a flattened wire can create problems. In Europe and Japan, profiled wires were used on the outer surface of the conductor to make it smooth and withstand very high voltages. In addition, to obviate any chance of failure, a layer of semiconducting tape was used to manufacture extrahigh-voltage oil- or gas-filled cables.

In the case of stranded and compacted conductor, a semiconducting layer as a screen is used from 6600 V and above to eliminate chances of failure due to any imperfection on the conductor surface. On XLPE-insulated cables, a semiconducting screen is applied by the extrusion process on the surface of a conductor giving a plain contour devoid of any imperfection. An outer semiconducting layer on insulation helps in distributing the electrical stress perpendicular to the axis of the cable core.

As stated earlier, any imperfection in the extruded semiconducting layer such as protrusion or pip on the surface, or an irregular, uneven contact between the insulation and the semiconducting layer, will develop high stress at the point, initiating failure over time.

Naturally, the semiconducting compound must fulfil very high quality standards. To produce a semiconducting compound, high-quality carbon black having smooth and extreme fine particles is required. Two types of carbon black are used to make a semiconducting compound. In one of the processes, carbon black is produced by burning mineral oil in a furnace called 'furnace black'. Another method is by a controlled pyrolysis of acetylene to produce 'acetylene black'. Normally, furnace black contains ionic molecules which may initiate the electrochemical tree formation. Acetylene black is more contamination free. Nowadays, high-quality furnace black having extremely low contamination level is also used to produce a semiconducting compound. To manufacture a semiconducting compound, the base polymer chosen is different from XLPE. A sufficient amount of carbon black is blended with the polymer to make it conductive and has a very smooth surface. Processing of the semiconducting compound is very tough. Compounding machinery is designed to maintain a homogeneous mix. Before cross-linking, additives are added. A very fine filtration is used to assure the smoothness of the final product. By extruding a sample in the form of a very thin film or tape, it is examined optically to determine the nature of protrusions (called pips) present within the material. As the presence of protrusion is detected, its dimensions and its spread within a given determined area are noted to be able to establish the total spread statistically. Before arriving at a conclusion, a large number of samples are examined. Obviously, it becomes essential to check all the ingredients for quality standard rigorously, especially carbon black. The interface between the insulation and the screen of an extruded sample is checked optically (immersing a cut sample in hot silicon oil under focused light) to detect any defects like protrusion, wavy formation, contamination, air trappings or any uneven contour and to ascertain the quality of the product. Thus, the semiconducting compound, as desired, should have a clean and smooth surface to provide an optimum interface.

2.2.3.8 Polyvinyl Chloride

PVC was the first item recognised as an insulating material for manufacturing low-voltage cables. Siemens first introduced PVC-insulated cables under the brand name Protodur. The basic structure of PVC is

$$\begin{array}{cccc} -CH_2- & CH- & CH_2- & CH- \\ & \downarrow & & \downarrow \\ & Cl & & Cl \end{array}$$

PVC is produced from a vinyl chloride monomer through polymerisation. In the initial stage, vinyl chloride is produced by reacting acetylene and anhydrous hydrogen chloride in the presence of mercuric chloride. Acetylene was produced by reacting calcium carbide with water. Calcium carbide was obtained utilising limestone. In the modern process, acetylene is obtained from ethylene, a cheap feedstock available from petroleum refineries. In the basic process, ethylene is chlorinated to produce 1,2-dichloroethene. The second step is a pyrolysis reaction which yields vinyl chloride and HCl:

$$CH{\equiv}CH + CH_2{=}CH_2 + Cl_2 \rightarrow 2CH_2{=}CHCl$$

Polymerisation of vinyl chloride is done to yield polyvinyl chloride. This can be done using four methods: (1) suspension, (2) bulk, (3) emulsion and (4) solution. The molecular weight of PVC is controlled during polymerisation by the reaction temperature. Different grades of polymers are specified by its '*K*' value. The higher the value of '*K*', the more difficult will be the processing of material. A resin with a higher '*K*' value is tough and brittle and has poor heat stability and cannot be used as such.

The resin has to be compounded with various ingredients for a particular end use.

2.2.3.8.1 Ingredients Used for Compounding PVC

2.2.3.8.1.1 Stabiliser The resin is sensitive to light and temperature and gets degraded on exposure to light and heat. To stabilise the product, lead stearate or barium and cadmium stearate are used. Barium and cadmium stearate prevent degradation against ultraviolet radiation and sunlight. There are organic stabilisers and antioxidants available, such as epoxidised resins, which are very effective as antioxidants. Presently, as per restricted practice of hazardous substances (RoHS) condition, the use of lead, barium and cadmium compounds is restricted. As a result, different types of organic antioxidants have been developed, which are free of inorganic elements.

2.2.3.8.1.2 Plasticiser The resin as such has to be blended with a plasticiser to make a homogeneous and flexible compound, imparting strength and flexibility. To reduce cost, secondary plasticisers are used in conjunction with primary ones such as chlorinated paraffin wax. The most popular plasticiser is dioctyl phthalate (DOP). For different applications, different plasticisers are used, sometimes in combination. Some of them are dibutyl phthalate, dioctyl adipate, tricresyl phosphate, tri-octyl phosphate, dioctyl sebacate, dioctyl succinate and so on. Presently, plasticisers are available in the market in a few hundreds. In this case too, the use of plasticisers having lead content is restricted. Plasticisers are to be selected considering the end use of the product and the quality requirement. Some of the specific requirements are as follows:

1. Fire-resisting properties
2. Low smoke and halogen-free quality
3. Heat-resisting properties
4. Cold temperature withstanding capabilities
5. Low dielectric loss compound
6. High flexing properties

Here, one or the other property, singularly or in combination with others, is to be incorporated, wherein, for the outer covering, some of the electrical characteristics may have to be sacrificed against tensile strength and elongation.

Presently, the demand for fire-resistant low smoke (FRLS) compounds has increased considerably. Some of these properties can be achieved through blending with antimony trioxide or aluminium trihydrate. Apart from this, a zero-halogen, low-smoke compound demand is preferred. Here, the compound is blended with EVA or other organic materials to obtain the desired properties. Nevertheless, different plasticisers, which are also being developed, are in use.

2.3 Fillers

Fillers impart better strength and make compounds competitive in the market. Fillers should be used selectively considering the type of compound to be produced. The fillers and their importance are listed as follows:

1. *French chalk*: Gives higher flexibility and better electrical properties.
2. *China clay*: Gives higher tensile and impact strength but increases the density of a compound. This type of compound is generally used for inner and outer sheath.
3. *Calcined clay*: Improves tensile strength, electrical properties and abrasion resistance.

4. *Calcium carbonate uncoated and coated*: This is a popular filler and gives good electrical properties, moderately increases density and allows easy flow during extrusion.

2.3.1 Carbon Black

A versatile and essential ingredient, it gives multifunctional qualities to the polymer. It increases tensile strength. Carbon-black-mixed compounds resist degradation due to ageing action of UV rays and sunlight.

Its reinforcement properties are excellent depending on the particle sizes. A lot of work is being done to study various phenomena and action of carbon-black-blended thermoplastics and elastomers. Those who are interested may study the technical literatures and publications on carbon black written by different authors.

In some cases, special fillers, like powdered mica and silica, are also used to increase fire- and heat-resisting characteristics. Aluminium trihydrate (Al_2O_3) is incorporated to reduce the halogen content during burning of the compound. Antimony trihydrate Sb_2O_3 is used to make a fire survival cable. But these compounds must be of extreme fine mesh; otherwise, the surface finish is impaired deteriorating tensile and elongation characteristics.

2.3.2 Lubricants

For smooth flow during extrusion and a glossy finish, paraffin wax and/or stearic acid or any metallic stearate is used. Too much of wax, however, leads to a blooming effect. Microcrystalline wax is used for better electrical properties, smooth finish and heat-withstanding and ozone-resisting capabilities.

2.3.3 Colourant

Different organic or inorganic colours are used for the identification of cores. Some of the colour do affect the insulation resistance values of the compound. Blended colour, if kept exposed to sunlight, gradually fades out with time. Hence normally, for outer sheathing, PVC should be black. Any other colour is not recommended. Some users prefer coloured outer sheaths but that will be at their own risk.

2.4 Other Chemicals

Different varieties of organic peptizers, UV-resistant chemicals, cold temperature withstanding ingredients, heat- and fire-resistant ingredients, flameproof materials and acid and low smoke additives are developed and used in compliance with RoHS regulations.

2.4.1 PVC Compounding

PVC compounding is carried out in a clean atmosphere. It is recommended that the resin be mixed with an antioxidant, lubricant, stabiliser and plasticiser and retained as stock for some time in order to allow the absorption of the plasticiser to be completed. Premixed powder is loaded with all other ingredients and mixed in a turbine mixture at about 70°C for about 20 min. The mixed resin with ingredients is then extruded through a twine screw extruder as filaments. During extrusion, the material is passed through the wire mesh at the exit side of the die head to screen all unwanted impurities and grits. Smooth filaments coming out of the extruder are then palletised, cooled and packed for shipping. The insulating compound must have a density of 1.35–1.38 and a dielectric constant of around 4–6 (max) and give a value of tan $\delta = 0.08$. For inner and outer sheathing, the density can be raised to 1.40 max by using suitable quantities of filler.

The extrusion of the PVC compound is considered on a kilogram per hour basis. To get more kilogram per hour, a highly filled compound will show the best result. But in terms of surface coverage, that is, kilogram per kilometre in cable covering, having a particular diameter will register relatively higher consumption.

If the diameter of a cable is 30 mm and is sheathed with a compound having a density of 1.40, then the quantum of compound required to cover 1 km of cable, for a density of 1.40 and a thickness of 1.80 mm, will be $W = \pi\,(30 + 1.8) \times 1.8 \times 1.40 = 251.75$ kg/km, and for a density of 1.45, the weight will be 260.74 kg/km, that is 9 kg more will be consumed per kilometre basis (approx. 3.58%). There will be some cost saving due to the inclusion of more filler, but this will get offset by the following:

1. As the output of the extruder depends on the density, such extruder output shall be higher by weight and lesser by volume. Hence, the coverage of the surface area shall be lower, the result being a higher material consumption at a lower line speed.
2. A highly filled compound being more viscous will generate more friction. This might increase wear and tear of screw and barrel, reducing the life of the extruder.
3. During extrusion, frictional heat generation will be higher, which at times may become uncontrollable even when heaters are put off. This will degrade the compound affecting quality.
4. At the initial stage, moisture absorption will be higher. The compound would be needed to be dried before use, incurring extra cost.
5. The evolution of gas will be more. This will generate porosity within the compound during extrusion.
6. Controlling of thickness/diameter within the specified limit will be difficult.

Points (1) and (2) may lead to the generation of more scrap.

It is, therefore, not advisable to sacrifice quality against an apparent lower value achieved by incorporating cheaper type fillers in substantial quantity. A specific parameter for the density of a compound (formulation) is needed to be established to gain maximum advantage, retaining the desired quality and lowering the processing cost, cost of maintenance and other associated unwanted expenses. At lower subzero temperatures, the compound becomes brittle unless specifically blended for low-temperature conditions. PVC dissolves in cyclohexanone, pyridine and furfuran. It gets softened in methyl ethyl ketone (MEK) and methyl isobutyl ketone (MIBK). The compound is not generally attacked by acid, alkalis or the surrounding affluent.

2.4.2 Reprocessing of PVC Compound

Reprocessing PVC scrap is a bold art. PVC being a thermoplastic material can be reprocessed by heating and brought back to its original form by cooling. The scrap formed by purging material from the extruder before setting the conductor and centring diameter should be collected in a clean container and placed underneath the extruder head. These materials should remain free from dust and contamination. The material stripped from cable scrap can be made free from all metal particles and dust, by chipping them in smaller pieces and then sieving them on a vibrator sieve to remove metal parts. In case materials are smudged with dust, they should be washed thoroughly and then chopped to pieces. Every burnt particle should be removed from chips and lumps received from the extruder head. Bigger lumps should also be chopped to smaller pieces and sieved. During sieving, magnetic particles can be removed by running a magnet over the shredded material. No material other than those removed from the cable, cable core and the extruder head are to be mixed for the purpose. Washed and chopped off materials are then fed into an extruder. One percent of DOP can be mixed with the chips. The material is then allowed to come out as thin rods passing through a wire mesh placed within the breaker plate. These rods are fairly clean and useable as outer sheathing on cables. To make the rods absolutely clean and homogeneous, chopped materials are fed on a mixing mill (same as rubber). The mill is heated to about 120°C while feeding the chips. One percent of DOP is mixed during milling. When a masticated homogeneous sheet is obtained, it is taken out and cooled. These sheets are chopped in a granulating machine and packed. The material can be directly fed into the extruder for cable sheathing. They can also be palletised if needed. Such material resemble virgin material. Out of these material, calendered sheets can be made for wrapping cables. It is important to remembered that no material other than the cable material is used for reprocessing. Unfortunately, nowadays, all types of collected materials, even burnt material from garbage, are mixed together and blended to form granules to be used as cable covering. Naturally, surface finish and other qualities are found to be lacking.

2.4.3 Special Polymer Compounds

Low smoke and fire-resisting properties are considered to ensure that in case of a fire breakout, personnel are not affected. Further damage is contained within a specified time limit. Nowadays, low-halogen or zero-halogen compounds are in great demand. PVC having chlorine content within itself evaluates HCL acid during burning, and though it helps in extinguishing fire, the gas is hazardous to health. Naturally, the test is to be conducted for the amount of acid gas generation in a compound. Compounds containing chlorine, bromine, etc. are now being considered toxic material and are used in a limited way when ventilation is not a problem and human habitation is very limited.

Though XLPE contains no halogen, it is not flame resistant. To make it fire resistant, polymers like EVA, silicon rubber or similar additives are blended. These additives are devoid of any toxic substance.

Nowadays, all power houses, substations and important installations need to be secured from fire hazards. As such, demands have increased for fire-resistant cables to be installed within tunnels of metro railways, metro stations, substations, airports, defence establishments (defence installations, warships, etc.), oil refineries (oil exploration sites, oil wells) and other important areas, where there should be no damage to human life when fire breaks out due to cable faults or short-circuit conditions. The emission of smoke and acid fumes should be kept under specified conditions to prevent health hazards and to keep the surroundings pollution free.

Naturally, insulating compounds are used to comply with such tests so as to ensure that smoke and acid fumes are not generated under such conditions and the compound extinguishes fire automatically within a given period of time by suppressing the flames.

In certain cases, after suppressing fire, the cable must remain in *live* condition for about 3–4 h, during which all the electrical systems should remain functional until alternative arrangements are made.

The choice of polymeric compounds and design parameters are being developed to comply with these regulations.

With the development of new technology, safety norms, choice of material, and design parameters, the testing of raw materials is being modified and changed from time to time. Here, effort is to be exerted to formulate or procure quality materials from reliable sources and to be tested rigorously at the raw material stage. It may be necessary to modify processing parameters considering material quality and the capacity of the machine.

Various blends of the PVC compound with fire, smoke and acid suppressing additives are available or are being formulated. PE blended with EVA or other thermosetting polymers is utilized to impart fire-resisting and low smoke properties. Silicon rubber is also being blended in different proportions to impart high-temperature-withstanding properties and fire-resisting qualities, where acid generation can become almost zero producing a zero-halogen compound. Today, many types of thermoplastic elastomers (TPEs) are developed, which can be extruded in a high-compression extruder where the cross-linking process is completely avoided.

2.4.4 Metallic Screen or Sheath

The insulation screen of a semiconducting material, though it establishes an equipotential surface, is not sufficient to divert induced voltage and currents to earth efficiently and does not have a low resistance path. Naturally, a conductive metal tape, such as a screening material, should be wrapped around the cable core touching the semiconductive layer. Copper or aluminium or lead tapes could be used as a screening material. In addition, this metal screen shall form a mechanical protective layer and a moisture barrier specifically for HV and EHV cables. A system of cables are up to 33 kV constructed as three cores laid up together. In this case, individual cores are screened by applying EC grade copper tape or wire, or a combination of both. Copper tapes are applied with a minimum of 10% overlap.

2.4.4.1 Lead

From early days, lead was used to cover paper-insulated low- and high-voltage cables. Lead, being soft and malleable, can be easily formed and bent. It has a melting point of 327.5°C. Taking advantage of its low melting point and softness, the material could be extruded as a seamless pipe over the cable to protect it from moisture and the surrounding affluent. The metal does not get corroded easily though its mechanical strength is poor. It could also act as a metallic screen to divert short-circuit current as required. The metal is susceptible to fatigueness and can get fractured due to recrystallisation, as it experiences constant vibration when laid along the side of a road or over a bridge, or when transported by a ship. Further, during installation of cables, repeated handling, bending and unbending, as well as thermal expansion and contraction, may cause fatigueness leading to failure of sheath. In order to prevent such incidences, lead is alloyed with a 0.4% tin and 0.2% antimony, known as lead alloy E. When severe vibration alongside a railway bridge or a railway track is experienced, or when the cable is manufactured as submarine cable and stress and strain are severe, lead alloy B is to be used. Lead alloy B contains 0.6%–0.8% antimony. As per Vereinigte Deutsche Electrotechnik (VDE) specification, Cu (0.03%–0.05%) and tellurium (0.04%) are mixed. In the United States, different types of alloys with a combination of lead/arsenic/tellurium/bismuth are used, which give good results. With the introduction of screw press, lead sheath can be extruded over longer cable lengths continuously. With the introduction of the RoHS EU directive, lead has been declared as a carcinogenic material, restricting its use. In fact, with the coming of XLPE insulation, lead sheathing for low-voltage and MV cables up to 33 kV is no longer protected by lead sheath. From 66 kV and above, its use as a sheathing material has not been withdrawn; same is the case for oil-filled cables.

With the introduction of the corrugated aluminium sheath, attention has now been shifted to aluminium-sheathed cable due to its better mechanical strength and electrical protection, particularly in the manufacturing of HV and EHV cables.

2.4.4.2 Aluminium

Aluminium, having better mechanical strength and low resistance, became attractive as a protective sheathing material for electric cables of all types and nature. The use of aluminium as the sheathing material found due attention during Second World War in Germany. Initially, the metal selected was 99.99% pure. Subsequently with the development of better extrusion press, a 99.5% pure metal could be used. A the melting point of metal was 600°C, initially it became difficult to press the metal in an extrusion press. With the introduction of induction preheating, metal billet could be heated at a temperature above 500°C. The billet was then fed into the press for extrusion under temperature and high pressure. Initially, the metal tube could not be extruded continually as the press was to be stopped for recharging. With the development of a double stroke feeding and recharging system, a continuous seamless tube could be extruded. The seamless plain tube can get flattened while bending and hence has to be corrugated for better handling.

Since the cost of the press is too high, it has been found that a seam-welded corrugated sheath can be produced without difficulty. In this case, welding should be done under an inert gas like oxygen-free argon or nitrogen as aluminium reacts with oxygen under heat and moist air.

The DC resistance of the sheath being low, it acts as a shielding material for the communication cable. For the EHV cable, it carries a total short-circuit current where temperature dissipation is quick. As the cable is mechanically strong, added mechanical protection in the form of an armour is not required. But as it is susceptible to corrosion, it must be protected by an anticorrosive layer such as a coat of bitumen layer followed by a thermoplastic covering.

2.4.4.3 Armour Materials

In order to impart mechanical strength without losing the flexibility of cables, steel strips, steel wires or steel tape is used as armour material to protect cables from damages, which may be caused by a sudden impact due to falling debris or stones and other hard materials. The armour is also utilised for pulling cables through tunnels or ducts during laying and installation. Armour metal also acts as earth connection for diverting short-circuit current. Hence, the resistivity of the armour material should be within a specified limit as per international or national specifications. Armouring must be protected from corrosion by applying a coating of zinc. Zinc coating is done by a hot dip galvanising process. The dimension of strip or wire or steel tape is selected from the given specification (IS 3975) or as per EC standard.

The wire and strip shall be galvanised. Heavily galvanised wires are used for subsea cable and other special types of cables where environmental condition is severe. These wires have the following features:

- Tensile strength shall be 30–45 kg/mm^2.
- Elongation shall be 4%.
- For mining cable seven-strand pliable armour is used for flexibility.

Nowadays, aluminium-coated armour wires are offered with better conductivity and corrosion resistance.

In the case of single-core LV, MV and HV cables, steel wires, strips or tapes, being magnetic material, absorb a certain amount of power to produce an appreciable amount of induced voltage and current within the armour. Hence, such those cables are protected by applying hard grade aluminium wire armour that is nonmagnetic in nature.

Armour materials, as strip or tape or wire, should be smooth without bars and kinks. The sides of the tape and strips should be smooth and free of bars and scales. Zinc coating must be smooth and uniform in thickness. A combination of aluminium wire/strip screen and steel tape armour is used to reduce the inductive voltage of overhead railway traction lines within signalling cables installed alongside railway tracks.

3

Wire Drawing: Copper and Aluminium

Transmission lines, whether for power or communication, require a conducting medium. A metal, or in general any metal, can be a conducting path. A good conducting path allows electrons to vibrate or oscillate more freely, pushing their neighbouring electron cluster forwards and backwards in a linear path. In case of a poor conductor, such movements are restricted because electrons are bonded within the lattice of a molecular structure.

To transmit power from one point to the other, longer lengths of conducting wires would be required. In earlier days, copper used to be the most preferred metal because its conductivity is very close to 100%. But with the passage of time and for economic reasons, aluminium gained momentum though it has a conductivity of 61%. Thus, copper or aluminium has to be drawn to smaller-diameter wires from 9.5 to 8 mm (copper) or 6 mm wire rods. The drawn wires must be sufficiently long and flexible to be able to be wound on a bobbin. During laying and installation, the wires are required to be bent and passed through constricted areas. Naturally, the thinner the wire, the easier is the installation procedure.

Electrolytic grade copper is obtained in large ingot forms from primary producers of ingots. The metal is then rolled in a rolling mill under red hot conditions to form wire rods. During this process, a black oxide coating forms on the surface of the hot copper rod. Before drawing to a thinner wire, this oxide film has to be removed by a pickling (H_2SO_4) and washing process. Naturally, a small amount of metal would get lost in the form of copper sulphate in the pickling bath. As specified, a cleaned dull red–coloured rod from the pickling bath has to be drawn as the wire. These coils are limited in length and need to be welded frequently, stopping the wire drawing machine. At times, the weld gets snapped due to slug deposition or the presence of oxide flakes at the end of the rod, which created production problems. In case of aluminium, hot ingots are rolled to form wire rods. Around the surface of these wire rods, a microfilm of oxide coating forms instantaneously. During wire drawing, this film gets ruptured, but a new oxide film forms immediately on the surface of the drawn wire. This is an inherent characteristic of the metal.

As the demand for electric cables and conductors increased, the requirement of drawn wires also went up. This required increasing production hours per wire drawing unit. This can substantially be achieved when the frequent stoppage time allowed for the welding of smaller wire rods can be recovered by replacing it with a much longer coil length. This enforced the idea of having a bright copper rod of longer length as well as that of aluminium. Copper coils, however, must be bright and free of oxide coating. The limitation of existing rolling mills and pickling tank could not meet with such requirements. In 1947, Ilario Properzi came up with the solution by inventing continuous casting and rolling (CCR) mill for the production of copper and aluminium rods without any interruption. The main consideration was to get the bright copper rod in continuous lengths without any cuprous oxide on the surface. In this process, the molten metal coming out of the furnace is transferred using a launder path fully covered to prevent heat loss and then poured into a grooved casting wheel on which a flat endless steel belt is kept rotating along with the wheel, covering 180° of the bottom part of the groove. The casting wheel is closed from both sides with steel sheets. The inner side of the wheel is cooled by spraying water at high pressure. The rotating steel belt is cooled by spraying water from outside. Hot cast bars emerging continuously from the top of the wheel are driven by rolls into the pre-pickling chamber.

During its passage through the air between the outlet of the casting wheel and the pre-pickling chamber, the bar, at about 950°C, is covered with a thick layer of oxide mainly comprising copper oxide (Cu_2O). This bar which is covered with oxide enters the pre-pickling chamber where the nozzles on pipes project liquid at a pressure of 2–3 bars on to the surface of the oxidised bar. This spraying causes breakage of the thick layer of oxide due to thermal shock. The bar remains for about a few seconds in the pre-pickling chamber. During this time, it is in contact with the liquid projected by the nozzles, and the vapour is produced when the liquid contacts the hot copper bar. The copper oxide is thus partially reduced and detached from the bar as the vapour penetrates the broken layer of oxide.

The bar then enters the descaling chamber in which jets of liquid under a high pressure of about 210 bars are projected against its surface. The flakes of oxide are removed from the bar and carried along with the liquid to the lower part of the chamber. Cleaned of its surface oxide, the bar then enters the rolling mill portion and passes into the first stand. During this short duration, descaling spraying under high pressure leads to a relatively slight cooling of the bar with the result that the bar which enters the rolling mill is still at a temperature enabling hot forming. During the passage of the bar into the mill and its transformation into a rod, the metal is continuously isolated from the outside atmosphere by the liquid sprayed onto the rolls from the distributors. The rod thus isolated from the oxidising atmosphere may be rolled without reoxidation, and the rolling pressures together with spraying with a cooling and lubricating liquid enable the elimination of any oxide which may still remain on the bar or the rod during the course of rolling. At the outlet of the mill, a bright rod free of the oxide layer is obtained, but at a temperature of up to 600°C, at which considerable reoxidation of copper is possible, the reoxidation being particularly rapid above 150°C. At the outlet of the rolling mill, the rod enters the cooling duct, bringing the temperature of the rod down to a value low enough for reoxidation to be possible. In practice, the outlet temperature of the rod should not be higher than 80°C.

Immediately after the introduction of the Properzi mill, improvement was made by adopting an up-cast process of manufacturing copper rods by the 'upward continuous casting' method, better known as the UPCAST system. The method was originally innovated and developed by Outokumpu in the late 1960s for the production of copper and copper alloy wire rods. At the UPCAST (Outokumpu) plant, high-purity copper cathodes are melted in a protective atmosphere in induction furnaces. The molten metal, transferred into holding and casting furnaces via atmosphere-controlled launders, is cast into wire rods with standard diameters of 8 mm (5/16″), 12.5 mm (1/2″) and 20 mm (3/4″) by the 'up-cast' continuous casting method. Silver-alloyed copper wire rods are also produced at this plant. In this case, copper wire rods can be produced in continuous lengths for days together. The deoxidation process is similar to the one discussed earlier. Here, the metal is drawn vertically from the pool of molten metal in viscous condition. The drawn soft rod is passed through a die enclosed in a tubular chamber, kept completely isolated. In this process, a number of rods can be drawn at a time. The entire plant is automatically programmable logic controller (PLC) controlled. In both cases, electrolytic cathodes are used where the purity of copper remains above 99.8% approximately.

Today, oxygen-free quality rods in bigger coils of 3000–5000 kg are available in the market (Figures 3.1 and 3.2).

FIGURE 3.1
Up-cast unit for manufacturing copper wire rod.

FIGURE 3.2
Coil of electrolytic bright copper wire rod.

In the case of aluminium too, the conventional rolling process was replaced by Properzi rolling mill, where wire rods could be produced continuously. The process is similar to that of copper, but in this case pickling and deoxidation systems are not required. Aluminium as such acquires a thin film of oxide immediately after it comes out of the casting wheel. This film protects the bar from further oxidation. During rolling, the oxide film breaks down into finer flakes, but as soon as the rod comes out of the final die, it automatically gets protected by a micro film of oxide which, however, is cannot become a macro film by the rapid cooling process.

Most of the large overhead conductor and power cable manufacturing firms have an in-house Properzi plant. Ingot is to be fed into a reverbatory-type melting furnace. However, the current practice is to use a vertical furnace. The normal reverbatory furnace consumes about 80–90 kg heavy oil/ton aluminium rod, and in the vertical energy-saving furnace, the energy consumption is about 55–65 kg heavy oil/ton aluminium rod.

The molten metal is tapped and taken out to be stored in a holding furnace. In the holding furnace, it is mixed with a deoxidising agent, making all the oxidised particles and impurities appear on the surface of the metal. A layer of charcoal is allowed to cover the surface to minimise the oxidising process. There would be two holding furnaces to be operated alternately in order to cast the rod continuously. The molten metal is poured on to a grooved casting wheel cooled sufficiently with water to solidify the metal into a cast bar. The effect of cooling is more important and actually controls the tensile strength of the wire at the final stage. Conventionally, the holding furnace was eliminated and molten metal was poured using a ladle directly from the melting furnace.

As cooling by spraying water is very critical therefore, during installation it is to be ensured that the correct individual spray-adjusting device

FIGURE 3.3
An overview of a Properzi mill showing the tip of the casting wheel in a pit below.

(on a typical 5 MT output mill) shall be such that during starting the flow of water can be adjusted as per requirement in such a way that at least 22 sprays out of 52 on the inner side of the wheel can be controlled and on outer periphery 20 sprays out of 48 and from the side of lateral sprays at least 5 and 15 out of total 32 can be in operation under controlled condition and is a must. The remainder nozzles can be fully opened and can be of non-adjustable type. For aluminium alloy production, it is essential that the condensation (steam) produced by the casting process does not enter the liquid alloy. It must not be close to the spot where the alloy enters the mould cavity.

The cast bar coming out of the wheel is led into the rolling mill to shape it into a wire rod continuously. The cast bar is shaped by three roll-forming dies, 13 sets (standard mill). At the final stage, an almost round form of wire rod is formed. This wire rod is wound on a take-up coiler or is taken in a basket-type take-up system. This rolled material will be free of any blowhole, scale, cut and dents to avoid any interruption during drawing of the wire (Figures 3.3 through 3.5).

The typical process flow of the casting process for aluminium is given in the following:

Feeding, melting and refining → (holding furnace) → filtration system → laundering → casting ladle → continuous casting (casting blank) → forward pulling → casting blank treatment unit (automatic shearer, ingot conveyer, straightening machine) → aluminium (aluminium alloy) bar induction heating → entering rolling mill (automatic feeding) → continuous rolling

↗ oil refilling bag and leading duct ↘

↘ aluminium alloy rod quenching device↗

→ backward hauling → rod orbital coiler → finished product

FIGURE 3.4
Three roll-forming dies within the rolling mill.

In early days, conductors were designed to suit specific requirements, which led to numerous sizes of wires and conductors. This created confusion during production where many sizes of wires were to be drawn and within short periods. Scrap percentage increased and the utilisation of production capacities became restricted, making the industrial scenario unviable. To avoid such chaotic situations, rationalisation was carried out by standardising conductor sizes and voltage systems throughout the world. Worldwide, various manufacturers of underground cables standardised the sizes for different cross sections in order to reduce inventories and lessen downtime.

FIGURE 3.5
Basket coiler unit for copper and aluminium wire.

3.1 Wire Drawing Process

A conductor can be of a single wire or can be formed by twisting a number of small wires to make it flexible. A smaller-diameter wire is processed from a primary wire rod with a diameter ranging from 6 to 9.5 mm, either of copper or aluminium. This technique is known as wire drawing. The machine employed is called rod breakdown machine (popularly known as RBD). The drawing process consists of pulling a larger-diameter wire through a series of drawing dies made of hard element such as tungsten carbide or industrial diamond. The design of a die is intricate and very important for smooth drawing and finish. A definite entry angle with proper landing length is to be calculated taking into account the characteristics of metal in order to obtain good production (Figure 3.6).

FIGURE 3.6
Coil of E.C. grade aluminium wire rod.

To start with, the end of the wire rod is tapered down by successively inserting it through a series of oscillating roller grooves with progressively decreasing dimensions. This equipment is called a pointer and stringer. The pointed end of the wire is threaded through a series of dies placed in their respective holding blocks. Simultaneously, the wire is wound on drawing capstans placed between die holding blocks. The case or die holding box should be of adjustable type (Figure 3.7).

FIGURE 3.7
Pointer and stringer.

To understand the actual technological significance of the wire drawing process, it is necessary to pay critical attention to the following aspects:

1. Role of the die
2. Constructional features of pulling capstan, rings or tyre
3. Function of the lubricant

The reduction in the diameter is achieved by pulling the wire through the orifice of several dies in succession. It may be questioned why the diameter cannot be reduced by passing through a single die. Why do we need several passes for drawing? How far can we reduce the number of passes, that is the number of dies to achieve the required diameter? On which parameters do the economy of production depend? These are questions to be considered, particularly the last one, in every phase of the production.

To get correct answers, the characteristics of the metal that plays an important role need to be clearly understood:

1. The specific density at the initial temperature during drawing and when temperature differs
2. Ductility
3. Tensile strength (modulus of elasticity, Poisson's ratio, etc.)
4. Elongation
5. Electrical resistivity upon which cable design and manufacturing parameters are fully dependent

Each of these factors, with their limiting values, has to be considered while drawing wires. The drawn wire should be uniform in diameter with a smooth surface. The finish of a wire during drawing actually confirms that all the parameters are well within the limit and are not allowed to exceed their threshold. High reduction speeds will stretch the wire beyond its tensile limit and cause frequent fractures and an uneven wavy surface. This approach will damage the die. A low reduction speed gives a good finish and a uniform diameter, but the production cost will be high. Similarly, speed plays an important role in the quality of produce as well. A higher speed will generate a quick rise in temperature due to high frictional force. This will cut grooves within the die orifice and also on the pulling capstan's surface rings. Here, lubricants play an important role.

3.1.1 The Die

Reduction in diameter means an overall reduction of the wire diameter from the initial to the final stage. On this basis, the percentage of area reduction can be decisively and accurately worked out in relation to the die design.

The percentage of area reduction is calculated as follows.

Let 'D' be the diameter of the wire before drawing and 'd' be the final diameter coming out of the drawing die. The volume per unit length before drawing – considering the unit length as 1 cm – is

$$\left(\frac{\pi D^2}{4}\right) \times 1\text{ cm} = \left(\frac{\pi D^2}{4}\right) CC \tag{3.1}$$

At the exit point, the same volume will be given by

$$\left(\frac{\pi d^2}{4}\right) \times (1 + \Delta L) CC \tag{3.2}$$

where

ΔL is the increment in length resulting from the drawing
d is the final diameter at the exit point

Hence,

$$\left(\frac{\pi d^2}{4}\right) \times (1 + \Delta L) \quad \text{or} \quad 1 + \Delta l \left(\frac{D}{d}\right)^2 \quad \text{or} \quad \Delta L = \left(\frac{D}{d}\right)^2 - \cdots \tag{3.3}$$

Thus, the length increment is proportional to the square of the diameter ratio.

Considering Poisson's ratio for copper as $\sigma = 0.35$, approximately, and taking into consideration the tensile strength of the wire to be 23 kg/mm^2 and elongation to be 32% on an average for soft annealed wire, the deformation in diameter is given by

$$\sigma = \frac{\Delta D / D}{\Delta L_1 / L_1} \tag{3.4}$$

where

σ is Poisson's ratio
ΔL_1 is the elongation at the rate of 32%
L_1 is the unit length (= 1 cm)

Hence,

$$0.35 = \frac{\Delta D / D}{0.32} \quad \text{or} \quad 0.35 \times 0.32 = \frac{\Delta D}{D} \quad \text{or} \quad 0.112 = \frac{\Delta D}{D} \tag{3.5}$$

Taking $D = 1$ cm as the initial diameter. The final diameter after drawing is

$$d = (D - 0.112D) = 0.888D \tag{3.6}$$

If $D = 1$ cm, then the wire diameter $d = 0.888$ cm.

Therefore, the reduction in area is

$$\left(\frac{\pi\times 1^2}{4}\right)-\left(\pi\times\frac{0.888^2}{4}\right)=(0.7853981-0.619321)$$

$$=0.166066=16.6\%\ (\text{apx.}) \tag{3.7}$$

and the increment in length is

$$\Delta L=\left[\frac{0.7853981}{0.7853981\ -\ (0.7853981\times 0.166)}\right]-1=0.1990$$

Hence,

$$0.1990\times 100=19.9,\ \text{that is } 19.9\%. \tag{3.8}$$

The area reduction is taken to be 20.6% as the standard value. Applying design modification, an area reduction as large as 22%–23% for copper and 26% for aluminium could be achieved. Modifying standard calculated practice, machine manufacturers are able to ensure good-quality production and higher productivity.

Practically, the increment in length is found to be lesser because of the apparent loss suffered due to frictional forces. The percentage of elongation tends to reduce somewhat, as the working hardness of metal increases in each stage passing through the dies.

Die angle plays an important role in the mechanism of wire drawing. Let us consider Figure 3.8.

Let α be the angle at the entry point of the wire within the die. The length of contact at the entry point by the wire is given by the length *'a'* and the landing length is *'b'* when the wire diameter is reduced to a value *'d'* at the exit end. Let the wire diameter at the entry point be *'D'*. While the wire is passing through the conical part of the die, it touches the sides AC and A_0C_0.

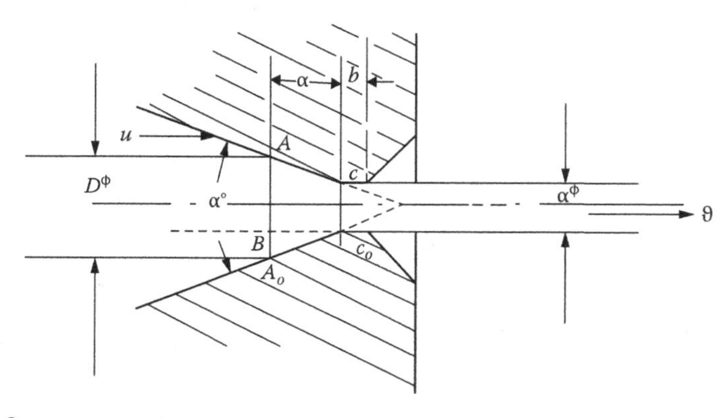

FIGURE 3.8
Schematic sectional diagram showing reduction of wire diameter through a wire drawing die.

With an area reduction of 16.6%, if the contact length is taken as $D/2$, then the angle would be $\alpha = 13°$ approximately. In actual practice, the angle is taken between 14° and 18° to achieve the best result. Here, the critical area will be the area covered by the exit point to the area actually reduced.

The area reduced ($= \pi D^2/4 - \pi d^2/4$) is the area at the exit point, that is $\pi d^2/4$. This means

$$D^2 - d^2 = d^2 \quad \text{or} \quad D^2 = 2d^2 \quad \text{or} \quad D = d(\sqrt{2}) = 1.414d \text{ apx.} \tag{3.9}$$

To avoid any breakage, when a frictionless die is used, D will be lesser than 1.414 days.

At a critical value, $AB = (D-d)/2$ or $AB = (D-0.7D)/2 = 0.15\,D$ approximately. Hence,

$$BC \tan\left(\frac{\alpha}{2}\right) = 0.15D;\ \text{since } BC = D,\ a = 17s° \text{ approximately.} \tag{3.10}$$

Therefore, when the contact length is equal to the diameter of the wire, the wire is likely to break, even if there is no friction. In practice, BC is taken almost ⅓D of the wire at the entry point. However, consider taking a little higher angle for drawing, say 18°. The area reduction and die angle can be more precisely calculated by the practical value by knowing the correct value of Poisson's ratio. In today's accepted system, considering all aspects, an area reduction of 20.6% has been selected as standard where an increment in length becomes approximately 25.94%.

In case of aluminium, Poisson's ratio is generally taken as $\sigma = 0.331$ = stress/strain when the area is unity. The stress per mm² varies from 10 to 15 kN (as given in specifications IS 8130 and international electrotechnical commission (IEC) 228).

Considering stress on an average as 12 kN, we get

$$\text{Strain (elongation)} = \frac{\text{Stress}}{0.331} = \frac{12}{0.331} = 36.25\% \tag{3.11}$$

Now

$$1.3625 \times \left(\frac{\pi d^2}{4}\right) = \frac{\pi D^2}{4}$$

where

D is the diameter before drawing
d is the diameter after drawing

Hence, the area reduction in percent is

$$= \left(\frac{D^2 - d^2}{1.3625}\right) \times 100 = 26.6\% \tag{3.12}$$

This heavy area reduction can be achieved in a bull block machine. In this machine, the drawing length can be stretched between 33% and 36% approximately. The angle of the drawing die is chosen between 18° and 20°. In practice, in order to achieve higher production of copper and aluminium wires and to rationalise the design of a wire drawing machine so that the same type of machine can be utilised for both metals, an area reduction of 20.6% is kept as the guaranteed value. However, the current trend in India is to achieve an area reduction of 22%–24% and to show a higher productivity without considering other aspects such as material quality and machine wear and tear. Even now the quality of indigenous machines cannot match the performance of Niehoff. They also do not consider the environmental condition of the country which determines if a high area reduction during processing is possible. What may be good and feasible for the European condition may not suit the Indian system.

The force of drawing also could be considered as follows. The mass within the annular space, between the diameters of inlet and outlet wires, will be drawn, pressed and reduced from diameter 'D' to 'd' by the application of force. If the starting diameter is unity, then the area is given by $\pi D^2/4$. When the area of the final diameter is taken lesser than 16.6%, then $\pi d^2/4 = 0.834 \times \pi D^2/4$ or $d^2 = 0.834D^2$ or $d = 0.913236D$, and the area of annular space is

$$S = \frac{\pi D^2}{4} - 0.834\frac{\pi D^2}{4} = \pi\frac{D^2}{4} \times 0.166 \quad \text{or} \quad S = 0.13D^2 \tag{3.13}$$

If we consider the diameter to be 1 cm, then $S = 0.13$ cm² = 13 mm².

Taking into account the tensile strength to be 23 kg/mm² on an average, the starting force of deformation will thus be $23 \times 0.35 = 8.75$ kg. For easy drawing, the force is increased by 1.5 times, that is 13.125 kg, which will be just near the limiting load, ½ of breaking load = 11.5 kg/mm².

Therefore, the pull required to deform the wire having 1 cm diameter for an area reduction of 16.6% will be $13 \times 11.5 = 149.5$ kg. This is to be multiplied by the length to be drawn per second to get the required force. In the actual case, the required force is higher than the calculated one. An additional force is applied to overcome the frictional resistance. From Figure 3.8, it is evident that when increasing the angle α the force should also be increased as area reduction increases and vice versa. But too low an angle α will increase the cost of drawing. As discussed, the preferred angle is 17°–18° for both copper and aluminium. It is important to choose a hard die material. The drawing angle and die orifice must be mirror polished and flawless. This can be verified with a profile projector. For large-sized wires die made of tungsten carbide is normally preferred. For finer wires, industrial diamond dies are used. Nowadays, diamond-powder-pressed dies are made with a particular orientation and are found to be very tough and durable. A good-quality die gives the best quality product and ensures longer life.

3.1.2 Constructional Features of a Drawing Ring

In the actual case, the length of the wire increases by 26% after every passage of the die for an area reduction of 20.6%. Based on this parameter, a table can be formed for a 9-die wire drawing machine, which includes the following:

1. Area reduction 20.6%
2. Length increase 26%
3. Starting diameter of wire rod of say 8.00 mm

It is interesting to note that the increment of length at the final stage can be found by the following formula.

If the number of drafts is 'n' and the area reduction is 'x'%, then the increment in length is given by $\Delta L = \{1/(1 - x/100)^n\}$ in metres. It is found that even with double lap, the diameter of the final pulling capstan becomes 1269.42 mm, when the speed of all the rings is kept constant. In this case, the dimension of the machine becomes too large. Threading and operation will be difficult and time consuming. To keep the machine within a reasonable dimension, the rings can be grouped in sets. In this case, the group consists of three rings to form a set. By increasing the speed of the second set (ring nos. 4, 5, 6) twice that of the first set (1, 2, 3) and four times that of the third set (ring nos. 7, 8, 9), the diameter of each of the capstans in the set can be kept the same. As shown in Table 3.5, capstan dimensions are 201, 253 and 318 mm per set and are same for all the three sets. Once this is fixed, other design parameters such as speed can be worked out. The previous calculation relates to the cone-type wire drawing machine similar to the Winget Syncro type as offered initially.

After threading the wire, the dies are placed within the respective holding blocks, normally positioned between the capstans. Every die holding block can be shifted forwards or backwards and sideways too. This facilitates adjustment of dies during operation.

1. When the centre position of the die changes due to the variation in the wire diameter
2. When the wire position on the drawing ring needs to be shifted sideways because of groove formation on the drawing rings

After the dies are positioned, the wire is lapped on the capstan with two or three turns. While lapping, should the length fall short between capstan blocks, it is compensated by drawing the wire by an inching process. Once stringing is completed, a coolant lubricant is allowed to be sprayed under pressure on dies and drawing rings to start the machine for regular production (Table 3.1).

It is seen that the diameter of the capstan rings does not take into account the calculated fractional quantum of the wire length, which may at times fall short or become longer than the capstan diameter, because the rings are made after considering practical manufacturing problems and to keep the system

TABLE 3.1
Calculated Diameter of Capstan Rings

Draft No.	Dia. of Wire at Exit End (mm) Apx.	Area of Wire (mm^2) Apx.	Length Increase at Each Stage (m) Apx.	Required Dia. on Capstan Ring (mm)	Actual Dia. Over Capstan Ring (mm)	Speed Ratio
Starting	8 mm	50.265	1.0000	Double lap	After rounding off	
1	7.1285	39.910	1.2595	200.45	201	1:1 Set-1
2	6.352	31.689	1.5863	252.46	253	1:1 Set-1
3	5.660	25.161	1.9980	318.00	318	1:1 Set-1
4	5.043	19.978	2.5165	400.52	201	2:1 Set-2
5	4.494	15.863	3.1695	504.44	253	2:1 Set-2
6	4.004	12.595	3.9920	635.34	318	2:1 Set-2
7	3.567	10.000	5.0279	800.21	201	4:1 Set-3
8	3.180	7.940	6.3326	1007.87	253	4:1 Set-3
9	2.833	6.305	7.9760	1269.42	318	4:1 Set-3

rationalised as far as dimensions are concerned. Naturally, this small amount of length variation needs to be accommodated while the machine is in operation; otherwise, after a certain period of time, this fractional amount of length variation will accumulate to a large quantum multiplying itself after every turn of the capstan ring. This phenomenon will create problems during production.

Normally, it was accepted that the bull block–type wire drawing machine is ideal for drawing aluminium wires. An area reduction up to 27% could be achieved in a bull block–type machine. Several blocks are placed in a row where individual blocks are driven separately by a motor. During drawing, the wire gets tightly wound on blocks (Figure 3.9).

FIGURE 3.9
Bull block–type wire drawing machine.

When passing wire from one block to another through a drawing die, unwinding is done by a flayer which guides the wire forwards. In this case, no downward slippage of wire takes place avoiding any entanglement. These machines operate in low speeds and are known as non-slip-type vertical bull block wire drawing machines. However, aluminium wire can also be drawn in a cone-type machine. In case the same machine is used for copper and aluminium, it would be necessary to clean the machine thoroughly by injecting a steam jet and a drawing lubricant. The lubricant for copper and aluminium is different and has to be kept in different tanks. However, it was found that in the cone-type machines aluminium cannot be drawn comfortably. Snapping of wires was experienced frequently. It is better to employ a tandem-type drawing machine for aluminium, which is universal, also allowing copper to be drawn comfortably. However, it is better to employ a separate machine for copper and aluminium. In a cone-type unit, the rings are grouped in three or four capstans, as stated earlier, with 9, 11 and 13 die compositions. The speed of each group is varied by the corresponding gear ratio combined with a different diameter of capstan in the group assembly so as to achieve the desired elongation (Figure 3.10).

In a tandem-type machine, capstans are arranged in a row, one after the other. All the capstans have an identical diameter, but the speed of each capstan is varied by combining an appropriate gear train ratio in such a manner that the speed of each capstan increases (1.26 times when an area reduction is 20.6%) progressively in order to match the increase

FIGURE 3.10
Cone-type wire drawing machine.

in wire length after each step of drawing. These machines are called slip-type tandem wire drawing machines.

As discussed, the diameter of capstans of each group in a cone-type machine is kept the same (Figure 3.11), whereas the diameter of all the capstans in a tandem-type machine is kept identical. In both drawing machines, there remains a fractional amount of variation between the length lapped on the capstan periphery and the length coming out of the die orifice. Apart from these, minute differences may occur in the die orifice. A small backlash in the gear train may cause further variation in the drawing length. This small amount of length variation is not allowed to multiply and create problems during wire drawing. In order to accommodate all these anomalies at each stage, a slipping effect is allowed to generate between the wire and the capstan giving a forward or backward movement of the wire over the capstan.

The lubricant facilitates this slip action, lubricating the system heavily, spraying forced oil and submerging the capstan rings partially in the oil bath below the drawing machine. During the rotational movement of the capstan, the lubricant is drawn from the bottom and allowed to form an oil cushion between the metal surface and the wire. The wire can thus slip easily on the periphery of the capstan surface without damage (rubbing effect). The lubricant not only helps the wire to slip easily but also takes away the heat generated during the drawing operation. It is always better to have a backward slip to retain proper tension on the wire. This is why these machines are called submerged slip–type wire drawing machine, cone or tandem type. The designer of a wire drawing machine must understand the full significance of the function of capstan and gear trains to allow minimum and smooth slippage. For this reason, the choice of metal, surface finish and meshing of gears is critical.

FIGURE 3.11
Tandem type wire rod breakdown machine.

3.1.3 Lubricating Oil

For both copper and aluminium, different grades of oil are available. Due to the elongations of different percentages for different metals and the down-time required to change the oil in case the same machine is employed, the current practice is to use different drawing machines for both metals to make them more efficient. During wire drawing, considerable heat is developed due to the following reasons:

1. A high frictional heat is generated within the die orifice as the wire passes through the same. At this point, the wire expands in volume as per the coefficient of volume expansion of the metal. This sudden expansion along with the frictional effect will tend to increase the drawing force while simultaneously causing a considerable rise in temperature at the die point. This excess quantum of heat must be taken away instantaneously to keep the system within the allowable temperature limit.
2. The slip occurring on the ring produces a rubbing effect on the ring surface and with the number of turns accelerates heat generation proportionately. In a dry state, this will produce a scratch mark on wires and the rings will also get damaged. This heat should be taken away by the flow of lubricant, allowing easy slippage of wires.

The lubricating oil passes with the wire into the die orifice during the drawing operation. It forms a very thin film around the wire surface. This film protects the wire from coming in contact with the inner surface of the die and allows the wire to pass smoothly. Here, the lubricating film experiences two types of pressure, one on the angular face at the reduction point and the other a longitudinal force parallel to the axis of the wire within the drawn length. This angular force can be resolved into two components, one which is perpendicular to the wire axis and the other which is parallel to the axis and gets combined with the already existing longitudinal force. The perpendicular component of the force does not act on the oil. As the speed increases, the longitudinal force increases manifold. This acts as a shearing stress trying to rupture the film of oil. Further, the heat developed within the die increases with speed. Naturally, the function of the lubricant is to retain the continuity of the film even under a very high stress and to take away the generated heat quickly from the interface of the wire and die orifice.

Second, the lubricant keeps the drawing rings wet all around and try to protect the wire and ring from the mutual rubbing action under pressure. If the lubricant is applied only from the top on the wire and rings, it will not reach the bottom of the wire gripping the ring. Wire at high pressure does not allow the oil to move between the interface of the metals. While two metals are allowed to move at different speeds with a slip on the wire producing the rubbing effect, the absence of oil between them will cause localized intense heat due to friction. Further, the ring and the wire will start to wear at contact point producing a rough surface and groove which will aggravate the situation. The surface of the wire will get damaged, producing fine metal cuttings. The surface of the ring will also give out fine particles, contaminating the oil. To obviate these adverse effects, the ring is kept submerged in an oil pool at the bottom part of the machine where a trough is kept fabricated to hold the lubricating oil. With the rotational movement of the capstans, the oil is carried upwards by the ring and allowed to be retained between the ring and the wire, preventing a relative rubbing effect of both the metal surfaces. At the same time, the rings are cooled by the flow of oil. Even then some fine dust particles are produced, which mix with oil and need to be filtered. The heated oil flowing out of the die case and ring is passed through a heat exchanger to bring down the temperature of oil before being recirculated.

It is necessary that the temperature of oil should be maintained at a constant level to retain the same viscosity of oil at every moment. If the viscosity of the oil is too high, it will not flow properly into the die orifice. The heat generated will increase gradually, keeping the die surface dry. The flow of oil will not be able to convey heat adequately, damaging the die and wire surface badly. A thin lubricant will dry out quickly under pressure between the die and wire surface and on the interfaces of the wire and the drawing rings. There the rubbing effect will damage both parts, though the heat generated would be eliminated quickly by the oil. Hence, it is essential that the temperature of the inlet oil and the temperature rise within the machine be kept under control to maintain constant oil viscosity. The measure of viscosity also determines the speed of the machine. Thus, every high-speed machine, apart from its constructional features, will be equipped to keep the viscosity of the oil at a fairly constant value.

The other benefit obtained from accumulating oil at the bottom part is that the metal dust carried by the oil during drawing gets settled at the bottom part as slug, which is filtered and taken out from time to time. Metal dust is formed due to the rubbing effect of wire in the die and on the capstan. If the die is of high quality and the ring is mirror polished, there will be decreased formation of dust. In the case the aluminium dust formation is very pronounced after emerging out of each die, an instantaneous micro film of aluminium oxide will be formed on the surface of the wire which gets broken down at each stage with every step of drawing, forming flakes of fine oxide, which remain suspended in oil. Up to a certain amount, these flakes help to retain the viscosity of the oil and the drawing aluminium wire, but too much of it will create a problem and hence will need filtration from time to time. Copper dust in oil, however, settles very quickly at the bottom and should not be allowed to be dragged on the rings from the bottom during the rotational movement of the capstan.

Even by taking all precautions, the groove formation on rings cannot be avoided, though it may be delayed. The deeper the groove becomes, the more will be the heat generated, as the area of the rubbing surface around the circumference of the wire increases. Precaution must be taken by hardening the ring metal. After hardening the surface is to be mirror polished to prolong the life of rings. Nowadays, hard polished ceramic rings are used to resist groove formation for a long time. Spraying of lubricant, which allows easy slippage of wire, also helps. But in any case, after a longer period, the phenomenon of groove formation can be seen, initially very faint but in time more pronounced. When a deep groove is formed, it should be wiped out by the grinding ring surface. Care should be taken to see that the depth of the groove on the ring does not fall far below a given level and becomes disproportionate in comparison to the other rings. If the dimension of the groove exceeds its given limit but the machine cannot be stopped due to a tight production schedule, then the position of the wire should be changed sideways by adjusting the die holder block. When too many grooves are formed, the ring should not be used without polishing. After polishing two or three times, if the diameter of the ring becomes disproportionate in comparison to others, then the ring should be changed. The limit of diameter variation can be determined on the basis of minimum or zero slip. This can be found by precise calculation. If the groove becomes deeper, the wires will tend to become loose. For this reason, sometimes a number of turns on the ring are to be increased, creating a problem.

Even though manufacturers may try to construct several wire drawing machines having the same constructional features, during performance each machine will have its own characteristic feature. It has to be remembered that there are many components that have to be incorporated to manufacture such a machine. Hence, it would be incorrect to think that every one of them will behave in the same manner. Further, it should be understood that environmental conditions play an important role during operation. The working efficiency of a machine will be different in Indian conditions than when operated in Europe. Also, the same machine may behave differently within different locations in India, like Kolkata and Jaipur. The occurrence of minute variations may not draw our attention, but such differences ultimately affect the performance and efficiency. In India, we always contemplate that when a machine is working in Madhya Pradesh without any problems, then why should it not behave in the same way in Rajasthan!

Regarding the point of wire drawing and die dimension, there is a close relation between the factors:

1. The temperature of the lubricant and die during processing
2. The pulling tension of the capstan ring and winding take-up system
3. The composition of the metal

The wire tends to expand within the die orifice after entering. After coming out of the die, it remains merged in hot lubricant. The recommended temperature for lubricant is 40°C ± 3°C. Considering the coefficient of expansion of copper or aluminium, there will be a variation in diameter. As an example, if we take the diameter of the wire as 3$^{\phi}$ mm and the coefficient of expansion as 0.1692×10^{-4} for copper and 0.23×10^{-4} for aluminium, then the respective increase in diameter will be 3.00228$^{\phi}$ mm for copper and 3.0031$^{\phi}$ for aluminium. Here, the virtual weight difference could have been 100 g/km for copper and 40 g/km for aluminium. Though there will be no actual increase in weight, the expansion in diameter will demand more force for pulling the wire, increasing the corresponding heat generation. On the other hand, considering the previous expansion, if the orifice of the die is lowered by the corresponding amount, it will cause problems after cooling when the diameter becomes lesser than specified. Now we know how the metal behaves during drawing and the reason for the quick generation of heat in the machine. In generally hot atmospheres, the range of temperature rise above ambient will be high. In such places, an efficient heat exchanger should be installed to keep the temperature rise within the limit and to achieve the required output from the machine.

While drawing and at the take-up stage, the wire is subjected to a constant pull by the capstans. This pull, that is the tension on the wire, must remain constant during operation. Even a slight variation or uneven winding will stretch the wire with a constant jerk, resulting in a nonuniform diameter with a wavy surface, which is undesirable. It is therefore necessary that the meshing of gears must be perfect and smooth. The take-up side also has to be synchronised perfectly. The thinner the wire, the more precise the machine dimensions and operational skill called for. Also to maintain the correct and uniform size across the length, a wire compensating device/dancer is provided, and it becomes almost mandatory to maintain the exact actual dimensions, more so in aluminium conductor steel reinforced (ACSR) conductors.

As for metal composition, if the same does not conform to the given international annealed copper standard (IACS) specification for copper and IEC for aluminium, the physical characteristics of the drawn wire will change, which will affect the output and efficiency of the machine.

The elongation of aluminium is lower than that of copper, but it does not acquire much working hardness by repeated drawing. High ductility allows aluminium to be compressed more than copper, and at times in a properly constructed machine, an area reduction up to 22% can be achieved.

In the course of drawing copper, the material acquires a working hardness. To draw fine wire, it is necessary to anneal the wire after one or two drawings in order to recover its softness and facilitate further processing. Copper wire can be drawn as fine as 0.0015 mm diameter. Further reduction in diameter can be obtained by taking special precautions. Aluminium wire, however, cannot be drawn below 0.30 mm diameter. An oxide film covering the surface of the aluminium wire becomes proportionately higher in comparison to the pure metal content. Naturally, the wire loses its strength and breaks frequently. At times, wire rods may have blowholes or air pockets within, the dimensions of which may vary. But even a small blowhole will produce piping within the drawn wire, causing frequent snapping as the wire dimension is reduced further. The blowhole is formed during improper casting of material at the primary stage. It also causes variation in resistance values. It is essential that the primary producers of metals must take precaution at the rolling stage to avoid the formation of blowholes within the metal casting.

It is not advisable to measure the resistance value of the wire immediately by directly taking a sample from the machine. The wire at the time remains hot. This should be cooled down to room temperature naturally. It is not advisable to cool the wire by dipping in water immediately after drawing from the machine. This will give a quenching effect, and the result will not be favourable.

The quality of a machine is judged on the basis of the following:

1. Power consumption
2. Running efficiency
3. Quality of output product
4. Productivity
5. Cost of maintenance

A machine designed properly and constructed with close tolerance will have minimal frictional force. Naturally, the unit will run smoothly. In this context, as explained earlier, factors such as the quality of the die, die angle, finishing of die and of drawing rings, quality of input material, proper lubricant and system of lubrication, temperature during performance and timely preventive maintenance will ensure quality and guaranteed output. When quality prevails, quantity is automatically achieved.

Without knowledge of the wire drawing technique and a proper method of testing, utilising a machine of poor construction (to avoid a little more investment) and thinking that anyone can draw wire, leads to inconsistency, inefficiency and poor quality, which ultimately results in losing all economic advantages.

3.1.4 Selection of Wire Dimension

Today, prices of copper and aluminium are steadily rising. The availability of copper is becoming a concern in the world market. It is therefore essential to control the wastage of metal/metals in every respect when high productivity has to be achieved with close tolerance. The content of metal(s) used should be towards the lower side, but it should not, at the same time, affect quality. This can be done by establishing a close relation amongst weight, diameter and resistance of individual wire size. To select an optimum size, the following discussion will be of importance.

TABLE 3.2

Variation of Weight as per Variation of 2.5 mm Wire Diameter

Area of Wire to Consider (mm²)	Diameter of Wire Nominal Value ± 0.02 (mm)	Weight of Copper (kg/km) (Density 8.89)	Weight of Aluminium (kg/km) (Density 2.703)	Difference in Weight of Copper from Nominal Value (kg/km)	Difference in Weight of Aluminium from Nominal Value (kg/km)
2.500	1.784	22.225	1.758		
2.556	1.804	22.723	6.909	(+)0.498	(+)0.1514
2.444	1.764	21.727	6.606	(−)0.498	(−)0.1514

Considering the size of a wire to be 2.5 mm^2 to be drawn with a tolerance of ±0.02 mm on diameter, the weight of the wire will be as given in Table 3.2.

Thus for a difference of 0.02 mm diameter, the difference in copper weight becomes 500 g/km and for aluminium it will be 150 g/km approx. A conductor made of 19 wires will thus consume more or less 19 × 500 g = 9500 g as per the selection of maximum or minimum tolerance limit. For 10 km of conductor, a difference of 95 kg will be experienced. Similarly for aluminium, the difference will be 150 × 19 × 10 = 28.5 kg.

While selecting the diameter, the conductivity of the metal must be taken into consideration and its proper selection be ensured. As per IACS for 100% conductivity of copper resistivity is fixed at 0.01724 at 20°C and for aluminium having 61% conductivity, resistivity is fixed at 0.02826 at 20°C. As per IEC 228 and BS 6360 and BIS 8130, the resistance has been specified for 2.5 mm^2 solid copper conductor at 7.41 Ω/km and for aluminium as 2.10 Ω/km. Considering this, a diameter table for aluminium can be arrived at based on the weight of the material for practical application. In deciding for the correct wire size, it should be kept in mind that the tolerance is more dependent on the specified resistance values. Therefore, theoretical values are used as a guide, but the actual weights of copper are dependent upon meeting the resistance value, rationalising the size and weight of a wire accordingly. One of the other factors in bunch copper conductor for reduction in overall weight is the size of the compacting die (Table 3.3).

TABLE 3.3

Comparison of Weight as per Resistivity & Resistance of 2.5 mm Wires

Condr.	Resistivity as per IACS (Ω/km) at 20°C	Resistance as per IACS (Ω/km) at 20°C	Resistance as Specified in IEC, BS,BIS (Ω/km) at 20°C	Resistivity as per IEC, BS,BIS Calculated at 20°C	Should Be Area as per IEC, BS, BIS (mm²)	Nom. Wt. as per IACS (kg/km)	Weight as per IEC, BS/ BIS (kg/km)
Copper	0.01724	6.89	7.41	0.01825	2.326	22.2250	20.6830
Alum.	0.02826	11.30	12.10	0.03025	2.335	6.7575	6.3129

A comparison of the standards IEC, BS and BIS indicates that the diameter of the wire should be 1.723 mm. Thereby, the difference in weight for copper would be 1.542 kg/km and for aluminium 0.446 kg/km lower than the values obtained when the weight is calculated as per IACS standard. These are substantial in quantity. In any case, if the design and production keep a close watch on the quality of the material, a saving can always be worked out based on the material parameter used without sacrificing the quality of the product. This is because at times the raw material coming from suppliers may have some unforeseen deviations that would be within the tolerance limit and would need to be taken into account. On the other hand, customers should not insist on the weight of the material provided in a conductor but should consider the electrical parameters that need to be fulfilled. All the specifications are based on electrical design parameters, and it is manufacturer's responsibility to satisfy the same as they try to rationalise the product values without deteriorating the quality. For this reason IEC 228 or BS 6360 or IS 8130 does not specify any wire diameter, but they are more concerned with resistance values. On the other hand, manufacturers have the responsibility of fulfilling conditions laid down in the specification. It will depend on their skill, ability and constant monitoring system to rationalise their production process for economical benefit and customer satisfaction.

3.1.5 Annealing

Every metal gets hardened when worked upon. When steel is cold drawn, lead is beaten, and copper is formed or beaten or drawn, they become tougher. The tensile strength increases, decreasing the value of elongation. During drawing or cold forming, atoms and molecular structures are stretched and find themselves randomly oriented. Naturally, they exert a physical resistance to the applied force. Further, the easy flow path of the electron gets disturbed, increasing the resistance value. This is very much pronounced in steel and more in copper. The high ductility of copper makes it possible to elongate the metal considerably without breaking. This property is utilised for cold forming and drawing of wire rods to thinner wires. When molten metal solidifies, it crystallises (as in the case of a salt from a saturated solution), but the metal crystals do not show any clearly defined facets. Yet, within the crystals, atoms are arranged in space in a 3D form. These metallic crystals of copper are ductile and can be distorted to a considerable extent. During drawing, at the beginning of deformation, the whole phase of the atom within the crystal slides relatively with one another, but later as the work progresses, the regular arrangement of the atom gets severely disturbed. This change brings about progressive increase in the hardness in the metal and resistance to deformation, diminishing ductility. In such a case, free movements of electrons are disturbed, increasing electrical resistivity. This phenomenon is known as work hardening.

The remedy for such work hardening is to rearrange the atoms to form new crystals, by heating the material sufficiently at high temperature. The energy given out by the metal during cold forming is gained by the heat supplied externally and uniformly in a closed system and thereafter cooling the metal gradually to atmospheric conditions. By alternate drawing, heating and cooling processes, metals can be formed to a very thin wire from a square or round billet, spanning many kilometres in length. This type of heat treatment is known as annealing. The recrystallisation temperature, which is influenced by many factors, ranges from 250°C to 600°C.

Care has to be taken to see that the metal does not get oxidized and tarnished black during the heating process. Normally in absolute dry air, the metal does not get oxidised, but at an elevated temperature under exposed conditions, oxide formation cannot be prevented and the oxide layer increases with increasing temperature. Sulphur also attacks copper, giving a greenish tinge to the surface. In the presence of moisture, sulphur and oxygen react very quickly. During annealing, these aspects have to be taken into account. The formation of oxide or sulphide on the surface will reduce the effective area of the metal, increasing its electrical resistance and creating problems during subsequent operation.

Presently, there are two types of procedures adopted for annealing copper wire. One is known as the batch process. The other one is continuous annealing by heating the wire electrically, taking advantage of its own resistance in tandem with the wire drawing unit. Otherwise, continuous annealing can be done separately as well.

As for batch process, the annealing furnace consists of two parts. The first part consists of a cylindrical pot lined with heat-insulating bricks. Heater coils are placed in half-round spiral grooves that are etched on the inside periphery of the pot, the terminal of which are taken out through hermetically sealed ceramic bushings. Normally, such furnaces are placed inside a pit below the floor level for easy loading and unloading of the annealing pots. The top of the heating pit is kept flushed with the floor level. The second part consists of annealing pots (several of them) made of high tensile steel and adequate thickness to withstand repeated heating and cooling cycle and the outside atmospheric pressure while they remain evacuated and under high vacuum. The top lids of the pots should remain hermetically sealed when in use. The lid is provided with an air inlet and an outlet pipe for evacuating air or filling the pot with inert dry gas like dry nitrogen or carbon dioxide. The gas must be free from moisture and oxygen. The lid has an airtight valve. Thermocouples are provided to measure the inside temperature during operational conditions. The extended rim of the lid covers the top of the heating pit to prevent leakage of heat from the heating chamber. Bobbins with copper wires are placed on a plate hanger which can be lifted or lowered down by a jib crane. The hanger with the bobbins is placed in the annealing pot. It is then closed by tightening the lid and is placed in the furnace pit which is also closed by the lid of the pot. After closing the pot with the lid, the air is evacuated by a high-vacuum pump and retained up to a value of 0.1–0.05 millibar constantly till the annealing cycle is completed. In other cases, nitrogen gas is introduced, which drives out the air completely. In both cases, it is ensured that the wire placed inside should not come in contact with the oxygen/sulphur present in the atmosphere. The temperature within the pot is allowed to rise to the desired degree and to be retained for a given period. The temperature for finer wires is to be maintained at 250°C and 400°C–450°C for higher sizes. The quantum of heat required depends on the time, temperature and quantity of the material. In ascertaining the quantum of heat required during annealing, the heat absorbed by the bobbin and the pot has to be taken into account along with the copper contents. Monitoring the consumption of energy is a vital part of batch annealing process in order to make the system economically viable.

The quantity of heat required to raise the temperature in the pot material (neglecting the radiation and other dissipation factors) is calculated as follows:

$$T_r\,°\mathrm{C} = (M \times S_1 + G \times S_2) \times (T_r - T_a) \text{ calories} \tag{3.14}$$

where

M is the mass of copper
G is the mass of total steel material
S_1 is the specific heat of copper 0.093
S_2 is the specific heat of steel material 0.1075
T_r is the temperature rise as required (°C)
T_a is the ambient temperature (°C)

The same amount of heat will be raised by the resistance heater of the heating pot:

Here,

$$H = \frac{i^2 R t}{J}$$

where

H is the quantum of heat
i is the current (amps)
t is the time (s)
J (Joules) (4.2)

or

$$H = \frac{V i t}{J} = 0.24 V i t \text{ calories}$$

where V is the voltage.

Therefore,

$$(M \times S_1 + G \times S_2) \times (T_r - T_a) = 0.24\ V\ i\ t$$

or

$$(M \times 0.093 + G \times 0.1075) \times (T_r - T_a) = 0.24\ V\ i\ t \tag{3.15}$$

With the help of this equation, the time for annealing of the copper material, temperature rise and all other factors can be worked out. Even the pot design and resistance values for the heaters can be determined. In this case, a factor for the dissipation of heat and surrounding condition must be considered while actual calculations are made.

After placing the bobbins in the pot and closing the lid, the pot is evacuated completely. Thereafter, heaters are switched on and the temperature is allowed to rise. The temperature rise in vacuum is very slow initially. The only way the heat wave spreads is by radiation. As the temperature seems to stabilise after a period of time, say from 6 to 8 h, the heat starts acting on the metal. Here lies the main problem of pot annealing. Heat acts on the upper layer of the spool and the sides and bottom layer touching the steel bobbins at the starting point, whereas in the middle portion, it is transferred through a conduction process and in layers. This is a very slow process. Before the temperature stabilises within the inner layer, the upper, lower and sides become overannealed. During further processing, forming a stranded conductor or insulating the wire is risky. The wire can stretch unevenly unless tension is monitored continuously, which becomes an impossible task. In order to obviate this problem, the temperature should be raised step by step for uniform heating and annealing of wires. The increment in temperature would be 100°C, 200°C, 275°C, 300°C, 325°C, 350°C and so on. The initial temperature difference in steps is larger, and as the temperature rises, the difference in steps is also reduced leaving a very small variation of temperature between the inner and outer layers. This type of annealing process is time consuming but allows the attainment of a uniform result throughout. This method was adopted by the author in one of the reputed cable factories in India. This is also true for the thinner sizes of wires where the winding is tight and dense. The time–temperature adjustment is to be made as per pot size, quantity of material, spool size, wire size, winding condition, etc., of the factory concerned. Accordingly, a processing specification is to be drawn. It is a wrong notion that by increasing the temperature rapidly to a higher range, the annealing time could be shortened. This action will lead to uneven results which will create problems at a later stage of processing. After heating is completed, spools are cooled down without bringing them in contact with the atmosphere. If this precaution is not taken, the copper wire will immediately tarnish, forming a layer of black copper oxide on the surface. The evacuated pot with the material inside takes a long time to cool even if the pot is sprayed with cold water from the outside. The cooling process can be shortened by introducing an inert gas free from oxygen, such as oxygen-free dry nitrogen, or carbon dioxide, into the pot and breaking the vacuum. Initially, the gas will absorb the heat and transfer it on the pot wall kept under the shower of cold water. The pot is opened after the temperature is brought down to 50°C. In order to get a fool-proof result and to obtain a bright annealed wire, the following are applied:

1. The pot should be hermetically sealed with a heat-insulating material.
2. Vacuum should be effective and complete.
3. The wire should be free from any lubricant on its surface.
4. Annealing temperature should be controlled properly.
5. Cooling should be done as per requirement.
6. The inert gas introduced should be completely dry and free of oxygen.

It is known that the batch-type annealing process is

1. Time consuming
2. Labour oriented
3. Complex

To make the annealing process easier and uniform all over the wire length, a continuous annealing process has been developed for larger- and smaller-diameter wires. This is done by using the wire resistance and taking the empirical formula i^2R as the basis of calculation. Let us consider the following:

1. Resistance of the length of the wire is 1 m and it is held between the two electrically heated contact pulleys.
2. The diameter of a 2.5 mm² copper wire is approximately 1.784 mm.
3. The voltage applied is 24 V DC (to be adjusted with the change in diameter, i.e. cross section of the wire).

In equation form

$$M\,S\,T = \frac{i^2 R\,t}{J} \tag{3.16}$$

where

M is the mass of material
S is the specific heat
T is the temperature rise
i is the current
R is the resistance
t is the time (s)
J = joule = 4.2

Mass of 1 m wire (=M) is = 2.5 × 8.89 = 22.225 g
Resistance of 1 m length (R) is = 0.00741 Ω/m
The temperature required for annealing is 450°C and ambient 27°C; hence, the rise in temperature is (450°C – 27°C) = 423°C = T
The speed of the wire is 9 m/s. Hence, 1 m is covered within 0.1111 s (t).
The specific heat of copper is 0.093 s
Therefore, the equation becomes $22.225 \times 0.093 \times 423 = i^2 \times 0.00741 \times 0.1111/4.2$
Hence, i = 2112 A approx., and the voltage is 15.65 V ($i \times R$)

The voltage is adjusted based on the diameter of the wire. Similarly, for a 3 mm wire diameter, the current requirement will be approximately 5700 A when the line speed is 8 m/s. This has been found to tally with the actual working system. Here, the voltage will be 14.5 V approximately. A table can be prepared for the production department listing the wire length between electrically heated contact pulleys, as given in the appendix of this chapter.

Example 3.1

Say there is already an annealer working in tandem with the intermediate wire drawing machine to anneal wires of 0.5–0.91 mm conductor for telecommunication cables. In this case, for a 0.91 mm diameter wire running at 20 m/s for a temperature rise up to 250°C, the current requirement will be 600 A approximately, and for a 0.5 mm diameter wire it will be 185 A approximately, having a contact length of 1 m (Figure 3.12). (In this case, the factor for dissipation of heat has not been taken into account.)

The annealer consists of a wire entry point, two induction heated contact rings having a steam chamber to protect the heated wire from coming in contact with the atmospheric oxygen, cooling ring attached to coolant tube, air blower to wipe out coolant and moisture, and a preheater and dancer to control tension. Contact rings are attached to a set of slip rings and contact fingers for collecting high current. Contact fingers are connected to the low-voltage (secondary) side of the current transformer through a heavy-duty bus bar. The primary side of the transformer is connected to the supply line. The wire coming in is passed over the heated contact rings. The wire to be annealed is heated by collecting the current from the contact ring utilising its own resistance values. While moving over the contact rings, the wire passes through a dry steam tube and remains isolated from the atmospheric oxygen. Emerging from the second ring, the wire is cooled gradually by a coolant water moving through a tube. At the exit point, the wiper and air blower remove all the moisture and let the wire pass over a preheater to the dancer. The body of the machine is kept isolated in order to remain neutral during working and to ensure the safety of working personnel. Nowadays, a heavy-duty annealer is also constructed to anneal higher-diameter wires up to 4.5–5.0 mm attached with a heavy-duty rod breakdown machine.

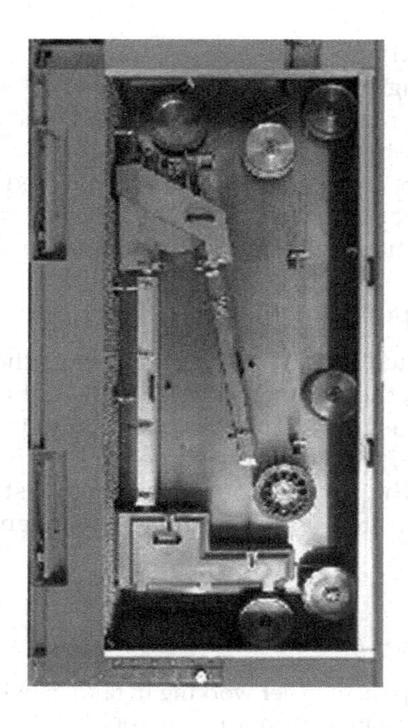

FIGURE 3.12
Continuous copper wire annealing machine to run in tandem with the RBD (rod break down) unit for aluminium alloy wires.

The advantages of an in-line annealer are as follows:

1. The annealing process is done on a continuous basis, requiring no waiting.
2. Intermediate handling is avoided, reducing labour cost.
3. Uniform elongation of wire is obtained, as constant temperature is allowed to work on the wire length throughout without any deviation.
4. Cooling of wire is done simultaneously.
5. Cause for the formation of the oxide layer is eliminated.
6. Power consumption is controlled as power can be switched *on* and *off* at the time of starting and stopping of the annealing process. Furthermore, the dissipation of power is very low.
7. The process can be automatically controlled and is not left to chance.

In India, until lately, batch annealing process was preferred for large-sized wires, as the initial apparent cost for an in-line annealer was high. But if one thinks rationally, the actual cost will be the same or lower in the long run, along with many advantages. But with the advancement of technology and very high competition, manufacturers are opting for an in-line annealer from both cost and quality points of view.

So far so good for the copper wire annealing process. Nowadays, there is demand for annealed aluminium wire. Customers think that if aluminium wires are annealed, it will also become a better conductor. Flexibility is possible, but by annealing, the conductivity of aluminium cannot be improved; rather, there is a risk of increasing the oxide layer on the metal which acts as insulation. If the metal needs softening, it can be done by heating. Proper precaution must be taken though. The metal is normally heated at the primary coil stage by placing in a box-type oven at about 200°C–250°C in a dry atmosphere. If the coil is heated in a moist atmosphere, a thick layer of oxide will form, and during drawing, a considerable amount of metal will get shaved as fine oxide particles contaminating the lubricant, while the life of the die will get shortened due to abrasive working friction. If annealing is done on spools, then the procedure adopted for copper to anneal under vacuum will be the best solution. Further, the temperature must be raised slowly to let the heat be absorbed gradually by the layer of wires. It is not advisable to anneal aluminium wire under normal atmospheric conditions, particularly when moisture is prevailing in the air. It is always better to request the supplier of primary rods to supply material in soft condition having almost zero hardness and specifying tensile values required at the initial stage so that after drawing the required hardness can be obtained. Aluminium normally does not show too much work-hardening during drawing. Primary producers can control the hardness or softness of the metal while rolling the rod in a Properzi mill by adjusting the metal temperature, rolling speed and cooling conditions. Using soft metal for a single wire or seven strands to make a 4 or 6 or 10 mm^2 conductor and for insulation of domestic wiring is very risky. In this case, the metal should have a tensile value in the range of 100–150 MN/m^2. Numerous developments have been made in the annealing of aluminium at the wire stage. Manufacturers have formulated methods by which they can draw an intermediate wire, anneal it and then redraw it to the required size. By this method, they are actually increasing the process time and adding one more process. But because of improvement of conductivity and reduction in diameter and weight due to higher compactness, the overall saving by weight is around 1%–1.5%, which is significant if there are large volumes in the pipeline.

3.1.5.1 All Aluminium Alloy Wire

Commercially, pure aluminium is weak and ductile. Electrical grades are even weaker. A large number of alloys can be formed with the addition of very small quantities of other elements in different combination and percentage. Elements like copper, manganese, silicon, chromium, zinc and magnesium and at times very small quantities of boron, titanium and zirconium are added to control the grain structure. After heat treatment and ageing process as per determined temperature and time, these alloys acquire a very high breaking strength, while the conductivity slightly decreases. The aluminium alloy contains 0.5% each of magnesium and silicon. The wire drawn

FIGURE 3.13
Bull blocks in tandem to draw an aluminium alloy wire.

from this alloy is used to manufacture the stranded conductor and support the overhead transmission conductor, replacing the steel core and messenger conductor for an aerial bunched cable.

A 9.5 mm alloy wire rod is drawn to a 5.5 mm diameter in the bull block machine (Figure 3.13). This is then put in a furnace to be heated at a temperature of 500°C–550°C (approximately) (the melting temperature of aluminium is around 630°C) for about 20–30 min. The red hot coil of wire is then dropped in cold water for quenching. The time for dropping the coil in water is very important (Figure 3.14).

The lesser the time, the better is the strength developed in the coil (2–4 s is ideal). For this reason, the water pit is put directly under the furnace.

FIGURE 3.14
Bottom drop solution treatment furnace.

FIGURE 3.15
Ageing oven to temper aluminium alloy wire.

The furnace has an opening at the bottom to allow the coils to run down directly into the water pit. After taking out, the coils are further drawn in a wire drawing machine to the required size. The wire is then tested for tensile strength to ascertain the ageing time and temperature that will be required to temper the wire to balance the tensile value and elongation (Figure 3.15).

This is necessary for the formation of a smooth conductor during stranding. The normal ageing temperature is around 200°C–250°C for about 8–10 h. These parameters are given as a guideline but need to be adjusted after a trial run taking into account material conditions and factory working conditions. Every spool must be tested for quality. Table 3.4 lists some of the values used during production.

TABLE 3.4

Grade of Alloy	Heat Treatment Temp. (°C)	Apx. Time of Heating (min)[a]	Quench[b]	Ageing Temp. (°C)	Time of Ageing
2014[c]	500–510	15–60	Water	170	10 h
2024	485–500	15–60	Cold water	Room	4 days[d]
6061	515–525	15–60	Cold water	155–160	18 h
7075	460–470	15–60	Cold water	120	24 h

[a] This depends on the size and amount of materials. In India, a 9.5 mm rod for a 2 ton coil weight is heated for 20–30 min max.

[b] This is done within a minimum time, say within 15–20 s, and the coil is dropped in water from the furnace.

[c] This is the type used by the Indian system as per IS 14255.

[d] More than 90% ageing is obtained on the first day of ageing.

3.1.6 Storage and Transport

Copper and aluminium are costly items and are the backbone of an electric cable. Therefore, it is necessary that proper precaution be taken for storing these items. The storage place should be dry and free of any fumes of acid, alkalis and moisture. The process of storage should be such that during loading and unloading, the surface of the metals do not get damaged or scratched. Hanger-type storing is ideal for metal coils. Transportation on a motorised flat car or hand driven is best instead of using a forklift truck, where there is a risk of damage. Lifting of the coil from the hanger can be done by a crane or monorail installed along the span of the storage line. Flat cars can be directly put behind the wire drawing machine for paying off the wire rod. All these costs for storage and transport ultimately bring about a saving, in the form of labour and damages, reducing unnecessary scrap formation.

3.1.7 Quality of Input Material and Drawn Wires

The quality of the raw metal received, and the finished product coming out of the drawing machine, has to be checked in terms of

1. Average diameter of wire rod and weight per metre.
2. Density of the metal.
3. Resistance or conductivity.
4. Check the surface (for scale formation, or oxide films, or impurities or porosity, etc.).
5. These preliminary checks will help determine the parameters for processing to follow.

The specifications for raw materials testing should be as per BIS standard IS 8130, BS 2627, BS 3988, BS 4109, BS 5714, BS 6360, BS 6791 or IEC 228.

During drawing a wire and after drawing, check for the following parameters:

1. Surface defects and diameter
2. Tensile and elongation (after annealing for copper, whereas for aluminium as it comes out of the wire drawing machine)
3. Resistance of the wire (to be measured after cooling to room temperature; the hot wire should not be measured)
4. Weight of the wire (correlate with resistance values)
5. Torsion test (for aluminium only)
6. Breaking load and resistance for all aluminium alloy (AAA) wire

TABLE 3.5

Wire Length Considered in the Process Stage (m)

Finished length delivered after final testing	1001
Take 2 m for final testing	1003
Take 1 m for outer sheath	1004
Take 0 m for armouring	1004
Take 2 m for inner sheath	1006
Take 1 m for laying up	1007
Take 1.5 m for insulation	1008.5
Take 2 m for stranding of conductor	1010.5
Take 1.5 m as bobbin scrap – for three cores (each wire)	3036

Wire drawing is the starting point for the manufacturing of cable. If the beginning is proper, subsequent process will follow without much problems. It is necessary to ascertain that the length drawn and wound on a spool is a multiple of such length that when the final cable length comes out, very little scrap is generated. The ideal condition is 'no scrap generation'. This is where the skill of the production chief lies, who formulates the factory process standards as per the available machine parameters. Production chief will direct and train operational personnel accordingly. The desired finished length is 1001 m. The last 1 m is for adjusting the counter metre's anomaly (though this is to be corrected after a few months' operation).

The quantum of length for each wire to be taken in a bobbin during wire drawing is being illustrated by taking the example of a three core 1100 V, 120 mm^2, PVC insulated, laid-up, extruded bedding, flat strip armoured and PVC sheathed cable.

In this case, the maximum number of kilometres is to be planned for production by clubbing several orders. Here, we assume only the length of 1 km cable (Table 3.5).

3.1.8 Further Consideration

For example, a 120 mm^2 conductor consists of 19/2.82 mm. This will give the weight of copper as 1066.8 kg/km and aluminium 324.36 kg/km. Now let us consider the area and weight on the basis of resistance values. As per IS 8130, the resistance of copper on the finished cable is 0.153 Ω/km and that of aluminium is 0.253 Ω/km.

Accordingly, the specific resistance of copper and aluminium becomes 0.01836 and 0.03036, respectively. Normally, specific resistance values as per the available material from the primary manufacturer are found to be 0.01750 and 0.02850, respectively. Even if we consider an increase of 4% due to stranding and laying allowance, the final values will be 0.01766 and 0.02920, respectively.

On this basis, the area is 115.42 mm^2 in both cases.

The weight of material will be for 1026.10 kg/km for copper, ensuring a saving of 40.7 kg/km, and 311.98 kg/km for aluminium, showing a saving of 12.38 kg/km, whereas in both cases, the final resistance values are in compliance with IS 8130. When the initial values are known, by measuring primary coils, the design parameters can be established. The wire diameter thus will be 2.78 mm. The weight of the wire per kilometre is 53.961 kg for copper and 16.4 kg for aluminium.

Nowadays, 630 diameter spools are being standardised for wire drawing machines. The dimensions of the spool are as follows:

Flange diameter 630 mm
Barrel diameter 275 mm
Travers inside 300 mm

The net weight of the material in a spool is given by the following general equation:

$$G = \frac{0.001 \times \pi(D^2 - d^2)b\,\gamma\,f}{4} \tag{3.17}$$

where
G is the weight of the material in the spool (kg)
D is the flange diameter of the spool (cm)
d is the barrel diameter of the spool (cm)
b is the width of the spool (cm)
γ is the specific weight of the wire (g/cm^3)
f is the filling factor of the spool generally taken as 0.85

Taking $\gamma = 8.89$ for copper and 2.703 for aluminium, the formula reduces to

$$G \text{ for copper} = 0.00593\,(D^2 - d^2)\,b = 571.548 \text{ kg} = 10{,}591 \text{ m}$$

$$G \text{ for aluminium} = 0.00180\,(D^2 - d^2)\,b = 173 \text{ kg} = 10{,}550 \text{ m}$$

For wire diameter of 2.78 mm
The length is almost the same as expected. The length required is to be in multiples of 3036 m to be taken in the spool. As such, the maximum length that can be accepted in a spool would be 9110 m, that is three times the length of 3036 m. This would be nearest to 10,591 m. The spool will not be able to accept four times of 3036 m as it will be too large a quantity. The weight of a 9110 m length cable is 491.58 kg for copper and 149.40 kg for aluminium to make a 1 km three-core finished cable. There will remain practically no scrap in the spool except perhaps a few metres within the tolerance of ±2 m. The percentage of scrap length, that is 36 m for 3036 m, is approximately 1.2%. However, this length is taken when all the wires considered laid are straight while forming the conductor. In actual practice, wires are stranded spirally. The length of each layer increases as per the lay ratio, the calculation

of which shall be shown in Chapter 4. An allowance of 1.2%–1.3% is taken on the positive side on the previous accommodate the lay ratio. If the design and production are properly planned, it is not difficult to achieve a final scrap figure below 1%. However, the following are important to obtain such a creditable performance:

1. Measurement of length must be perfect. If necessary, counter-checking of length can be incorporated to compare two counter metres simultaneously.
2. The second method is control by weight. If the weight of the empty spool is known, then by knowing the net weight of the material in relation to its length, an automatic stop can be actuated as soon as the selected weight is reached. The system should have initial and final setting pot that has to be controlled electronically (through the PLC).
3. Diameter monitoring must be accurate. An increase or decrease in wire diameter by 0.001 mm will give either a positive or a negative length in the spool. The weight of a 2.781 mm wire will have weight for 1 km length at exactly 54 kg. Therefore 491.58 kg weight will be for a length of 9103.30 m having a short fall of 6.7 m from the mean value. On the other hand a 0.001 mm reduction in diameter will give a wire length of apx. 6.45 m in excess of mean value. Naturally it is imperative that the wire drawing machine must run very smoothly without any vibration having a precision control system.
4. Die selection should be such that the opening of the die diameter selected gives maximum tonnage with minimum variation in ± tolerance limit. This has to be standardised after a few days run.

If the input speed of the wire is 1 m/s, then the final length coming out of the last die will be $L = \{1/(1 - X/100)\}^n$ m, when the area reduction is assumed to be X% and the number of drafts to be 'n'. When $X = 20.6\%$ and $n = 13$ and the input speed is 1 m/s, then $L = 20$ m/s. The weight of the wire produced per second, when the outlet diameter is in millimeter, will be

$$W = \left(\frac{\pi d^2}{4}\right) \cdot \left\{\frac{1}{1 - X/100}\right\}^n \cdot \gamma \text{ g/s} \tag{3.18}$$

By knowing the initial and final speed, a timer can be set to get the exact length on the spool. Similarly, the output diameter 'd' of the wire is calculated by the following formula:

$$d = D\sqrt{(1 - X/100)^n} \tag{3.19}$$

3.2 Wire Drawing Machine

Some of the functional requirements and parameters for drawing wire have already been mentioned in the previous discussions. As indicated, there are two types of machines being constructed. One is the cone type and the other is the tandem type (Figure 3.16).

In a cone-type machine, drawing capstans are arranged in groups, where the diameters are increased proportionately on the basis of area reduction. Each group is assembled on a shaft and kept supported on the bearing placed in a bearing housing. Housings are sealed by oil seals on both sides to prevent leakage of oil from the gearbox side to the drawing side so that the oils do not get mixed up. Corresponding to each drawing cone assembly, one set of idler cone assembly is kept mounted on the wall of the machine. The die holder box is placed between drawing blocks and idler blocks. On the capstan and die holder box, a lubricant is sprayed by a pump through pipes and nozzles. The bottom part of the machine is provided with a tray to hold the lubricant in which capstan rings are allowed to remain partially submerged. On the back side of the machine, in a separate compartment, a train of gears are kept attached to the shaft of the capstan block and idler block. Gears are of the same ratio as the increased lengths and are accepted by the corresponding larger-diameter capstan in succession. The unit is supplied with an inching mode and running mode along with emergency switches. The top lid can be openly supported on the hinge. The machine is compact and sturdy in construction and requires less space. A spooler or a coiler is put online to collect finished wire. The spooler can be of pintle type or shaft

FIGURE 3.16
Cone-type wire drawing machine.

type as per the customer's requirement. The main drive motor is connected to the drive pulley. Nowadays, a DC motor is preferred, or the drive can have an EC coupling for a smooth start-up. The spooler can have a separate drive synchronised with the main unit. The machine will run smoothly without jerk or vibration. These machines are old-fashioned and maintenance can be somewhat cumbersome. Normally, these machines are constructed as per customer's choice; hence, no two machines are found to be identical in construction and performance. However, the basic principle remains the same.

In the case of a tandem-type wire drawing unit, pulling capstans are arranged in a row and mounted on shafts. One end of the shaft is mounted on a bearing housing for a smooth rotational movement. Shafts are attached to sets of gear placed in a separate compartment completely isolated. An automatic lubrication spraying device is provided by a pump as soon as the machine starts operating. The gears are made of hard chrome alloy steel which is ground polished and hardened. Capstans are provided with rings of high chromium high carbon (HCHC) steel, again ground polished and hardened. Adjustable die holder boxes are provided between the capstans. Lubricant spraying nozzles are put on the die holder box. The finishing die normally is to be rotating type to achieve bright and smooth finish (Figure 3.17).

The ratio of gears is such that the speed of each capstan keeps increasing proportionately to the percentage of wire elongation vis-à-vis area reduction. Thus, the gear set can be arranged to get an area reduction of 20.6% where the lengthening factor will be 26% approximately, that is 1.26 times the previous block.

The starting sets of Gear teeth has to be 63/50. In between capstans an idler gear of 50 teeth to be set in. This gear is coupled with a gear set of 63 teeth to form a duplex system. This 63 teeth gear will again drive a gear

FIGURE 3.17
Tandem-type wire drawing machine.

of 50 teeth to increase the speed of the next forward capstan by 1.26 times. As such, the configuration of the driver driven with idler gear becomes 63/50/50–63/50/50–63/50/50 and so on. The gear distance and module can be calculated to determine the centre distance of shafts and the power to be transmitted. The bottom part of the machine is formed like a tray for holding and circulating the lubricating oil. A filter is provided to remove the sludge of fine metals. A heat exchanger is provided to cool down the temperature of oil to 25°C before entering the machine. The main drive motor is connected to the drive pulley. The machine will then run smoothly without jerk or vibration. Normally, DC or AC with inverter drive may be used for smooth start-up and constant speed maintenance. Emergency, start, stop and safety switches are also incorporated.

The type of take-up system depends on the choice of the customer. This also depends on the production facility available. A basket-type take-up or pintle-type bobbin holding high-speed take-up units is to be selected as required. A spooler or a basket coiler is to be put online to take up the finished wire. The spooler can be pintle type or shaft type as per the customer's requirement and can have separate drives synchronised with the main unit (Figure 3.18).

Bobbin is kept in position by clamping between two spindles and the tail stock and self-centring facilities for driving pins, automatic safety lock system on tailstock and pneumatic ejectors for the full reels. High-precision linear wire traverse, by a ball nut screw mechanism driven by AC synchronous motor, is recommended. A pre-selectable stroke of the traverse, with automatic fine adjustment, for fitting the actual width of the reel is incorporated. The pre-selectable winding pitch, with automatic adjustment of the traverse speed to the actual RPM of the spindle, is an essential feature of the state-of-the-art unit. To accept the wire coming from the annealer, a dancer is incorporated.

FIGURE 3.18
Spooler unit to accept drawn wire.

Fully automatic reels' changeover cycle should include the following:

- Wire catching and cutting on full reels
- Wire latching on empty reels
- Full reels' download
- Empty reels' upload

The intermediate and fine wire drawing machine constructions are similar in principle. When an in-line annealer is provided, a dancer in between is a must to control the tension of the wire. The principle of a working annealer has been discussed earlier. With the incorporation of an annealer, the speed of the machine comes down to a certain extent, but this is compensated by the gain in time and doing away with the intermediate handling of spools. Nowadays, machines are built to draw multiple wires within a single unit at a time, in order to increase the output and reduce manpower. In this case, the cost balance needs to be worked out thoroughly to determine the economic advantage for a particular factory.

Wire drawing is the very starting point for manufacturing of conductors and cables, whether domestic, low tension or high tension, extrahigh-voltage cable, AAC, ACSR, AAAC or enamelled wire, or even communication or high-frequency cables, for either instrumentation or control. One can lose or gain from this start process. Quality maintenance and quantity control both start from here. It is the heart of the electric supply system and needs to be given utmost importance during processing.

3.2.1 Segmental Wire to Make a Smooth Conductor

It has become a practice to compact stranded conductors passing through die or shaping rollers. This is to reduce the diameter and smoothen the outer surface of the conductor simultaneously. In case of medium-voltage and high-voltage cables, a smooth outer surface facilitates in reducing electrical stress on the surface of the conductor. Compacting makes the outer surface somewhat smooth, but due to pressure and drawing force, the edges of wires are flattened and become rough and uneven. These edges develop stress at the corner of the wire and at times can become vulnerable. To obviate such a potential high-stress point, segmental smooth wires are drawn and stranded over the inner round contour, thereby eliminating the danger.

This method avoids sharp edges on the outer wires and possible damage which may be caused by the compacting pressure of the inner layers while also ensuring that the flexibility of normally stranded conductor is retained.

This practice was introduced long before in the United Kingdom and Europe for EHV cables and also for overhead conductors. With the development of high-strength steels and a new type of central core material for overhead conductors, this type of compacted trapezoidal conductors allows to pack more aluminium per unit length, decreasing losses.

3.2.2 Design Consideration

3.2.2.1 Calculation of Segmental Wire Shapes

The segmental wire shapes required to form a smooth conductor exterior can be calculated mathematically. In the following, the solution employed by the theory is first propounded and then a design procedure based on this is set out.

The following symbols are used throughout the theoretical design calculations:

R is the inner radius of the segmental layer
Z is the outer radius of the segmental layer
t is the thickness of the segmental layer
α is the lay angle of the segmental layer
α_1 is the lay angle of the inner surface of the segment
α_2 is the lay angle of the outer surface of the segment
l_1 is the lay ratio of the inner surface of the segmental layer
l_2 is the lay ratio of the outer surface of the segmental layer
P_1 is the inner radius of the segment
P_2 is the outer radius of the segment
G_1 is the clearance of the inner surface of the segment
G_2 is the clearance of the outer surface of the segment
C_1 is the inner chord of the segment
C_2 is the outer chord of the segment
r is the radius of segment corners
A is the sum of segment areas
θ_1 is half the angle subtended by C_1 at P_1
θ_2 is half the angle subtended by C_2 at P_2
L is the lay length
L_1 is the lay length of line at right angle to the segmental wire
L_2 is the lay length of the segmental wire

1. *Theory.* Consider a layer of 'N' segments stranded round a core of radius 'R'. Let the radial thickness of the layer be 't 'and the lay angle 'α'. Figure 3.19 depicts a stranded segmental wire.

   Now consider a section through Figure 3.19 at X–X' and the axis of the segment at right angle. This segment is shown in Figure 3.20. The wire that would be covered by a longitudinal straight line of the length is

$$\text{Length} = N\left(1+\frac{L_1}{L_2}\right)$$

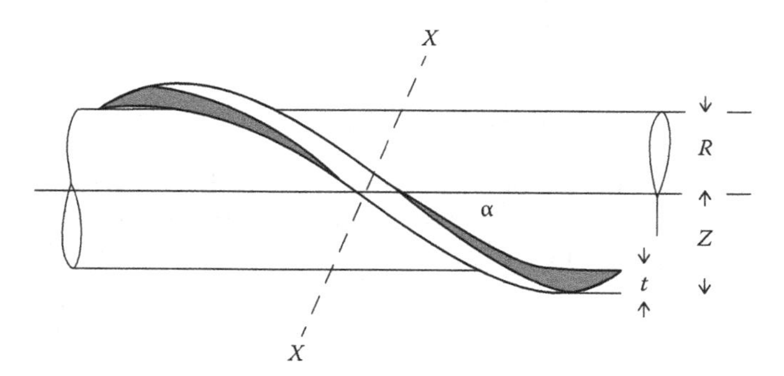

FIGURE 3.19
Segmental wire stranded around a cylindrical core of a conductor.

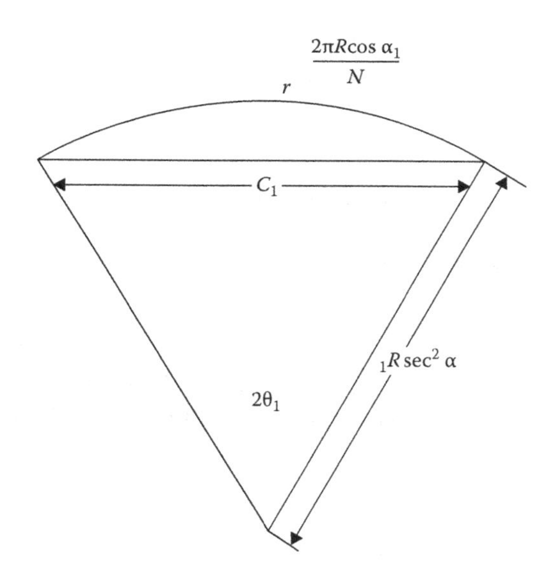

FIGURE 3.20
Inner arc of one segment.

Generalising from Figure 3.20,

$$\tan \alpha \text{ (lay angle)} = \frac{2\pi R}{L}$$

or

$$L_1 = \frac{2\pi R}{\tan(90-\alpha)} = 2\pi R \tan \alpha$$

and

$$L_2 = \frac{2\pi R}{\tan \alpha}$$

Therefore, the number of wires crossed by the line = $N(1 + tan^2\alpha)$

Assuming no clearance between the wires, the length of the inner arc of the segmental wire is

$$\text{Length of the helix} = 2\pi R \frac{\sec \alpha_1}{N\left(1 + \tan^2\alpha_1\right)} = \frac{2\pi R}{N\sec \alpha_1} = 2\pi R \frac{\cos \alpha_1}{N}$$

Figure 3.20 shows the inner curve of the segment of known radius and arc length, from which

$$\theta_1 = \frac{\pi R \cos \alpha_1}{NR \sec^2 \alpha_1} = \frac{\pi}{N \sec^3 \alpha} \text{ rad}$$

Therefore, the inner chord of segment C_1 is given by

$$C_1 = 2R \sec^2 \alpha_1 \sin \theta_1$$

Similarly for the outer chord,

$$C_2 = 2(R + t) \sec^2 \alpha_2 \sin \theta_2$$

2. *Relationship of the segmental layer radii and the total segmental area.* Consider a cross section through the stranded cable as shown in Figure 3.21.

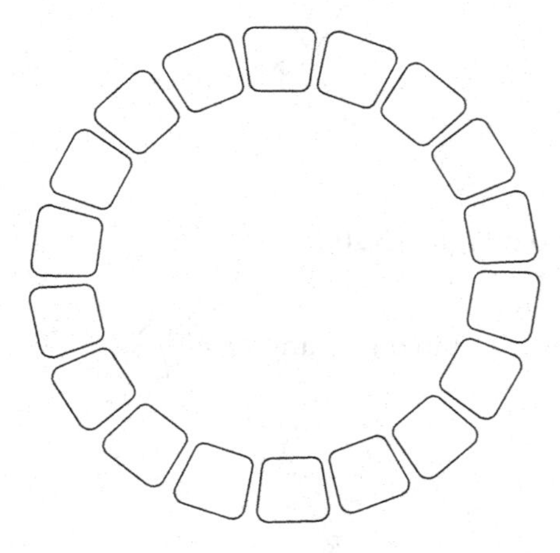

FIGURE 3.21
Cross section of a segmental layer.

3. *The area of the annulus is* $= \pi(Z^2 - R^2)$. If 'A' is the total of the individual wire areas, then since the wire has a lay length α, the total sectional area of the wires when cut perpendicular to the conductor axis will be $A \sec \alpha$.

   The remaining area of the annulus consists of the gap due to the clearance and the space left by the rounded corners of the segments. The space left by these rounded corners is almost the same as the area left by a circle of radius 'r' drawn inside a square of side $2r$. This is $(4 - \pi)r^2 = 0.858r^2$. The area left by the radius corners of 'N' wires, cut with an angle of α_1, is $0.858\, Nr^2 \sec \alpha$.

   The clearance between the segments has a constant angle. Therefore, the fraction of the annular area taken by the clearance is $NG_1 \sec \alpha / 2\pi R$.

   NOTE: Although 'α' is used here in place of the strictly accurate 'α_1', this does not introduce any significant error. We can therefore write

$$\pi\left(Z^2 - R^2\right) = \frac{\left(A + 0.858Nr^2\right)\sec \alpha}{\left(1 - \dfrac{NG_1 \sec \alpha}{2\pi R}\right)}$$

$$Z = \sqrt{\left(\frac{(A + 0.858Nr^2)\sec \alpha}{\pi\left(1 - \dfrac{NG_1 \sec \alpha}{2\pi R}\right)} + R^2\right)}$$

   The dimensions of the designed segments are shown in Figure 3.22.

4. *Area of the segment.* The corners of the segment shown in Figure 3.4 have a radius of 'r'. A formula for the area can be derived as follows:

$$\text{Area of trapezium } QPYX = AB \times \left(\frac{C_1 + C_2}{2}\right)$$

$$= \left[t + P_1(1 - \cos\theta_1) - P_2(1 - \cos\theta_2)\right]\left(\frac{C_1 + C_2}{2}\right)$$

5. *Area between the chords.* QP and arc QS_2P = area of sector OQS_2P – area $OQP = P_2^2\theta_2 - P_2 \sin\theta_2 P_2 \cos\theta_2 = P_2^2(\theta_2 - \sin\theta_2 \cos\theta_2)$; the area lost due to radius corner $S = 0.858\, r^2$. Therefore, the segment area is

$$\left[t + P_1(1 - \cos\theta_1) - P_2(1 - \cos\theta_2)\right] \times \left(\frac{C_1 + C_2}{2}\right)$$
$$+ P_2^2(\theta_2 - \sin\theta_2 \cos\theta_2) - P_1^2(\theta_1 - \sin\theta_1 \cos\theta_1) - 0.858r^2$$

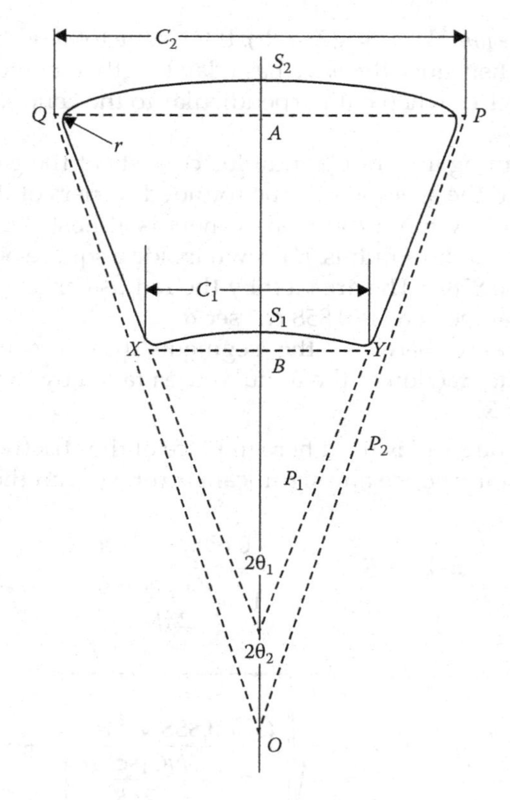

FIGURE 3.22
Shape of the drawn segment.

6. *Lay ratio.* In cable design, the lay ratio is defined as the lay length of a layer of wires divided by the mean diameter of the layer. It is therefore

$$\tan\alpha = \frac{\pi}{\text{Lay ratio}} = \frac{\pi D}{\text{Lay length}}$$

7. *Design procedure*
   a. Select values for the constants R, N, t (and hence α), G_1 and r.
   b. Select a value for A or Z where the aim of the design is a given overall diameter or a given total area.
   c. If designing for a given overall diameter works through the following formulae

$$l_1 = l \times \left(\frac{Z+R}{2R}\right)(\text{hence } \alpha_1)$$

$$l_2 = l \times \left(\frac{Z+R}{2Z} \right) (\text{hence } \alpha_2)$$

$$P_1 = R \sec^2 \alpha_1$$

$$P_2 = Z \sec^2 \alpha_2$$

$$G_2 = G_1 \times \frac{Z}{R} (\text{to give a constant angular clearance})$$

$$\theta_1 = \left(\frac{\pi}{N \sec^3 \alpha_1} \right) \text{rad}$$

$$\theta_2 = \left(\frac{\pi}{N \sec^3 \alpha_2} \right) \text{rad}$$

$$C_1 = 2P_1 \sin \theta_1 - G_1$$

$$C_2 = 2P_2 \sin \theta_2 - G_2$$

$$t = Z - R$$

All the segmental dimensions are now known.

d. Check the dimensional area using the following formula:

$$\text{Area} = \{t + p_1(1 - \cos \theta_1) - P_2(1 - \cos \theta_2)\} \left(\frac{C_1 + C_2}{2} \right) + P_2^2(\theta_2 - \sin \theta_2 \cos \theta_2) - P_1^2(\theta_1 - \sin \theta_1 \cos \theta_1) - 0.858r^2$$

e. If designing for a given area, first determine Z from the following formula:

$$Z = \sqrt{\left\{ \frac{\left(A + 0.858Nr^2\right) \sec \alpha}{\pi \left(1 - \left(\frac{NG_1 \sec \alpha}{2\pi R} \right) \right)} + R^2 \right\}}$$

Then proceed as detailed in (c).

Example 3.2

Showing actual calculation of profiled wire used to manufacture 400 mm² conductor, here a 300 mm² compacted conductor is selected as inner core. On this conductor 100 mm² area is to be built with profiled sections (Table 3.6):

TABLE 3.6

Calculated Details of the Profile Wire

Showing Outer Layer Profiled Wire Details to Construct a 400 mm² Conductor Over a 300 mm² Round Stranded Center Core	Units and Symbols	Dimensional Details	Units, Values of Angles & Referred Angle
Diameter of inner conductor is taken to be 300 mm² compacted	mm	20.5	
Final conductor to be made 400 mm²			
Area to be added	mm²	100	
Number of profiled wires selected	N	24	
Notations	m	16	
Lay length of the profiled wire	L	354.24	mm
Inner radius of the segmental wire	R	10.25	mm
Outer radius of segmental wires	Z	11.89	mm
Thickness of segmental wires	t	1.64	mm
Lay angle of segmental layers	α	11°10′	0.196
Lay angle of the inner surface of segment	α_1	9°34′	0.168
Lay angle of the outer surface of segment	α_2	12°45′	0.227
Lay ratio of inner surface of segmental layer	l_1	18.66	
Lay ratio of outer surface of segmental layer	l_2	13.87	
Inner radius of the segment	P_1	10.54	
Outer radius of the segment	P_2	12.50	
Clearance of inner surface of segment	G_1	0.05	
Clearance of outer surface of segment	G_2	0.058	
Inner chord of the segment	C_1	2.30	
Outer chord of the segment	C_2	2.77	
Radius of the segment corner	r	0.3	
Sum of the total segment area	A	100	
Half the angle subtended by C_1 at P_1	θ_1	0.13	7°10′
Half the angle subtended by C_2 at P_2	θ_2	0.12	6°56′
Lay length	L	354.24	
Lay length of the line at right angle to the segmental wire	L_1	382.58	
Lay length of the segmental wire	L_2	329.81	
Number of segments	N	24	
Area of the segment	mm²	4.12	
Total area of all segments (to be 100 mm² as required); actual	mm²	98.97	
Outer diameter of the conductor with segmented wires	mm	23.78	
Outer diameter of conductor stranded compacted 90%	mm	23.79	Calculated

3.2.3 Profiled Wire Drawing

As shown earlier, once the design parameter of the profiled wire is established, the wire can be drawn using a normal 13-die wire rod breakdown drawing. The die should be made of a tungsten carbide material. The die should be profiled in a spark erosion machine and should be flawless. The profiled wire can be drawn from a 9.5 mm wire rod. Shaping should be done during the last four stages. Take-up should be done with a proper guide system to keep the profiled wire always on the same plane. The wires can be accepted on a 630 mm bobbin DIN standard.

The other method, though not very popular but still used, is forming of trap wires just before the stranding by the use of rollers. In this case, a round wire is drawn and allowed to change shape before stranding in the stranding machine itself. All these methods are employed depending on specific requirements.

4

Conductor Formation–Stranding: Theory and Practice

Power that is transmitted from a supply station to the point of consumption is the product of a predetermined amount of current and power voltage and the phase angle between the current and voltage. If the power is transferred in a three-phase circuit, then it is three times that of the phase. This power is carried through a bare or insulated conductor of a network. Here, the conductor forms a vital part of the system.

The amount of power to be transmitted through a conductor depends on the following:

1. The area of the conductor to be selected, based on the conductivity of the metal which carries the given amount of current at a predetermined voltage. It is analogous to a water pipe that would carry a given amount of water at a certain pressure.
2. The choice of the conductor which should not only ensure smooth flow of current but also withstand short-circuit current, which at times would be very high. A short-circuit current causes an instant rise in temperature. If the area of the conductor is not sufficient enough to endure the combined stress developed, it may sustain severe damage and disrupt power supply. Therefore, both the current-carrying capacity and the short-circuit capacity of the conductor are important, whether it is designed for underground cable or an overhead line.
3. The voltage drop to be allowed for a particular length of a circuit. The conductor area has to be calculated accordingly.
4. Underground cables for which other external factors like thermal resistivity of surrounding materials and environmental and installation conditions are important. Naturally, the current-carrying capacity vis-à-vis the conductor area has to be selected taking into account all these relevant factors.

The diameter of the conductor is an important factor in designing medium voltage (MV), high voltage (HV) and extra high voltage (EHV) cables.

A small-diameter conductor will develop high stress in the MV and EHV systems. This can be shown by the following equation:

$$\text{Stress} = \frac{V}{r \log_e \frac{R}{r}} \tag{4.1}$$

where
V is the working voltage
r is the radius of the conductor below insulation
R is the radius of the conductor over insulation

This shows that stress is inversely proportional to r. If r is small, V becomes large. Naturally, after a certain voltage, the conductor diameter cannot be made too small and needs to be restricted at a particular dimension for a given voltage.

A conductor is formed by twisting a number of wires (called stranding) arranged in a regular fashion or at random, or a combination of both, to transmit a definite amount of power through a given length from a substation or transformer end to a subscriber point having a given cross-sectional area. It is the live portion of the cable that acts as the artery of the system and the power supply station is its heart. There are many forms of conductors being designed and manufactured and are still being developed taking into consideration the required characteristics of transmission parameters. For a small amount of current, a single solid wire can be used, but to carry a large amount, the cross-sectional area of the conductor should be correspondingly increased. It has been found that above 10 mm^2, a solid single conductor becomes stiff. At one time, a single solid conductor out of an aluminium wire rod was made by rolling in a Properzi mill and was used to manufacture cables. In such a case, the following disadvantages were observed:

1. To form a solid conductor above 10 mm^2, heavy troll rod–type drawing machines or Properzi rolling mills were employed.
2. To control dimensional parameters, precise processing tools were used, which were costly.
3. A conductor thus formed became stiff, and flexibility was less. Processing bobbins and dispatch drums were made larger, which called for a heavy and larger-capacity pay-off and take-up unit vis-à-vis higher investment. The bending radius during installation was large. Packing drums required larger bending radii and thus a larger size of the drum and cost.
4. Skin effect was higher resulting in an increase in AC resistance.

5. Special jointing ferrules and tools were required. The dimension at the joint was large. Contact resistance created the problem of heating, thereby increasing power loss.
6. The cable could not be made in longer lengths.

To overcome these difficulties in one go, larger-size conductors were made by stranding a number of smaller-diameter wires. (Wire drawing details have been elaborated in Chapter 3.)

The optimum area for a conductor was determined based on international standard (IEC 60228) and the Bureau of Indian Standard (BIS 8130). A guideline on the minimum number of wires has also been indicated. Naturally, it is left to the skill of manufacturers to choose the wire size and rationalise their production parameter in order to gain maximum economic advantage.

Wire conductors need to be stranded, whereas a solid conductor would eliminate intermediate processes and do away with extra equipment and labour costs. It is desirable to examine this aspect closely.

A solid-type conductor would develop undesirable internal stress and strain during bending.

A bent section of a solid round or sector conductor *A–B*, as shown in Figure 4.1, is considered. In this form, assume that the solid section has been formed by many thin sections of filaments stacked together as a bundle. If the solid section is bent as shown in Figure 4.1, each of these filaments will experience unequal stress and strain. The top portion of the filament will elongate more, whereas the bottom parts will get compressed. The cross section of both ends will try to become perpendicular to the middle axis or the neutral plane *A–B*. If the portion of a metal is kept in this position for long, it will have a permanent set. While unbending, the equilibrium of the set will be disturbed and a reverse stress will come to play. The cross section will also deform. Ultimately, cracks and scales will start forming on the top surface. The bottom surface will get compressed, creating uneven undulation. This constraint is eliminated by the stranding process.

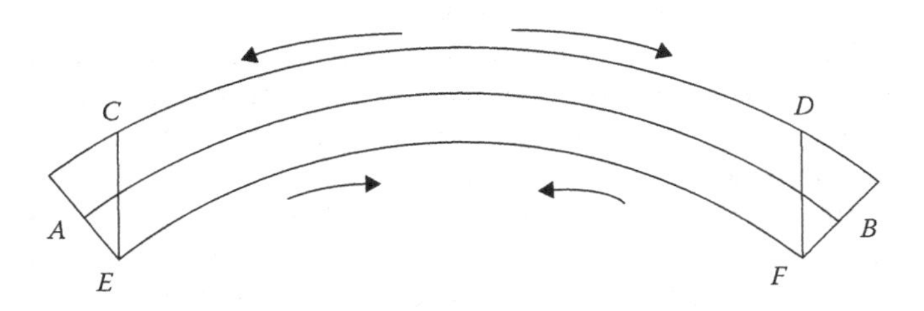

FIGURE 4.1
Forces acting on a bent section of metal strip.

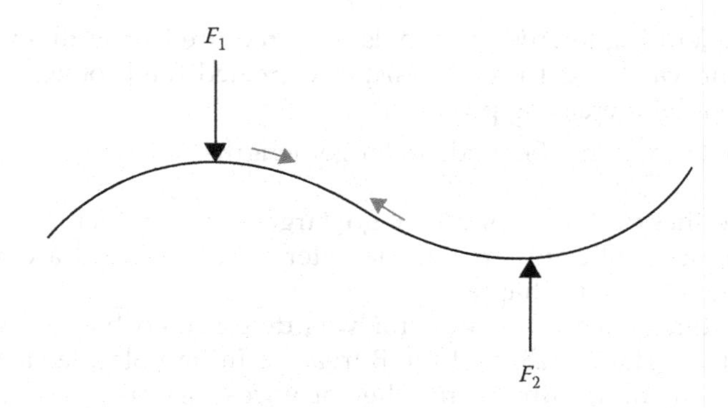

FIGURE 4.2
Forces acting on individual wire during stranding.

While stranding wires, in general, we get a long and open spiral form. During bending, a force comes to play on the conductor (Figure 4.2) trying to displace the position of the wire. Considering that a force F_2 is acting on a particular position of a wire, immediately an opposite force starts acting on point F_1 counteracting F_2 bringing an equilibrium and compensating the bending length. This is the reason why a stranded conductor can withstand repeated bending and unbending without wires getting displaced. For the same reason, multicore insulated conductors are laid together spirally to form a cable core. Wrapped armour tapes, plastic tapes, tapes of fibrous materials, or armour wires are also laid spirally to accommodate bending and unbending stress and strain. During winding and unwinding on bobbins or drums, sliding back of wires or conductors is not experienced. An additional advantage is gained particularly in the case of an aluminium-stranded conductor. Aluminium wire, after drawing, acquires a very thin oxide film on its surface which partially acts as insulating media. For this reason, when current flows through a stranded aluminium conductor, it divides in parallel paths to flow along the axis of the individual wire and not along the axis of the conductor as a whole. The resultant effect is a slight lowering of the effective resistance value. (Formation of oxide film renders jointing difficult, hence, before jointing conductor needs to be treated with special flux to remove the oxide layer in order to get proper electrical contact with the wires.) However, benefits thus obtained because of the parallel path mostly get compensated by the increase in effective length during stranding. However, some fractional amount of benefit can be extracted by designing the conductor intelligently. Thereby even if a small amount of reduction in diameter is achieved, the same will bring down the consumption of raw materials to some extent during subsequent processes.

Against all these advantages, a torsional force comes into play within the wires during stranding in a rigid-type stranding machine, where bobbins containing wires are kept in a fixed position. During rotational movement, carriage wires are twisted in a regular form. The axis of each wire rotates

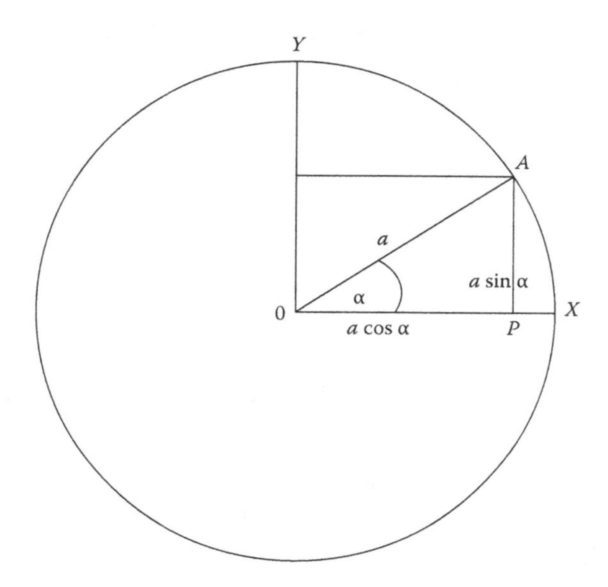

FIGURE 4.3
Torsional force acting on a wire during stranding.

360° around its centre along a helical path, with each revolution of the bobbin. Naturally, an inherent torsion develops with each turn. If this torsion is retained within the conductor, it may cause snaking and bird caging, damaging the cable. The amount of torsion that develops can be calculated theoretically as explained later.

Consider a point A (Figure 4.3) moving along a circle with a constant angular speed. The circle itself, which remains perpendicular to the axis of the cylindrical path, moves forwards along the axis of the conductor at a speed v. The equation of the helix can be in the form

$$\begin{aligned} &OP = X = a\cos\alpha \\ &\text{Now } AP = Y = a\sin\alpha \text{ and } Z = v\,t \end{aligned} \tag{4.2}$$

where

- Z is the axis of the cylinder formed in time t
- α is the angle subtended by the wire perpendicular to the axis of the cylinder
- v is the speed of the capstan

In this equation, Z is the axis of the cylinder formed by the progressive movement of the helix moving forwards. The point on helix A turns through an angle α in time t. When the angle makes a full turn ($\alpha = 2\pi$), it covers the complete lay length h along the direction of the cylinder. In this case, Equation 4.2 may be written as

$$x = a\cos\alpha; \quad y = a\sin\alpha \quad \text{and} \quad z = \frac{h\alpha}{2\pi} = b\alpha \tag{4.3}$$

where

$$b = \frac{h}{2\pi}$$

$$a = \frac{D+d}{2} = \frac{D'}{2}$$

where
D is the diameter of the centre core
d is the diameter of the wire moving helically on the outer layer of the conductor

Here, the helix is a space curve. It has both curvature and torsion. The curvature indicates the degree of deviation of a curved line at a point from the direction of the relevant tangent. The value of ρ_1, which is the reciprocal of curvature k, is defined as the radius of curvature. The torsion indicates the degree of deviation of a space curve at any point from the corresponding osculating plane. (An osculating plane is a plane that passes through the tangent line and the principal normal of the curve.)

Thus, the torsion is characterised by the angle formed by two adjacent osculating planes.

When the radius of curvature is given by ρ_1 then:

$$\rho_1 = \frac{a}{\cos^2\varphi} = \frac{D+d}{2\cos^2\varphi} \tag{4.4}$$

where φ is the lead angle of helix.

The torsion of the wire per unit length of the helix is given by the following equation:

$$T = \frac{1}{\rho_2} = \frac{b}{a^2+b^2} \tag{4.5}$$

or

$$T = \frac{\dfrac{h}{2\pi}}{a^2 + \left(\dfrac{h}{2\pi}\right)^2} = \frac{2\pi h}{(2\pi a)^2 + h^2} \tag{4.6}$$

Since $h = 2\pi\, a \tan \varphi$, it follows that

$$T = \frac{(2\pi)^2 a \tan\varphi}{(2\pi a)^2 + h^2} = \frac{(2\pi a)^2 \tan\varphi}{[(2\pi a)^2 + h^2]a}$$

Since

$$\frac{2\pi a}{\sqrt{(2\pi a)^2 + h^2}} = \cos\varphi,$$

$$T = \frac{\cos^2\varphi \tan\varphi}{a} = \frac{\sin 2\varphi}{2a} = \frac{\sin 2\varphi}{D} \tag{4.7}$$

As such, torsion is proportional to the angle φ. The maximum value is obtained when $\varphi = 45°$.

Of special interest is the torsion of a complete turn, that is the relevant length of a helix. It is given by

$$S = 2\pi\sqrt{a^2 + b^2}$$

and the torsion for this length is

$$T = \frac{b}{a^2 + b^2} \cdot 2\pi\sqrt{a^2 + b^2} = \frac{2\pi b}{\sqrt{a^2 + b^2}} \tag{4.8}$$

Since $2\pi b = h$, it follows that

$$T = \frac{h}{\sqrt{\left(a^2 + \left(\frac{h}{2\pi}\right)^2\right)}} = \frac{2\pi h}{\sqrt{(2\pi a^2) + h^2}} \tag{4.9}$$

Taking the value as

$$\frac{h}{\sqrt{(2\pi a^2 + h^2)}} = \sin\varphi$$

the torsion per length is obtained as

$$T = 2\pi \sin\varphi \tag{4.10}$$

Here, torsion is expressed in radians per one turn, whereas torsion per unit length is expressed in radians per centimetre.

To obtain the torsion per unit length of a helix for each turn of the twisting of carriage, the previous equation is divided by 2π to obtain

$$T = \sin \varphi$$

T is generally measured in cm^{-1}. This is a dimensionless quantity and is expressed in a number of turns as referred to the length of one turn of helix.

It shows that when twisting is done without a back twist, there is a complete turn in the twisting of the wire. Theoretically, only for an infinitely large lay length, that is $\sin \varphi = 1$, the maximum value of the torsion per unit length will correspond to angle $\varphi = 45°$.

To release the wire from torsional force, untwisting is being done by rotating the cradles in the opposite direction and by actuating the anti-twist device. In this case however, a small residual torsion is retained within the wire which can be computed on a per one turn revolution basis of the carriage, that is the length of lay. It can be seen that the torsion per unit length due to untwisting is

$$T_1 = \frac{2\pi}{\sqrt{(2\pi a)^2 + h^2}} \tag{4.11}$$

Since

$$\frac{2\pi a}{\sqrt{(2\pi a)^2 + h^2}} = \cos \varphi$$

it follows that

$$T_1 = \frac{\cos \varphi}{a} = \frac{2\cos \varphi}{D'} \tag{4.12}$$

The residual torsion exists when twisting is done with a back twisting device

$$T_0 = T_1 - T = \frac{(2\cos \varphi)}{D'} - \frac{(\sin 2\varphi)}{D'}$$

or when expressed in degrees

$$T_0 = 180° \frac{2\cos \varphi - \sin 2\varphi}{\pi D'} \tag{4.13}$$

where $D' = D + d$.

If residual torsion is computed in revolutions per one complete turn of the cage, then

$$T_0 - T_1 - T = 1 - \sin \varphi \tag{4.14}$$

It has been seen that the value of torsion for twisting with a back twist is very small. However, when the angle φ is very small, torsion can become appreciable even when twisting is done with the back twist system. The value of φ at which such phenomenon is to occur can be determined from the following equation:

$$\frac{\sin 2\varphi}{D'} = \frac{2\cos\varphi - \sin 2\varphi}{D'} \tag{4.15}$$

Solving this equation, we get the minimum value for $\varphi = 30°$. Below this value, a 360° twist even with the back twisting system becomes impractical. Therefore, small wires, insulated or bare, can be twisted with a partial back twist when the angle selected is very small.

Now we consider the twisting of wires which have a definite dimension (diameter). For twisting without a back twist, the wire should always contact the centre with the same general matrix. While the wire undergoes the twisting only in one direction to form a helix, the axis of the wire shall form an arc of the radius:

$$\rho = \frac{D+d}{2\cos^2\varphi}$$

The internal layers of the wires immediately adjacent to the previous layer form an arc of the radius

$$\rho_{int} = \frac{D}{2\cos^2\varphi}$$

And the external layers make an arc of the radius

$$\rho_{ext} = \frac{D+2d}{2\cos^2\varphi}$$

Thus for twisting without a back twist, the external layer is stretched and the internal layer compressed. Every wire is deformed along the entire length not only by torsion but by pure bending as well. This deformation usually takes place in the elastic and plastic zone and is the main cause of layer disruption.

It is important to know that torsional deformation takes place in every layer of the wire. The equation given previously is to be considered as the average values for the entire cross section, because they correspond to the torsion of the wire axis. The curved shape of the wire is caused by a shear deformation, as the shearing force acts continuously on the wire, rotating about its axis and at the same time moving along its length.

Considering all this, it can be inferred that

1. When twisting without a back twist is considered, one has to take into account the angle of φ of helix, because torsion depends on its value.
2. Even then, residual torsion rises to its maximum when back twist is applied during twisting when φ is 45°.
3. Since lay length *m* is equal to or less than 1.8, torsion due to twisting is not large enough, and hence a back twisting device will not be necessary.
4. During a back twist as applied during stranding (twisting), residual torsion being less, whatever deformation curve observed is on account of the shearing force.
5. Torsional force due to twisting can be eliminated by the force of compacting.

Initially, the a conductor was formed on a machine with a back twist arrangement driven by a gear train, whose speed reduced with each cage used. Here, only round non-compacted conductors are used. Naturally, these machines are ideal for avoiding internal torsion. In the course of time, it was seen that by compacting, the internal stress can be minimised if not altogether eliminated (point No. 5). Naturally, a high-speed, rigid-type machine came into existence that gave a better production rate by allowing the diameter to be reduced from 3% of the main pitch circle diameter (PCD) in the case of a compressed conductor without impairing the resistance value and flexing characteristics. As the technology advanced, a compaction of 92% was achieved, and sometimes even better. The reduction in the diameter lowered the consumption of materials during subsequent operations. To get the best results in terms of diameter and resistance values, a combination of intermediate annealing, drawing and stranding process was adopted to achieve lower dimensions for the conductor.

The area of the conductor is selected considering the following factors:

1. The amount of current it has to carry. Simultaneously, a short-circuit rating also needs to be considered where the condition is critical.
2. The expansion and contraction due to fluctuation of heat generated during operation and due to change in atmospheric conditions.
3. The formation is sufficiently flexible to withstand twists and turns and mechanical abuses during operation and installation.

In all cases, service conditions are to be defined clearly.

Conductors are designed in many forms as per the type of cables and their usage. Just for the sake of interest, a few types are mentioned here which will give a fair idea of varieties that are generally required:

1. For domestic building, wire cable conductors are made flexible to accept twists and sharp bending during installation within a building complex and in constricted areas. These conductors are smaller in size and are formed by bunching a number of fine wires together in a single or double twist bunching unit. They are highly flexible in nature. These conductors are preferably made of bunch or solid copper. However, aluminium conductors have limited use due to frequent failures. Some efforts have also been made to develop copper-clad aluminium conductors.
2. For mining, cables such as drill cables, coal cutter cables, shuttle car cables, excavator cables and miner's lamp cables are being made. The conductors are made flexible by bunching and stranding a large number of fine wires. Cables are used in a very constricted area where bending requirements are severe. The cables are subjected severe stress and strain. These conductors have to be insulated and sheathed with vulcanised natural rubber or elastomeric compounds and blended with sulphur as a curing agent. Sulphur reacts with bare copper wires and deteriorates their electrical characteristics. For several decades, natural rubber and butyl rubber ruled as insulating material for rubber cables. During the early 1960s, ethylene propylene rubber (elastomer) (EPR) or ethylene propylene diene monomer (EPDM) as a new insulant replaced the traditional insulating material with peroxide as a curing agent. The advantage of this new material is that it has higher thermal properties, which makes their use possible not only in MV cables but also in EHV cables above 66 kV. Naturally, wires are to be coated with tin to protect them from the attack of sulphur and other corrosive chemicals used in manufacturing rubber compounds. Further, while in use the cable can get damaged because of the highly abusive surroundings damaging the protective coverings. Conductors of damaged cables become exposed to the polluted atmosphere. If bare copper remains exposed in this condition, the conductor will get corroded, making it dysfunctional. For this reason, tinning of copper wire becomes an essential part of manufacturing.
3. Similarly, cables for electrically operated overhead trolley (Crane) (EOT) cranes, laughing cranes at port area and cables for mobile equipments are also made by bunching and stranding fine tinned copper wires.
4. Railway locomotive cables and internal loco control wiring cables are made with fine tinned copper wires bunched and stranded.

These cables have to be installed within a restricted space where bending, twisting and turning are sharp. Winding and unwinding are done repeatedly under severe stress and vibration. Sometimes, depending upon the local specifications, locomotive cables are made with bare copper wrapped with a separator tape and insulated with normal peroxide cure EPDM rubber and chloro sulphonated polythene (CSP) sheathed.

5. There are cables made with single solid conductors of smaller cross-sectional area, such as 1.5 and 2.5 mm², to manufacture signalling and control cables of multicore type.
6. For instrumentation cable of 7 × 0.3 mm, tinned copper wires stranded in regular form are used.
7. For fixed installation type of cables, wires are stranded in regular concentric form layer by layer, where each layer is laid in opposite direction from the previous one. Nowadays, rope lay is also considered by twisting concentric layers in the same direction. These conductors can remain non-compacted, compacted or shaped in sector form or oval shape. By compacting, the conductors' inherent stress within the wires could be eliminated and the diameter reduced.
8. For high voltages and higher sizes, above 1000 mm², conductors are formed by being laid in several segments separated by a thin layer of insulating tape to minimise the skin effect. The number of segments, either 4, 5 or 6, is the designer's choice depending upon the requirements and infrastructure available. These conductors are called Milliken conductors.

For high-voltage and extrahigh-voltage oil and gas pressure cables, conductors are formed as a hollow tube by stranding flat segmental or profiled wires in order to

1. Minimise the skin effect and circulate oil or gas through the tube for cooling purposes.
2. Make single seamless extruded hollow conductors for very-high-frequency (VHF) and ultrahigh-frequency (UHF) cables as well as for wave guides. Some of these tubes are made by seam-welding continuously and are corrugated to impart flexibility without distorting transmission characteristics.

There are conductors made for overhead high-voltage transmission lines called all-aluminium conductor (AAC), all-aluminium alloy conductor (AAAC), aluminium conductor steel reinforced (ACSR), Al-59 and aluminium alloy conductor steel reinforced (AACSR).

Currently with the issues of ROW and with the increase in demand of power transfer in the existing line, a new generation of conductors called high-temperature low sag (HTLS) have been introduced in the market. There are different technologies available, such as aluminium conductor steel supported (ACSS), GAP type, super thermal alloy invar reinforced and composite carbon core conductors. These conductors can operate at higher temperatures, sometimes up to 210°C. These conductors are capable of transferring double the power without any increase in sag and are best fit for reuse and for new lines. In case of increase in load, instead of making a new line with normal conductors and having a huge capital investment, it is recommended that these conductors be used, which shall result in overall CAPEX saving of more than 26%. These technologies are gradually gaining popularity worldwide.

Conductors come in various sizes and with different cross sections.

It is imperative to understand the technique of forming a conductor. While stranding a conductor, one has to obtain optimum results both in terms of electrical and mechanical characteristics. Stranding of a conductor is done in a regular configuration in a concentric manner. It follows a given mathematical pattern as shown in Figure 4.4. There is a centre wire around which only six wires of the same diameter can be placed.

It is interesting to note that if another layer of wires having the same diameter is placed on this layer and twisted around, the number of wires on this layer will increase by 6 making it 12 in total. Thus, if the number of wires at the centre is 1, then the number of wires over the centre wire as stranded will be 6, and the number of wires on top of this given layer will increase again by another 6 wires, following the sequence 1 + 6 + 12 + 18 + … . This can be established mathematically.

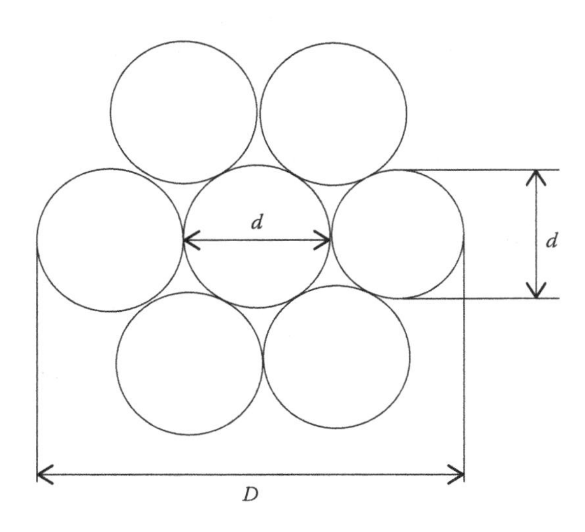

FIGURE 4.4
7 – wire strand.

Let d be the diameter of all the wires. If there is a single wire in the centre, then on top of the central one, say n number of wires can be accommodated and are given by

$$n = \frac{\pi(d+d)}{d} = 2\pi = 6.283 \text{ (as per Figure 4.4)}$$

where n is the number of wires and is equal to six, since the number of wires cannot be a fraction.

Thus, it can be shown that there will always be an increment of six wires on each successive outer layer from the previous inner layer. It also shows that if these six wires are laid parallel without twisting, there shall remain a total gap of $0.283d$ on the given layer.

If this space is distributed equally amongst all the six wires, then the gap between each wire shall be $0.283/6 = 0.047d$ approx. This proportionate gap will remain throughout in each layer between two parallel adjacent wires. This is the gap actually to be filled in by the stranding process, that is by twisting in a regular helical pattern around the corresponding centre at an angle. When the stranded conductor is sectioned at the right angle of its axis, the cross section of the wires on the outer layer(s) is viewed elliptical in shape (Figure 4.5).

The major axis of the ellipse is given by

$$d' = \frac{d}{\sin\alpha}$$

Considering $\sin\alpha$ is the angle of twist, that is the angle between the axis of the twisted wire to the perpendicular plane of the axis of the conductor,

$$\sin\alpha = \frac{d}{d'} = \frac{d}{d+0.047d} = \frac{1}{1.047} = 0.9551 \text{ or } \alpha = 72^\circ 46'\ldots \tag{4.16}$$

Now the straight length of one complete turn of helix (crest to crest) is denoted by h called the lay length, and then $\tan\alpha = h/\pi\, dm$, where dm is the

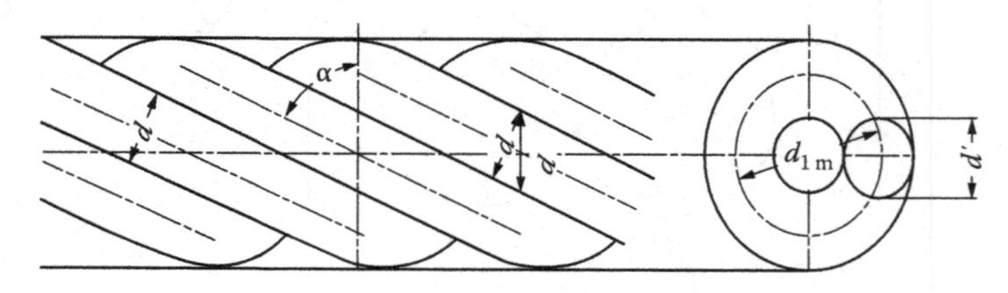

FIGURE 4.5
Lay angle of stranded wire.

mean diameter of the stranded conductor, as shown in Figure 4.5. Now let us consider that $h = m\ dm$, that is $h = m$ times the mean diameter of the stranded conductor. Then

$$\tan\alpha = \frac{h}{\pi d_m} = \frac{m d_m}{\pi d_m} = \frac{m}{\pi} \quad \text{or} \quad m = \pi \tan\alpha \tag{4.17}$$

With $\alpha = 72°\ 46'$, we get $m = 10.128$, that is $h = 10.128 \times dm$.

This shows that while stranding, the minimum lay length should never fall below 10 times the *pitch diameter* of the conductor *dm*. Below this value, the result would be an overriding of wires. Normally, for non-compacted conductor, the lay length is selected at 16–18 times the mean diameter of the conductor, depending upon the size, and for compacted ones it can be selected up to 20 times the mean diameter of the conductor, unless otherwise specified by the customer. In case of AAC, AAAC and ACSR, as well as for aerial bunch cables (ABCs), the lay length has been specified to be kept within 12–14 times the conductor diameter (as per BIS 398).

On stranding, gap formations are found between the wires when viewed through the cross section of the conductor (Figure 4.6).

These are air gaps formed between wires. Naturally, the area covered within the diameter of the conductor is not totally filled in by metal and is the sum of the area covered by metal and air gaps. The total area is called the geometrical area. The percentage of the ratio between the actual area covered by metal and the total geometrical area is known as the filling factor denoted by *f*.

Thus, let *D* be the diameter of the stranded conductor and *n* the number of wires with diameter *d*.

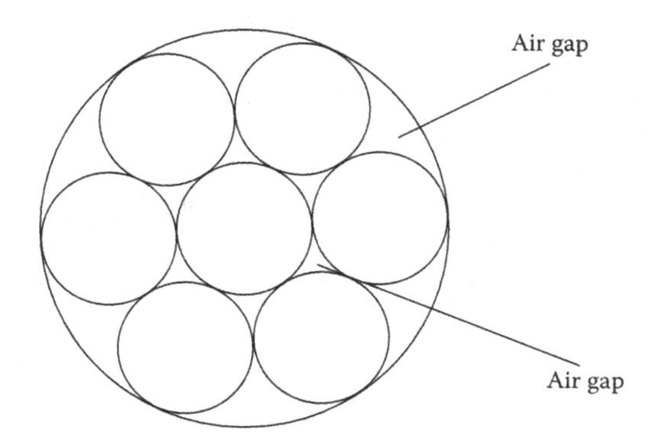

FIGURE 4.6
Stranded conductor cross section showing air gap.

Then the ratio is

$$f = \frac{n\pi d^2/4}{\pi D^2/4} = n\left(\frac{d}{D}\right)^2 \quad (4.18)$$

In case of a seven-wire strand, the diameter of the stranded conductor is $D = 3d$, and the ratio of the metal area of wires to that of the geometrical area of the stranded conductor is

$$\frac{7d^2}{(3d)^2} = \frac{7}{9} = 0.778$$

which means that if the total geometrical area is taken as 100, then the metal content would occupy 77.8% of the geometrical area. The rest of the area would be filled by air. This factor is given by $f = 77.8/100 = 0.778$ and is known as the filling factor. With the increasing number of layers, the value of the filling factor approaches 0.75, i.e., 75%. Here, the endeavour of every designer would be to achieve as much higher filling factor as possible, which is possible by taking right wire sizes as well as higher compacting %.

The filling factor thus is defined as

$$f = \frac{\text{Area of metal content} \times 100}{\text{Geometrical area of the stranded conductor}} (\%)$$

NOTE: It is interesting to know that by compacting a conductor, mutual capacitance can be reduced as can be seen from the following example:

The diameter of round non-compacted stranded 120 mm² conductor is approx. 14.0 mm. If the conductor is compacted to gain 93% filling factor, then the geometrical area becomes $120/0.93 = 129\ \text{mm}^2$ and the diameter is $\sqrt{129 \times 4/\pi} = 12.82$ apx. The diameter is reduced by 9%.

Taking insulation thickness of 1100 V grade as the 1.4 mm diameter of non-compacted conductors becomes 16.8 mm and the compacted one shall be 15.6 mm, respectively. Taking $\varepsilon_r = 5$ for polyvinyl chloride (PVC), we get capacitance values to be calculated as per the following formula:

$$c = \frac{2\pi\epsilon_0\,\epsilon_r}{\ln(R/r_0)}\ \text{F/m}$$

For conductor diameter $r_0 = 14.0$ mm and insulated diameter $R = 16.8$ mm, and $\epsilon_0 = 8.85 \times 10^{-12}$, the value becomes 1.345 μF/km, and in the case of a compacted conductor, the value becomes 1.246 μF/km. In the latter case, power loss is likely to be lesser. This is also true for sector- and oval-shaped conductors. This is an indication that lower power losses can be achieved by compacting the conductor.

A stranded conductor is formed by twisting wires around a centre wire while the centre wire remains straight. It is observed that around one number of centre wire, six other wires can be brought over, provided that all the wires are of the same diameter and those on successive layers also go on increasing by six wires. The configuration is also true when the centre wire consists of two or three or four twisted wires. In these cases also the number of wires twisted over the given centre shall increase by six wires from the adjacent inner layer. During stranding, wires are laid spirally. The effective length of the wire increases proportionately to the lay length. A definite relation exists between the mean diameter, lay length and the effective length of the wire, which can be seen in Figure 4.7.

Here, *dm* is the mean diameter of the stranded conductor, *h* is the lay length, *L* is the the actual length of the wire when straightened and α is the the lay angle.

From Figure 4.7, we get $h = L \sin \alpha$. Hence, the increment in length is

$$L - h = \left\{ \frac{h}{\sin \alpha} - h \right\} = h \left\{ \left(\frac{1}{\sin \alpha} \right) - 1 \right\} \tag{4.19}$$

Now we have seen that $\tan \alpha = m/\pi$.

By selecting the value of *m*, the lay length is to be chosen which is *m* times the mean conductor diameter (this is given by $m \times dm$, where $m > 10$). From this, we can calculate the increment in length on the top layer of the wire as in percent of the straight length:

The percentage of the increment is

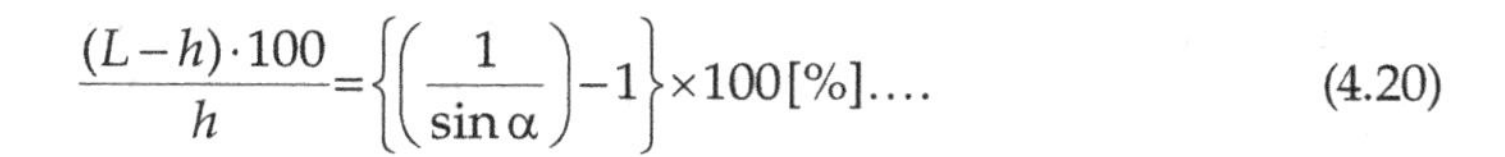

$$\frac{(L - h) \cdot 100}{h} = \left\{ \left(\frac{1}{\sin \alpha} \right) - 1 \right\} \times 100\,[\%] \ldots \tag{4.20}$$

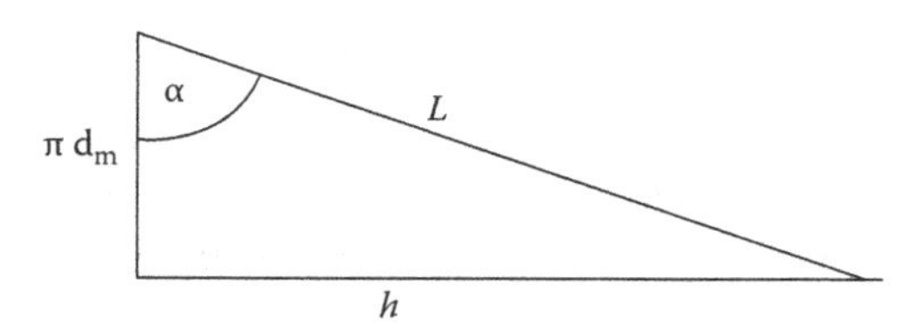

FIGURE 4.7
Lay length in relation to PCD.

Example 4.1

If $h = 100$ *m* and $m = 16$, then we get $\tan \alpha = 16/\pi$ or $\alpha = 78.89°$.

Therefore, the percentage of increment in length is $= \{(1/\sin 78.89°) - 1\} \times 100 = 1.94\%$ apx. Thus, for 1000 m stranded conductor length, the actual length of the wire on top of the centre wire, when taken in straight form, will be 1019.4 m.

It can be seen from this formula that the increment solely depends on factor *m* during stranding. By fixing the value of *m*, the exact length of the wire required per layer per 1000 m can be calculated. This is useful to plan the quantum of wire that is to be taken on a bobbin for stranding a conductor of a particular length without keeping any leftover. This means that the quantity of scrap will be zero. In practice, however, it does not happen, but this action will definitely help in minimising scrap to its lowest value when planning is meticulously done. Practical experience has proved that the longer the lay length, the better the circulation of the conductor. But the lay length is a factor of size and is part of process adapted for insulation.

By stranding, an effective cross-sectional area of the conductor increases. This can be seen by cutting the conductor at the right angle plane of the axis. The cross-sectional area of the stranded wires in each layer becomes elliptical, except the centre wire (Figure 4.5).

The major axis of the area of the wire is given by $d/\sin \alpha$ and the minor axis is *d*.

d is the diameter of the wire. The increment in area is

$$\Delta a = \left\{ \frac{\pi d}{2} \times \frac{d}{2 \sin \alpha} \right\} - \frac{\pi d^2}{4} = \frac{\pi d^2}{4} \left(\frac{1}{\sin \alpha} - 1 \right) \tag{4.21}$$

When $m = 16$ and $\alpha = 78.89°$, $\sin \alpha = 0.981$.

From Equation 4.21, taking into account $d = 1$, the increment in area is also seen to be 1.94% apx. Naturally, the weight of the conductor will increase proportionately.

Example 4.2

Considering a conductor having an area of 120 mm² made of 19 strands, each wire diameter when considered parallel to the axis, leaving no gap, becomes 2.836 mm approx. If we take the lay length factor as $m = 16$, then $\alpha = 78.89°$ and $\sin \alpha = 0.981$. The length in each layer will uniformly become longer by 19.4 m for each 1000 m wire length. Apart from the centre wire, all other wires will be laid helically. Thus, the assembled final length of 19 wires will form a conductor of 1000 m. Now the length of each helically laid wire when straightened becomes 1019.4 m, and for all the 19 wires, the total length will be 1000 + 1019.4 × 18 = 19,349 m = 19.349 km.

The area of each wire is 6.3158 mm². As such the weight of copper wire becomes 56.15 kg/km (density 8.89). Hence, the rgw weight of a stranded 120 mm² copper conductor becomes 1086.408 kg/km, whereas a solid 120 mm² conductor will weigh 1066.80 kg/km. The increment in weight is 19.608 kg/km. This is 1.84% of the total conductor weight. The lower value obtained against 1.94 is due to the fact that the centre wire remained straight and did not have any incremental value. There was an increase only on the 18 wires which were twisted around the centre wire.

A similar result is obtained when cross-sectional increment is considered. The wire diameter at 2.836 is taken as the minor axis, whereas the major axis is 2.891. Therefore, the area is $(\pi/4) \times 2.836 \times 2.891 \times 18 + 6.3158 = 122.225$ mm² and the weight of copper is $122.225 \times 8.89 = 1086.6$ kg/km approx., which agrees with the previous value.

Further, the lower the value of m, the higher will be the input of material, and the maximum value is obtained when $m = 10$. A chart can be built up giving the length of the wire against the corresponding values of m. This helps to plan the length to be taken in a bobbin for stranding a conductor of a certain length at a certain value of m which will help to theoretically obtain zero scrap on the length taken in the bobbin. In practice, the scrap value never comes to zero but can be brought down to a minimum level by selecting a zero scrap programme based on the proper selection of wires for each layer as described in Table 4.1.

By selecting m, the corresponding value of α can be obtained. Knowing the value of α, an increase in length l can be computed.

The following table shows the percentage of increase in length against certain specified lay lengths ($h = mD_m$).

TABLE 4.1

Percentage of Increase in Length of Wire in Relation to Lay Ratio

mD_m = Lay Length	Length Increase in Percentage (Δl %)	mD_m = Lay Length	Length Increase in Percentage (Δl %)
5 D_m	18	15 D_m	2.2
6 D_m	13	16 D_m	1.9
7 D_m	9.5	17 D_m	1.7
8 D_m	7.5	18 D_m	1.5
9 D_m	5.5	20 D_m	1.2
10 D_m	4.8	25 D_m	0.7
11 D_m	4	30 D_m	0.5
12 D_m	3.4	40 D_m	0.3
13 D_m	3	50 D_m	0.2
14 D_m	2.5		

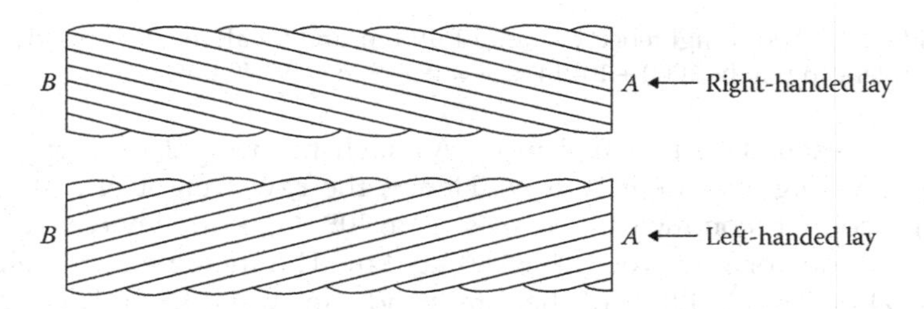

FIGURE 4.8
Direction of lay in a stranded conductor.

In a stranded conductor, the twisting of wires is done layer by layer in a concentric form while each layer is laid helically on the inner layer. Normally, successive layers are stranded in opposite directions from the previous inner layers. If one layer has a left-handed lay, the top layer on the same is stranded with a right-handed lay, thus continuing in an alternate direction one above the other. If we try to lay each layer in the same direction, there are chances of the conductor opening out, which is not recommended, and the manufacturer is compelled to follow standard specifications. Figure 4.8 shows the direction of lay. A right-handed lay occurs when one looks from the capstan side towards the rotational direction of the final cage. The cage rotates in a right-handed direction to give the right-handed lay. The opposite is the case for the left-handed lay.

The direction of the lay remains the same when viewed from the *A* end or *B* end of the strand, as shown in Figure 4.8. Alternate layers spiralling in alternate directions bind the conductor tightly and prevent any bird-caging action during winding and unwinding. During armouring, it helps keep the conductor in a tight position.

Initially, the electrical cable conductor design was dependent on the voltage system and the quantum of power that could be transmitted through a transmission line, that is cable. For safety reasons, urban and industrial distribution lines were made of insulated conductors and cables.

However, emphasis was given to increase the voltage range, keeping the conductor size as per selected parameters. Quality, which included the breakdown voltage of available insulating materials, in those days was also not satisfactory. To ensure safe transmission parameters, conductors were thus constructed round, either solid or stranded (Table 4.2).

The problems faced were as follows:

1. For a three-phase, multicore, round, insulated conductor, a large amount of interstice filling material had to be inserted to make the laid-up cable core round. Figure 4.9 illustrates a cable with round conductors having round fillers within interstice spaces. Still there are voids that need to be filled. Actually, bunched paper/jute or non-hygroscopic bunched polypropylene threads were used to fill

TABLE 4.2

Wire Diameter Determines the Area of Interstice Space in a Stranded Conductor

Number of Wires or Cores in a Stranded Conductor or Cable	Diameter of Round Interstice Filler	Diameter after Stranding or Laying Up	Area of Outer Interstice Filler	Area of Inner Interstice Filler	Total Area of Interstice Filler
2	$0.67d$	$2d$		$0.785d^2$	$1.57d^2$
3	$0.15d$ (centre) $0.48d$ (outer)	$2.155d$	$0.04d^2$	$0.417d^2$	$1.291d^2$
4	$0.41d$	$2.414d$	$0.215d^2$	$0.305d^2$	$1.434d^2$

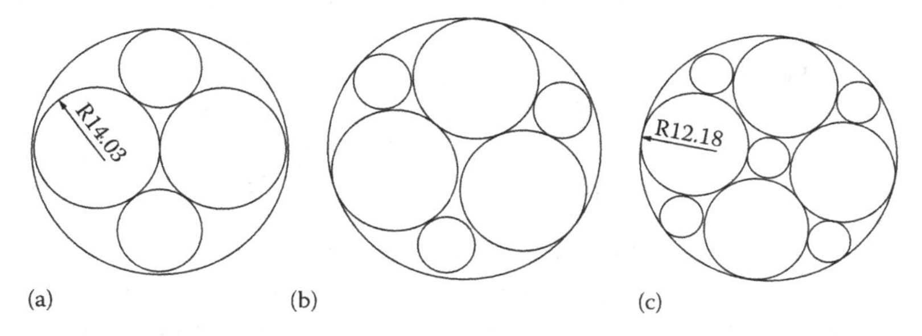

FIGURE 4.9
Cables made with a round conductor with interstice round fillers: (a) two-core, (b) three-core and (c) four-core.

in the total space, and the laid-up diameter of cable became large. For cross linked polythene (XLPE) and PVC cables, PVC fillers were extensively used.

Consequently, the consumption of outer protective materials increased, increasing the diameter and the weight and cost of the cable. Also cables of loner length could not be made.

2. This increased packing, transportation, installation and jointing costs.
3. These problems were considerably solved by making sector-shaped conductors which reduced the diameter, input raw material consumption and overall cost of packing, transportation, installation and jointing, since these cables could be made in longer lengths.
4. In the lower-voltage range such as from 1000/1100 to 3000/3300 V paper-insulated cables, shaped conductors were used. From 6000/6600 V and above 66 kV paper-insulated oil- or gas-filled cables, conductors were made elliptical at times. Above 66 kV, all conductors were round and constructed with profiled strips, making the outer surface smooth. The inner side of such conductors was made hollow allowing the flow of oil or gas.

5. The initial reaction to thermoplastic insulation was that the insulation will not be able to withstand voltage ratings if conductors are made sector shaped or elliptical; this was based on the following
   a. Lower thermal endurance property
   b. Lower dielectric strength compared to paper insulation
6. With the development of superior-quality cross-linked polythene as insulation, continuous catenary vulcaniser (CCV) line extrusion and dry curing system, such limitations to use sector and elliptical conductors can now be overcome.
7. Low tension (LT) PVC cables were made with shaped conductors leading to a lower geometrical area, which led to saving of space and material.
8. In the late 1970s, attempts to popularise three-layer 11 kV XLPE cables and sector-shaped cables were met with limited success. The main reason was the difficulty to address uniformity of the electrical stress in the sector corners. This challenge was not observed in circular cables. The ease in manufacturing circular cables by the three-layer extruded process made them more popular.
9. However, looking at the current competition in the market, there is a need to readdress the issue by using XLPE insulation sector-shaped conductors to produce cables in the 10,000/11,000 V range and 30,000/33,000 V cable with elliptical-shaped conductor has been felt. These cables are more compact. At present, the main issue is to obtain the right design and shape (sector/ellipse) that can be extruded and insulated in the extrusion lines successfully by using a super clean quality XLPE compound. In the near future, the issues related to elliptical conductors up to 33 kV will be fully eliminated, and cables with elliptical conductors will offer a cost-effective solution with reduced material costs and distribution losses.

Initially, attempts were made to construct a shaped conductor for a 1000/1100 V cable. Most of them were based on the concept of area measurement. The first attempt was made by Norman C. Davis of General Cable Corporation in 1928. The next patent was filed in 1933. But the concept of compaction came much later. In 1960, compacted sector conductor was designed by the author along with Herr Friesenhagen of Felten and Guilleaume, Kȍln, Germany, on the basis of precise geometrical calculation. This concept puts forward generalised formulae covering all types of sectors with the possibility to compact a conductor as required.

Constructing rolls for sector-shaped conductors is given by the formulae, as shown later. These calculations apply to all types and angles of rolls and

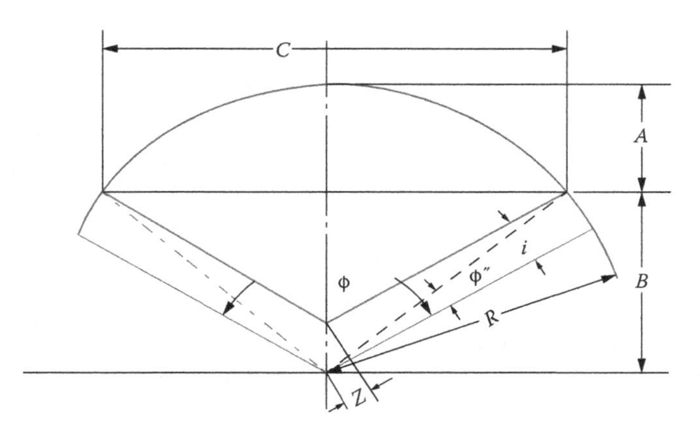

FIGURE 4.10
Schematic diagram of a sector conductor-design parameters.

as per the choice of filling factors. These formulae are fairly accurate and are being used for manufacturing sector conductors.

The general formula for sector conductors is as shown below (see Figure 4.10):

Q is the effective area of the conductor
f is the filling factor to be taken
$Q/f = F$ is the area of the conductor without insulation
i is the thickness of insulation

To design the sector-forming rollers, the dimensions of R, C, A, B notations must be calculated.

Referring Figure 4.10 the total area of the sector with insulation without rounding the corners is = S

Now $S = \pi R^2 \varnothing / 360° = \pi R^2 / n$ when $n = 360°/\varnothing$ = the number of sectors and $\varnothing$ is the angle of the sector.

Now

$$S = \frac{\pi R^2}{n} = \frac{Q}{f} + 2i(R - Z) + iZ \quad \text{since } Z = i\cot\frac{\varnothing}{2}$$

$$= \frac{Q}{f} + 2iR - iZ$$

Therefore,

$$\frac{\pi R^2}{n} = \frac{Q}{f} + 2iR - i^2\cot\frac{\varnothing}{2}$$

or

$$\frac{\pi R^2}{n} - \frac{Q}{f} - 2iR + i^2 \cot\frac{\varnothing}{2} = 0$$

or

$$R^2 - \frac{2n}{\pi} iR - \left(\frac{nQ}{\pi f} - \frac{n}{\pi} i^2 \cot\frac{\varnothing}{2} \right) = 0$$

This is a quadratic equation.

Solving this equation for positive root, the following value of R is obtained:

$$R = \frac{ni}{\pi} + \sqrt{\left\{ \left(\frac{ni}{\pi} \right)^2 + \frac{n}{\pi} \left(\frac{Q}{f} - i^2 \cot\frac{\varnothing}{2} \right) \right\}} \tag{4.22}$$

From here it follows that

$$A = R\left[1 - \left\{ \cos\frac{\varnothing}{2} - 57.29^\circ \left(\frac{i}{R} \right) \right\} \right] \tag{4.23}$$

$$C = 2R\left\{ \sin\frac{\varnothing}{2} - 57.29^\circ \left(\frac{i}{R} \right) \right\} \tag{4.24}$$

$$B = \left(\frac{C}{2} \right) \cot\frac{\varnothing}{2} \tag{4.25}$$

This sector conductors thus formulated should be used for 1000/1100 to 3000/3300 V cables, where stress levels are not appreciable. Such conductors were found to produce a finished cable with a round contour rationalising raw material consumption. A filling factor of up to 92% has been achieved in this design system.

NOTE: During extrusion of thermoplastic compounds, minute undulations on the surface of insulated conductor cannot be completely eliminated, and hence the following should be considered.

While designing the shaping roller, it is advisable to reduce the angle Ø by ½° or 1°. This allows the insulated conductors adjust mutually to form a perfect round contour while assembling (Figure 4.11).

In this case, the shaping rollers have collars to avoid any sideways displacement during operation, as was experienced earlier when collars were absent. Due to sideways shifting of rollers, the apex position gets displaced making core assembling difficult. The shape of the cable also gets distorted.

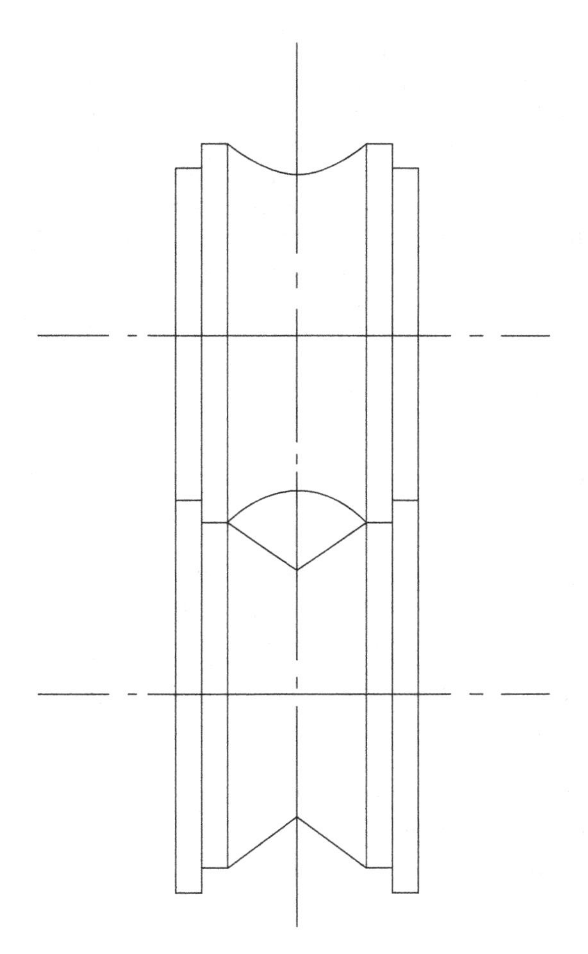

FIGURE 4.11
Shaping roller for sector conductors.

The rollers are designed as per the formula given earlier. The surface of the rollers must be ground mirror polished to get accurate dimensions and an optimal shape for the conductor. A clearance of 0.05 mm is preferable at the meeting point of two rollers for a smooth rolling action. Rollers are mounted on bearings to avoid any dragging effect. The roller should be made out of EN-24 steel and tempered (case hardening).

While designing a cable, the main considerations are the voltage gradient and maximum stress that will develop during operation and during a short-circuit condition. Let us consider the following points.

With the introduction of quality XLPE insulation compounds, an average breakdown voltage of 20 kV/mm and an impulse breakdown voltage of 55–60 kV/mm have been achieved. Compounds are manufactured with a dielectric constant ranging from 2.3 to 2.5 (comparable to impregnated paper insulation). In combination with high-quality semiconductive materials, including supersmooth compounds, it would be possible to design and manufacture medium-voltage cables with shaped (11,000 V) and elliptical conductors (33,000 V). These designs will reduce the cable diameters considerably vis-à-vis the consumption of raw materials.

In a shaped conductor, the stress voltage can be brought down to a minimum level by increasing the radius of the curvature at the edges. In an elliptical conductor, the stress voltage at the edges of the major axis can be easily controlled as the radius of curvature is fairly large. Further, the application of a smooth semiconducting layer would help in the even distribution of the stress voltage. This layer will also have the required properties to withstand an impulse voltage.

4.1 Sector-Shaped Conductor for an 11,000 V cable

The sector-shaped conductor should be used for an 11,000 V cable. As per calculations shown later, the stress voltage developed can be contained by the thickness of the XLPE insulation as specified in the standard specifications and by rounding off the sector edges (Figure 4.12).

To get the value of R when computing the design parameters of the conductor, the following assumptions are made:

t is the thickness of insulation (mm)

Q/f is the area of the sector (mm²)

Q is the effective area of the conductor

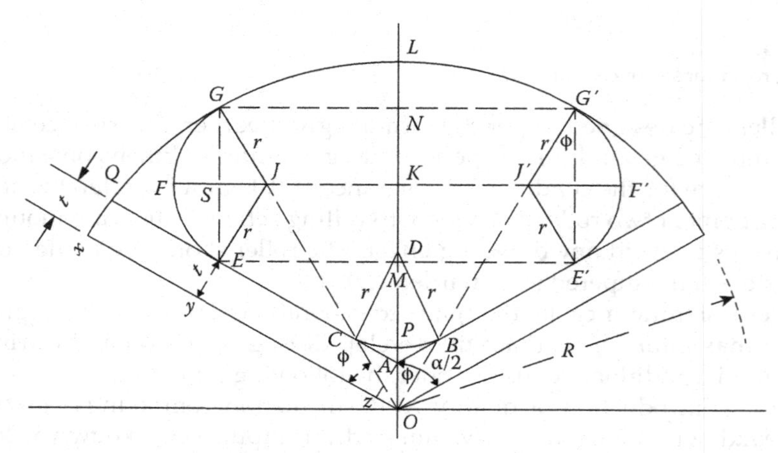

FIGURE 4.12
Shaped conductor with rounding of edges for 11 kV cables.

f is the filling factor, as defined

R is the radius of rounding at the edges

The cable is of three cores; hence the sector angle of 120° Q/f is considered a geometrical area of the conductor with filling factor of 88%. The geometrical area becomes S = 136.36 mm², approximately.

Once the value of 'R is determined, all other parameters of the sector can be worked out to construct the shaping roller for manufacturing a conductor.

It would be necessary to calculate the sector area without insulation. In order to initiate the process of calculation, the following assumptions are made taking into account the constructional feature of the sector:

1. $\angle EJG = 120°$
2. $\angle FJG = 60°$
3. $\angle JGE = 30°$
4. $\angle BDC = 60°$

This is a three-core cable.

Let r be the radius of the rounding at the edges.

$\phi = 30°$ $JG = JE = CD = DB = S'G' = J'E' = r$, by construction.

Now to get the conductor area, the following are calculated and added together:

1. Area of sectors $GFESG$ (two sides)
2. Parallelogram $GG'E'F$ area
3. Sector area of $GG'LG$
4. Triangle EAE' – area of $CABp$ (gap area)

Calculate the following:

1. Given $OG = R$, $JG = r$,
   $\angle EJG = 120°$ now the area of the sector is

$$EFGSE = \frac{\pi r 120^\circ}{180^\circ} \times \frac{r}{2} - \frac{2r \sin 60^\circ \{r - (r - r\cos 60^\circ)\}}{2} \tag{4.26}$$

or

$$\frac{\pi r^2}{3} - \frac{2r^2 \sin 60^\circ \{1 - (1 - \cos\ 60^\circ)\}}{2}$$

or

$$\frac{\pi r^2}{3} - \frac{2r^2 \sin 60^\circ \times \cos 60^\circ}{2} \quad \text{or} \quad 1.0472r^2 - 0.433r^2 \tag{4.27}$$

or $EFGSE = 0.6142r^2$

Now for two sides, the sector area is = $\mathbf{1.2284r^2}$. (4.28)

2. Area $GG'F'F$: $GN = R \sin 30°$ and $GG' = 2R \sin 30° = 2 \times 0.5R = R$

$$GE = 2r\sin 60^\circ = 1.7320r. \text{ Hence, area } GG'F'F = \underline{1.7320Rr}. \quad (4.29)$$

3. The area of the triangle AEE' is

$$EM = GN = R\sin 30^\circ \text{ and } EE' = 2R\sin 30^\circ = R$$
$$AM = EM\tan 30^\circ = R\sin 30^\circ(\tan 30^\circ)$$
$$\text{Therefore, } AM = R\times 0.5\times 0.57735 = 0.2887R$$
$$\text{Now, the area of a triangle is } 0.2887R^2. \quad (4.30)$$

a. From this area of triangle AEE' area $CABp$ to be deducted:
The area of $CABp$ is obtained as follows:
   i. Area of $DCAB$ – Area of sector $DCpB$
   Area of $DCAB$: $DB = r$ and $AB = r \tan 30°$
   Hence, the area of $DCAB = r^2 \tan 30° = 0.57735r^2$
   ii. Area of sector $DCpB$: $\dfrac{\pi r^2 60}{360} = \dfrac{\pi r^2}{6} = 0.5236r^2$
   iii. Hence, the area $CABp = (0.57735 - 0.5236)r^2 = 0.05105r^2$ (4.31)
   Hence, the area of the triangle AEE' minus the area $CABp$ is $0.2887R^2 - 0.05105r^2$....

4. Area of the top sector $GLG'NG$:
We know that $GG' = R$.
Hence, the area is given by

$$\frac{\pi R60^\circ}{180}\times\frac{R}{2}-\frac{R\{R-(R-R\cos 30^\circ)\}}{2}$$

$$=\frac{\pi R^2}{6}-\frac{R^2\{1-(1-0866)\}}{2}=0.5236R^2-0.433R^2$$

$$=0.0906R^2 \quad (4.32)$$

Now, the total area of the conductor is given by

$$\frac{Q}{f}=0.0906R^2+0.2887R^2-0.05105r^2+1.7320Rr+1.2284r^2 \quad (4.33)$$

or

$$0.3793R^2 + 1.7320Rr + \left(1.17735r^2 - \frac{Q}{f}\right) = 0 \tag{4.34}$$

This is a quadratic equation:

Taking the positive root of R, the solution becomes

$$R = \frac{-1.7320r + \sqrt{\left((1.7320r)^2 - 4 \times 0.3793 \times \left(1.17735r^2 - \frac{Q}{f}\right)\right)}}{2 \times 0.3793} \tag{4.35}$$

To design the pair of sector-shaped stranding rollers, the following are the formulae to be used:

1. R = Radius. Refer Equation 4.35. (4.35)
2. Width: $FF' = (R + r)$ (4.36)
3. Lower half: $PK = 0.2887R + 0.7113r$ (4.37)
4. Top half: $KL = 0.134R + 0.866r$ (4.38)

The value of Q is taken from the standard conductor area, f is the filling factor as per design system and r is determined by the designer of the cable.

Example 4.3

Taking $r = 3$ mm, $Q = 120$ mm² and $f = 0.9$, we get $R = 12.40$ mm.

Now taking cables with 11,000 V range, the insulation thickness is 3.8 mm as per IS 7098-(part-II) specification. We accept 4.00 mm as the average thickness with a manufacturing allowance of +0.2 mm. Further, the inner and outer semiconducting layers are 0.6 mm each. Hence the total thickness becomes 5.2 mm. A 2 × 0.05 mm copper screen thickness is also considered to build up a total thickness t of 5.3 mm. Hence the laid-up diameter of the cable will be $D = 2 \times (12.40 + 5.3) = 35.40$ mm. This is an ideal condition. A multiplying factor of 1.15 is taken to accommodate the feathering effect and laying gaps to establish the final laid-up diameter of the cable. This then becomes 40.71 mm. Against this, the laid-up diameter of a 120 mm² × 3 core × 11 kV round conductor cable is 50.14 mm. A reduction of (50.14 – 40.71) = 9.43 mm is gained.

This can be substantiated by considering one single insulated round conductor area that would be equal to the area of three individual insulated round conductors of a particular cable.

Let us again take the case of a round conductor of 120 mm² with a filling factor of 0.9 and a diameter of 13.03 mm. The combined thickness of the insulation and the copper tape is 5.3 mm.

Hence, the diameter of the conductor with insulation and copper tape screen becomes 13.03 + 2 × 5.3 = 23.63 mm. The area of three such cores is = 438.55 × 3 = 1315.65 mm² (approx.).

The diameter as calculated from the area is now 40.93 mm against 40.71 mm.

Basing on the above example, a comparative table on diameters for round- and sector-shaped laid up cable can be worked out for all types of 11 kV cable conductors.

The maximum value of the potential gradient or the field stress occurs at the surface of the cable conductor when $r = r_o$ for a single-core cable. Then

$$E_{\max} = \frac{V_o}{r_o \ln(R/r_o)} \text{ kV/mm} \tag{4.39}$$

In three-core cables with segmental conductors, the electric stress is computed at the three points and is of considerable importance, such as *a*, *n* and *m*, as per Figure 4.13. This being a belted cable of three core shaped conductor, the radius R_{seg} of the curvature of the shaped cable conductor can be considered as a radius *R* of a round single-core cable conductor. Thus the radius of belt insulation along with the insulation of the conductor, gives

$$E_a = \frac{V_{ph}}{R_{seg} \ln\left(\dfrac{R_{seg} + \Delta + \Delta_1}{R_{seg}}\right)} \text{ kV/mm} \tag{4.40}$$

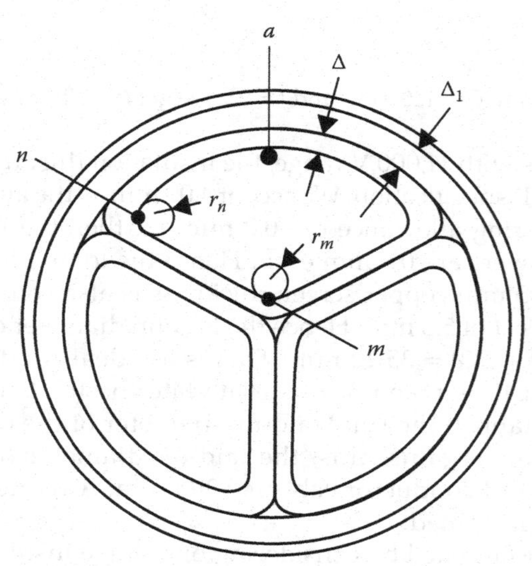

FIGURE 4.13
Development of electrical stress in a shaped conductor.

The electric stress at the centre point m is given by the radius of the segment r_m and the thickness of insulation is taken to be equal to the distance from point m of the centre of the cable. The stress at the point is given by

$$E_m = \frac{V_{ph}}{r_m \ln\left(\frac{r_m + 1.555\Delta}{r_m}\right)} \text{kV/mm} \tag{4.41}$$

The stress voltage at the corner points n is computed as follows, considering the edges as two parallel cylinders, in which case r_o is substituted by r_n and R by $(r_o + \Delta)$ and is given by

$$E_n = \frac{V_1\sqrt{\left[\dfrac{\dfrac{r_n+\Delta}{r_n}+1}{\dfrac{r_n+\Delta}{r_n}-1}\right]}}{2r_n \ln\left\{\dfrac{r_n+\Delta}{r_n} + \sqrt{\left[\left(\dfrac{r_n+\Delta}{r_n}\right)^2 - 1\right]}\right\}} \text{kV/mm} \tag{4.42}$$

But when individual segmental (shaped) cores are screened, in which case no belt insulation is applied, the electric stress at point a is given as follows:

$$E_a = \frac{V_{ph}}{R_{seg} \ln\left(\frac{R_{seg} + \Delta}{R_{seg}}\right)} \text{kV/mm} \tag{4.43}$$

The electric stress at points m and n is also given by

$$E_{m,n} = \frac{V_{ph}}{r \ln\left(\frac{r + \Delta}{r}\right)} \text{kV/mm} \tag{4.44}$$

In this case, when $\Delta = 4$ mm, $R = 17.31$ mm and $r = 3$ mm, and $V_{ph} = 11{,}000$ V, then

$$E_a = 3{,}055 \text{ V/mm} \quad \text{and} \quad E_{m,n} = 4{,}328 \text{ V/mm}.$$

At an impulse voltage of 80 kV, E_{imp} stress at points m and n will be $E_{imp} =$ 31.5 kV/mm, remaining within the safe limit as for XLPE insulation.

Considering these factors, British Insulated Callender Cables Ltd., BICC (UK) (BICC), United Kingdom, produced sector-shaped 11,000 V cables with

XLPE insulation. It is quoted that a specification has already been established for a 6/10 kV three-core cable which comprises three shaped solid aluminium conductors insulated with XLPE and screened with a strippable semiconducting layer. A stranded conductor can also be formed when the inner extruded semiconducting layer is applied to even out the wavy surface.

4.2 Elliptical/Oval-Shaped Conductor for a 33,000 V Cable

The radii of the edges of sector-shaped conductors could develop a rather higher stress at the operating voltage of 30,000/33,000 V. Considering this technical limitation, elliptical conductors were used.

Since the quality of XLPE and dry curing system has improved the stability and life span of medium- and high-voltage cables, it can now be used in designing 30/33 kV cables with elliptical conductors. This will reduce the diameter and raw material consumption to a large extent, enhancing the all-round financial benefit.

Design consideration of elliptical conductor (Figure 4.14)

The equation of the ellipse is

$$\frac{x^2}{a^2}+\frac{y^2}{b^2}=1 \tag{4.45}$$

Major axis = $2a$
Minor axis = $2b$

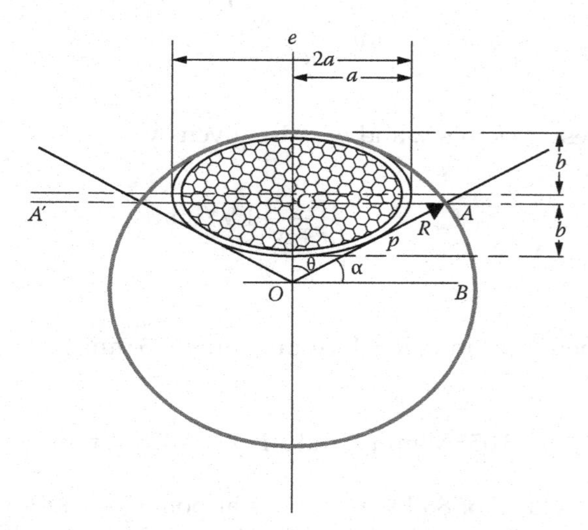

FIGURE 4.14
A schematic diagram of elliptical conductor design parameters.

e = eccentricity of the ellipse
O = the centre of the laid-up cable diameter
R = radius of the circle
OA = tangent touching the ellipse
C = meeting point of the major and minor axes at right angles (centre of the ellipse)
OB = radius of the big circle
Q = area of the conductor
S = area of the insulated conductor
$\angle BOA = \alpha$= angle subtended by tangent with the radius of the circle OB
$\angle COA = \theta = (90° - \alpha)$
Further
f = filling factor of the conductor taken as 0.92
$\tan \alpha = m = 33°$ approx.
Point P is where the tangent touches the ellipse (x, y coordinate)
Now

1. Taking the conductor size as 300 mm^2.
2. Filling factor is 0.92.
3. The geometrical area of a conductor is 326.09 mm^2.
4. The diameter of the round conductor is 20.38 mm.
5. Insulation thickness for 33 kV with allowance taken to be 9.00 mm.
6. Inner semicon thickness 0.6 mm.
7. Outer semicon thickness 0.6 mm.
8. The diameter of insulated round conductor is 40.78 mm.
9. The area of an insulated round conductor is 1305.88 mm^2.
10. The area of an insulated elliptical conductor thus will be 1305.88 mm^2 = πab.
11. The angle of the tangent m = 33°; now $m = b/a$ or $b = 0.65a$.

    On this basis, the value of a = 25.29 and value of b = 16.44.

    The equation of the tangent is $y = mx + \sqrt{a^2m^2 + b^2} = mx + c$.
    When $x = 0$ and $y = 0$, $c = 0$; hence, $m = -b/a$,

    as $c = \sqrt{a^2m^2 + b^2}$ and taking positive values.

12. The value of c obtained is 23.25 mm.
13. Therefore, the radius of the laid-up core (OA) is 39.68.
14. The laid-up diameter of all three insulated elliptical cores is 79.37 mm.
15. Now the laid-up diameter of the insulated round conductor is 87.87.

16. The diameter reduction thus gained is 8.50 mm.
17. The a_1 of the metallic elliptical part of the conductor is 15.09 mm.
18. The b_1 of the metallic elliptical part of the conductor is 6.24 mm.
19. Therefore, the area of the stranded conductor of the metallic elliptical part is $= \pi a_1 b_1 = 295.82\ \text{mm}^2$.
20. Eccentricity $e = 0.91$.

By making conductors elliptical or oval shaped, the overall size of the cable is decreased. With the three-cable core screened and laid up to form a round cable core, the electric stresses corresponding to points A and B are calculated by the following equations with reference to Figure 4.15:

$$E_A = \frac{V_{ph}}{\left[r_A \ln\left(\frac{r_A + \Delta_A}{r_A} \right) \right]} \text{kV/mm} \tag{4.46}$$

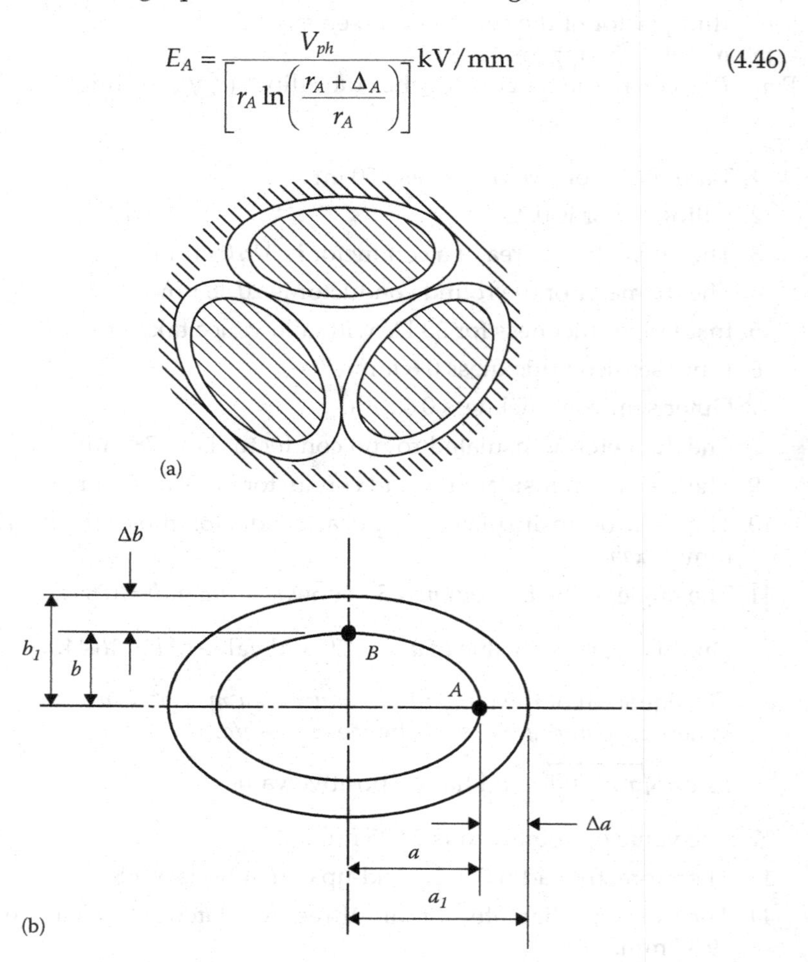

FIGURE 4.15
Points of development of electrical stress in an elliptical conductor.

$$E_B = \frac{V_{ph}}{\left[r_B \ln\left(\frac{r_B + \Delta_B}{r_B} \right) \right]} \text{kV/mm} \tag{4.47}$$

Now $r_A = b^2/a$ is the radius of the curvature at point A and $\Delta_a = a_1 - a$ is the insulation thickness at A. $r_B = a^2/b$ is the radius of curvature at point B and $b_1 - b$ is the insulation thickness at B.

Thus, the stress value at point A on the conductor E_A at operating voltage 33,000 V is 8519 V/mm and at E_B is 4126 V/mm. The impulse level stress is generated as per British recommendation (at 194 kV):

$$E_A = 50 \text{ kV/mm} \quad \text{and} \quad E_B = 24.3 \text{ kV/mm}$$

And as per IEC recommendation (170 kV),

$$E_A = 43.9 \text{ kV/mm} \quad \text{and} \quad E_B = 21.3 \text{ kV/mm}$$

Now, an XLPE compound with a breakdown voltage of 55 kV has been developed. Hence, when the cable is manufactured with proper care, there is no fear of failure of such cables during operation.

Example 4.4

A 300 mm² × 3 core × 33 kV cable is made of an elliptical conductor. The insulation thickness of the 33 KV cable is 8.8 mm (XLPE cable). With a filling factor of 0.9, the diameter of the bare conductor is 20.60 mm and the diameter of the insulated conductor in a circular shape is 38.20 mm. The area of circular conductor becomes 1146 mm², which will also be the area of the insulated elliptical conductor.

Thus, πab = 1146 mm²; now taking $b = 0.70a$, a becomes 22.82 mm and b becomes 16.00 mm

To get the value of c, coordinates of points x and y are determined where the tangent (equal to the radius of the laid-up diameter) touches the ellipse. The angle formed by the tangent (radius) with the y-axis is 60° and with the x-axis is 30° = (90° − 60°). Hence, tan 30° = 0.57735.

The equation for the tangent is

$$y = mx + c = mx + \sqrt{a^2 + b^2} \tag{4.48}$$

and $c = \sqrt{a^2m^2 + b^2}$ and also C = 20.73 mm.

Also

$$X = -\frac{a^2 m}{\sqrt{a^2m^2 + b^2}} \tag{4.49}$$

And

$$Y = \frac{b^2}{\sqrt{a^2m^2 + b^2}} \text{ is any point where the tangent touches the ellipse} \quad (4.50)$$

Hence, the laid-up diameter of a three-core elliptical cable will be

$$D = 2R = (b + c) \times 2 \quad (4.51)$$

Hence, $R = 20.73 + 16.00 = 36.73$ mm; the diameter of the cable $= 2R = 73.46$ mm approx. The diameter of a round insulated conductor is 38.20.

Reduction in laid-up diameter becomes $38.20 \times 2.155 - 73.46 = 8.86$ mm; in the actual case, reduction becomes approximately 8–8.5 mm.

On this basis, the dimensions of the conductor become $a_1 = 14.02$ mm and $b_1 = 7.20$ mm. Approximately $a_1 = 14.00$ mm and $b_1 = 7.20$ mm, and the geometrical area becomes 316.60 mm². In this case, the ratio $b_1/a_1 = 0.513$ and $e = 0.714$.

Medium- and high-voltage cable conductors are round, and so the electrical stress is minimised substantially. Even then the wavy form of individual wire produces localised electrical stress over its surface (due to a smaller radius of curvature) in a complex form. This wavy contour is polished to a great extent by compacting the conductor and making the surface more even and smooth, filling interstice spaces to a large extent.

Nowadays, shaped and round compacted conductor manufacturing has become common practice, where the stranded conductor is passed through shaped rollers and the round conductor through compacting die, thereby eliminating the torsional force to a great extent. The diameter of the conductor is reduced in order to save raw materials. A compact circular cable could thus be produced controlling the overall dimension. During compacting, work hardening is observed within the metal, more so in copper. Further, a stretching effect comes into play on wires which may reduce the overall cross-sectional area. To compensate this and to obtain the required resistance values, a slightly higher wire size is considered based on work experience. Care is taken to keep the weight and resistance within the permissible limit of tolerance. In order to gain a smooth, round and compact insulated laid-up multicore cable for conductors sized 185 mm² and above, it is desirable to rotate shaping and compacting roller on its axis allowing the formation of a long pre-spiral lay. During the assembling of insulated 2-, 3-, 4-, and 3½-core cables in a laying-up unit, pre-spiral lay helps the conductors to adjust mutually in their twisted position without experiencing any torsional deformation. Nowadays, conductors are compacted during shaping and rolling, where the pre-spiral system is done away with. While assembling insulated cores with a long lay ratio, the torsional force on insulated conductors was found to be negligible.

FIGURE 4.16
Pre-spiral compacting roller assembly.

In Figure 4.16, compacting rollers have plane contact surface for rolling action. Experience has shown that in time, holding shafts of rollers wear out, making the rollers shift sideways or tilted, distorting the shape of the conductor. In such cases, the shifting of the conductor apex becomes a common phenomenon. While laying up, such distorted cores create a bulging effect, damaging insulation. To eliminate this problem, compacting rollers are designed with sideway collars, as shown in Figure 4.16. This would prevent sideway movements of the rollers, ensuring smooth production.

The area and type of a conductor are selected as per the transmission parameters of a network. The surroundings, laying and installation conditions influence the temperature rating vis-à-vis the heat dissipation characteristics of a cable in the form of line loss. This heat is developed mainly within the conductor in the form of I^2R. If I (current) is kept constant, the variable value of R (resistance) would be responsible for the fluctuation in heat generation within the conductors. Naturally, it follows that the resistance of a conductor plays a vital role in determining the transmission loss. When current flows steadily without fluctuation through a conductor, it is termed as DC system. The resistance offered by the conductor is constant throughout a given temperature. In the case of a three-phase supply, current fluctuates alternately, changing the magnetic flux density with time. In an AC supply, a periodic change in magnetic flux generates self-inductance within the conductor, retarding the flow of current within the middle part. With an increase in cross-sectional area, the flow of current will tend to shift further around the periphery of the conductor because of the increased magnetic flux. In such cases, the current density forms an annular ring inside the periphery of the conductor. Thus, when current flow is restricted, the value

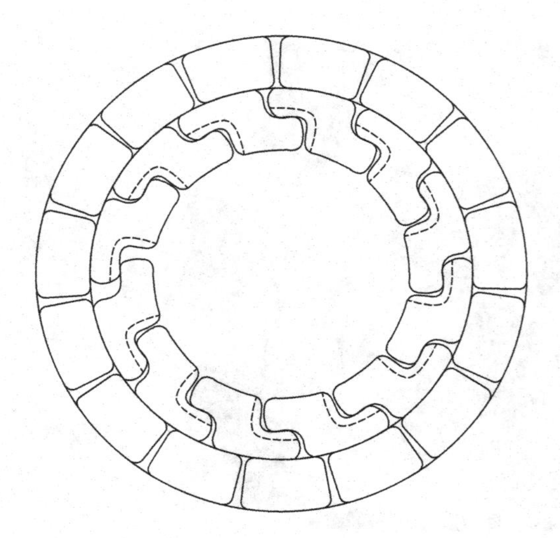

FIGURE 4.17
Hollow conductor made of profiled wires (for oil filled HV and EHV cables).

of resistance will increase considerably. The inner section of the metal in such cases remains inactive. This phenomenon is called the skin effect. At 50 Hz, a skin effect below 150 mm^2 is not much and can be neglected. As the cross-sectional area increases, the current density also rises considerably whereby the skin effect becomes more pronounced.

To tackle the problem of skin effect, two design concepts were adopted. In the case of paper, insulated and impregnated cables, the conductor is made hollow (Figure 4.17) to save metal content when the diameter becomes large. This type of conductor is made for oil- and gas-cooled extrahigh-voltage cables of larger cross-sectional area. A special advantage that is obtained by this design is that the conductor can be cooled by passing oil or gas under pressure through the central hollow pipe-like structure and a large amount of heat to maintain the current density as required. Utilising this formation, oil- and gas-filled cables were designed to transmit a large amount of power through a single circuit. There are various types of hollow conductors being manufactured. In one case, wires are stranded around a spiral metal tube in a normal way. Here, the disadvantages are that the wires, even after compaction, produce minor cleavages that are hazardous for very-high-voltage systems. The most acceptable construction developed was by stranding profiled curved strips that interlocked. The conductor formed by this process is more flexible. No internal support is required, and the tube becomes perfectly round with a smooth contour (Figure 4.17).

To manufacture such conductors, special profiled stranding machines are developed. These types of conductors are used for manufacturing oil- and gas-filled cables, ranging from 110 to 500 to 750 kV, or more (for power supply more than 1000 MVA).

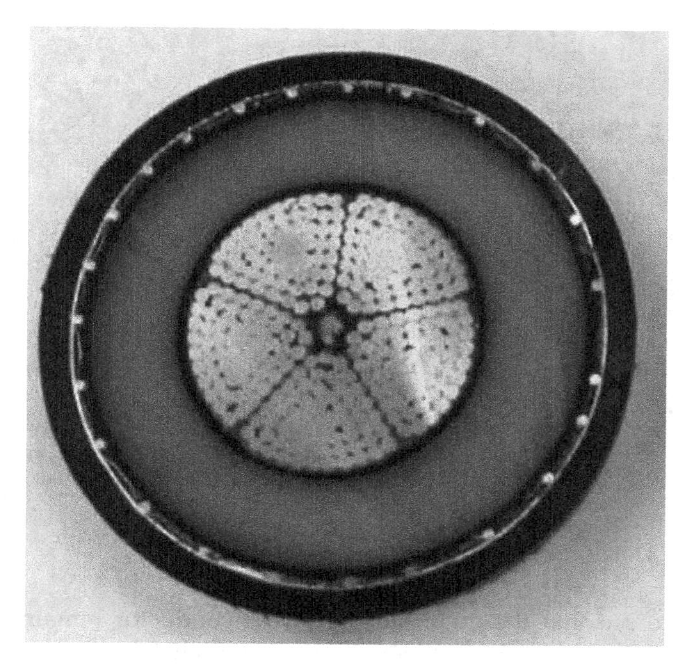

FIGURE 4.18
Milliken conductor.

Special profiled conductors are also used for overhead transmission line conductors, and one Belgium utility, Elia, used more than 90% of their conductors for AAAC, using special profile layer conductors made as part of their specification with critical parameter measure in such wires being the drag coefficient.

With the introduction of super clean XLPE insulating material, cables of up to 500 kV could be produced in a dry cure vulcanising system in VCV and CCV lines.

With the introduction of solid dielectric, hollow conductors became unsuitable. Further, a particular insulating gas or oil may react with the polymeric insulating material. In this case, the conductor with a large cross-sectional area is developed by splitting it in several segments of stranded sectors as a Milliken conductor (Figure 4.18) and by assembling them together to form a round construction.

This construction minimises the skin effect by dividing the current equally through each segment separately. To form an individual compacted sector, a conventional stranding machine with a shaped roller is employed. These sectors are then wrapped with one layer of semiconducting synthetic tape in a taping machine provided in-line with stranding machines. Or, during assembling, one separator tape is introduced longitudinally between two sectors to keep each one isolated from the other (Figure 4.19). It is, however,

FIGURE 4.19
Assembling of a Milliken conductor.

important to see that the outer surface of the conductor remains smooth after compacting and laying and should be absolutely free from any kind of defect. In some cases, enamelled wires are also stranded together to form a sector of the Milliken conductor. In this case, the skin effect is minimised to a very large extent as the current flows through the wires individually, further reducing the value of resistance. For aluminium conductor, enamelling is not required. The thin oxide film forming on the surface of the wire/strip acts as a natural barrier between wires and conductor segments.

It is evident that designing and manufacturing of a conductor must be done with utmost care and precision taking into consideration its nature of performance. If the conductor is made with a knowledge of the all-round quality requirements, it is sure to achieve the best of competitive and economical balance. One of the precautions taken for such conductors is that they are lapped with a paper tape to protect from dust and it is automatically removed before entering the dust-free floor in a VCV or CCV line.

It has become a practice to compact stranded conductors passing through die or shaping rollers. This is to reduce the diameter and smoothen the outer surface of the conductor in one go. In case of medium- and high-voltage cables, a smooth outer surface reduces the electrical stress on the surface of the conductor. Compacting does make the outer surface smooth, but due to pressure and drawing force, the edges of wires get flattened and become rough and uneven. These edges develop stress at the corners and at times can become vulnerable. To obviate such a potential high-stress point, segmental smooth wires are drawn and stranded over the inner round contour, thereby eliminating the danger.

This method avoids sharp edges on the outer wires and possible damage which may be caused to the inner layers by the compacting pressure, ensuring that the flexibility of the stranded conductor is retained. The segmental wire shapes required to form a smooth exterior can be calculated mathematically, as elaborated in Chapter 3 (also refer to Figure 4.17).

The stranding machine is designed to form a conductor by twisting wires layer by layer, with a definite configuration. Carriages are built separately to accommodate bobbins, where the first carriage accommodates 6 bobbins, the second carriage 12 bobbins, the third carriage 18 bobbins and so on. The largest stranding machine made so far accommodated 127 = (1 + 6 + 12 + 18 + 24 + 30 + 36) bobbins and 6 carriages. Most of the modern high-speed machines are designed with a fully automatic bobbin loading system and a PLC-driven software. Each carriage consists of a central shaft in which bobbins are placed on forks. Forks are placed in a star formation in the centre shaft or accommodated between parallel plates placed at 90° or 60° angles, depending on the size of the bobbin and carriage. Bobbin holdings can be of shaft type or pintle type with a proper locking arrangement. Carriages are placed on a central bearing in the back. The front side of each carriage is supported by an under roller of adequate strength. At the end of the central shaft of each carriage, a guide plate is provided to guide the wires at equal angles. Wires are placed with equal spacing through the guide plate. Bobbins are provided with a proper tensioning device so that the movement of the wires is not too loose or tight. Modern machines are provided with sensors to monitor tension automatically. In case of any wire break during operation, the machine will stop immediately. At the front side of each carriage, an adjustable die holder box along with the stand for sector-forming rollers is provided. The last front carriage will have the sector-forming roller with a rotating pre-spiralling system (if required).

For low-voltage, 2-, 3-, 3½- and 4-core sector-shaped cables, it is better to make higher-size compacted conductors from 185 mm^2 and above, with pre-spiral twists. This actually reduces the diameter of a laid-up cable core after insulation. All the cores will get adjusted in position within the specified lay, forming an almost round contour. This not only saves material but also does not allow any feathering effect due to torsional stress that develops during laying operation. Using a drum twister, the lay length can be adjusted as required, through the PLC system, so that pre-spiralling will cause no problem, as was experienced with the conventional 1 + 3-core laying-up machine. In this case, pre-spiral lay lengths were specified to match the position of cores during assembling. In order to increase production, nowadays, even in a rigid stranding machine, no pre-spiralling system is provided with the compacting head.

Figure 4.20 shows a conventional stranding machine. During operation, wires are led through guide holes provided with hard polished bushings (*A* and *B*) to allow for easy movement of wires without any scratch marks or damage. Wires are led through a guide plate, forming an equal angle with each other. Figure 4.20 shows the obnoxious angular position of wires when no guide plate is used. The length and tension of the wire are found to be unequal due to an uneven angular position. A stranded conductor in this case will come out with

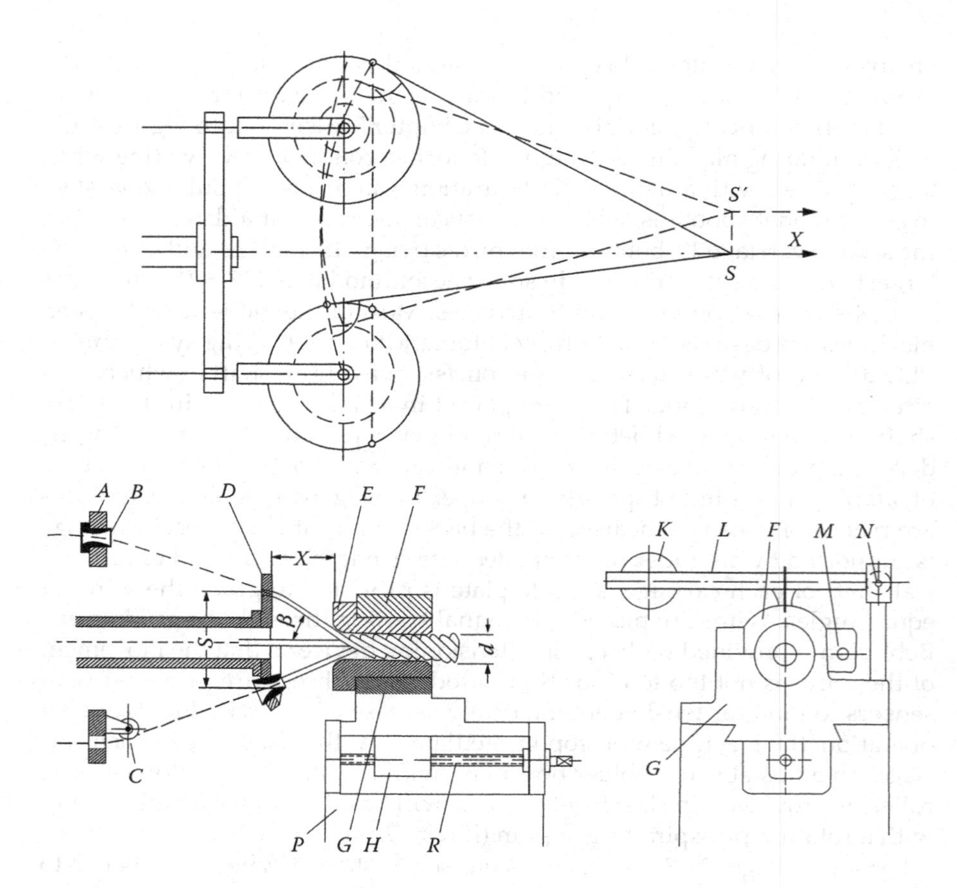

FIGURE 4.20
Picture showing the importance of a lay plate during stranding and laying of cables and adjustable die holder stand. A-Wire guide nipple holder; B-Wire guide nipple; C-Wire guide roller; D-Front guide plate; E-Stranding die; F-Gripping clamp for keeping the stranding die fixed; G-Die holder Box; H-Screw nut kept fixed at the bottom of die holder Box; K-Counter weight; L-Lever rod to hold the counter weight; M-Hinge for die gripping clamp 'F'; N-Hinge for the fulcrum 'L'; P-The die holder Box holding stand; R-Screw for moving the die holder box forward and backward; S-Middle axis of the machine and of stranded conductor; X-Distance between the guide plate and stranding die; β-Wire leading angle from guide plate to die (comfortable angle for stranding is 30°); 'd'-Diameter of stranded conductor.

an uneven twist having a wavy formation. The position of the wires will be displaced. The wire having a higher tension will cut through the strand becoming almost straight, whereas other wires having uneven and loose tension will sit awkwardly in spiral formation, leading to bird caging and snaking of the conductor. Hence, it is essential to have a front guide plate (*D*), as in Figure 4.20a, with holes at an equal distance, and lead all the wires at an equal angle. The guide plate also helps in maintaining uniform tension. The construction of guide plates can be different but the purpose remains the same. Wire guiding

FIGURE 4.21
Rigid-type stranding machine with automatic batch loading system.

holes or guide rollers must be very smooth. The die holder box is adjustable (*G*) and can move forward and backward having a clamping arrangement (*K*) to exert pressure as a lever connected to the hinge (*N*).

Tension of all the wires must be correctly adjusted. This will lead to an equal and uniform pulling of wires while passing through the die. Any slackness will lead to overriding and bird caging of the conductor. Carriages along with capstans are driven by a single drive shaft attached to the main motor through a reduction gearbox. Each carriage is provided with a direction change gearbox for left or right rotational movement. A lay change gearbox is coupled with a capstan. The change is actuated by an operating lever to arrange the gear combination when changing the lay length. The rotational speed of each carriage is set differently to bring about a given lay combination on the successive inner layer. Recently, machines have been designed to have individual carriages driven individually, as against the conventional practice of common shaft and gearbox. This gives flexibility when adjusting the lay ratios of each layer. Nowadays, a DC motor is preferred in order to operate the system smoothly. A lay chart of the machine with a gear combination is provided by the manufacturer. Automatic batch loading of bobbins is provided with modern units (Figure 4.21).

Conductor pulling capstans are of dual type provided with semicircular or shaped grooves (Figure 4.22).

The take-up system is electrically operated for up and down and lateral movements to accommodate different types of process drums with traversing system. There are various types of take-up units being developed to cater to requirements of different customers having different choices (Figure 4.23).

High-speed tubular stranding machines are employed to make 7-wire or 19-wire strands and can have compacting and shaped rollers (Figure 4.24).

FIGURE 4.22
Stranding machine with guide plates, die holder box and dual groove–type capstan.

FIGURE 4.23
With take-up unit.

In these units, the wires are passed through small rings attached to the rotor which is the outer shell of the unit. Bobbins are placed on yokes hanging between two pedestals and remain at a fixed position. The wires passing through the rings experience a twist at the beginning of the first entry point in the ring of the rotor. While coming out before entering the guide plate, the twist gets untwisted, releasing torsion in the wire. These machines are used to produce torsion-free AAC, AAAC and ACSR conductors, as well as steel central cores. To keep the wires and conductor completely torsion-free and straight, machines can be provided with pre-formers and post-formers as well. Nowadays, rotors are made of carbon fibre to make them lightweight. The rotor of such units can reach a speed of up to 1200–1500 rpm.

When higher-size conductors of alloy wire and stainless steel wire are produced for defence and for special applications, it is necessary to keep the

FIGURE 4.24
Tubular stranding machine.

conductor torsion-free. In such applications, stranding machines with anti-twist devices are employed. For stranding multiwire bunched conductors (tinned or bare) for wind power and other cables for special applications, such units yield best performance (Figure 4.25).

The speed of the machine calculated is shown hereafter: h is the lay length, V is the conductor line speed and S is the carriage revolution per minute. This means that the lay length is the length of the conductor and moves forwards with one revolution of the carriage. The speed of the carriage is to be specified by the manufacturer. For rotational movement, the centrifugal force developed is given by mv^2/r where m is the total mass of the carriage along with the materials contained in the bobbins, v is the rotational speed of the carriage and r is the radius of the disc/yoke holding the bobbins. The machine is provided with all safety devices and interlocking systems.

FIGURE 4.25
Wire stranding machine with a back twist arrangement.

The smoothness of the outer surface of a conductor is very important. The stress developed on the surface of the conductor is given by

$$E = \frac{V}{r\ln(R/r)} \tag{4.52}$$

where

r is the diameter of the conductor (when plain)
R is the diameter over insulation
V is the operating voltage
E is the stress on conductor

If the conductor is plain and smooth (i.e. on a plain solid rod), stress can be contained to a lower value. But if any cut mark or scratch or protrusion occurs on the surface, then r is considered as the dimension of the sharp edge. In case r becomes too small, the value of E increases considerably.

Example 4.5

The conductor diameter is 19.0 mm and insulation thickness 1.4 mm for 1100 V. The sharp point is 0.2 mm and the value of E is 2645 V.

When the operating voltage is high, such small protrusions can initiate a failure.

It is also essential to know and calculate interstice space (area) on the outer periphery of a stranded conductor. This can be accurately calculated by the following formula for all types and sizes of wires and stranded conductors having a concentric round form:

$$\left[\frac{\pi D^2}{4} - \left\{\frac{\pi(D-d)^2}{4} + n\frac{\pi r^2}{2}\right\}\right] \tag{4.53}$$

D is the diameter of non-compacted stranded conductor
d is the diameter of the wire
r is the radius of wire
n is the number of wires on the outer layer of the conductor

This formula is fairly accurate and can be used for all numbers of n for round and non-compacted conductors. It is not valid for compacted round and sector conductors.

It has to be understood clearly that while constructing and manufacturing a conductor, all precautions should be taken to avoid any damage, marks or protrusions. The surface should be smooth and uniform.

The production of a stranding machine can be computed as follows:
When v = line speed per minute
In such a case:
S = rotational movement of the final carriage per minute.
L = lay length (for non-compacted conductors shall be 14 times the mean diameter [Dm] of the conductor coming out of the last carriage; and for compacted conductor, it shall be 20 times the mean diameter of the conductor coming out of the last carriage). Also the lay length is given by = v/S i.e.

$$\frac{\textit{Line speed } (v)}{\textit{Bobbin holding rotational speed of carriage } (S)} (\textit{per minute})$$

Considering the linear speed of the conductor and rotational speed of the front carriage, production on 8 hrs a shift basis at 60% efficiency is being calculated as follows:
Taking RPM of the front carriage as 120 and conductor diameter Dm = 12 mm
Production per shift in meters would be = (20 × 12 × 120 × 60 × 8 × 0.6 mm)/1000 = 8294.4 m/shift
(Here L = 240 mm). This 240 mm is again the length drawn by the Capstan for one revolution of carriage. Hence the line speed per minute is

$$240 \times 120/1000 = 28.80 \text{ m/min}$$

For compacted conductors, Dm will be taken as per the calculated diameter of the round conductor.
The following are the qualities of the conductor that should be checked during the process to avoid rejection and undue scrap generation:

1. Before stranding wire diameter, the surface and tension on the machine are checked.
2. The lay length and direction of lay are checked and noted.
3. For a round conductor, the diameter at every stage and the height of sector-shaped conductors are controlled.
4. Few metres after starting point of the final stage stranding process of the conductor, two metres to be cut and taken to check the resistance value and for weight measurement. This is done after conditioning the conductor at room temperature. Once these values are found to be in compliance with design parameters regular production could start (resistance and weight should not be measured when the conductor is in hot condition. The conductor also should not be cooled by spraying water to hasten the measuring process to save machine time for production).

5. The surface of the conductor is smooth and bright, free from burrs, scratches, scale formations and cut marks, and there are no sharp edges.
6. After final winding is taken on the drum as per standard length or multiple thereof, the outer surface is covered with polythene sheet to avoid deposition of dirt, etc.
7. Completed drums are marked with proper label for easy identification and are placed in rows flange to flange or sideways so that no damages occur during storage.

Nowadays, development efforts are focused on manufacturing conductors with water and moisture barrier. In such a case, the water barrier compound in powder form, or water swellable tape, is applied between stranded layers. This calls for modification on the stranding machine. This is becoming an increasing trend and manufacturers are to comply with this demand, particularly for MV, HV and EHV cables.

4.3 Manufacturing of Conductor with Profiled Wire

4.3.1 Stranding

The profiled wires are placed on the front cage of the stranding machine (profiled wire manufacturing is elaborated in Figure 3.23). Lay gear is adjusted as per calculation. The wires are to be guided through a front guiding plate and led through a cone-type front guide system as is done in the case of flat strip armour wires. The front of the cage is to be modified accordingly. Care has to be taken to ensure wires do not turn obliquely. The profiled wire guiding system thus must be modified accordingly. The profiled wire will set plainly in place while passing through the die fixed on the die holder box. One should never use a split die. A bell mouth die is suitable for this purpose.

1. To manufacture such profiled wires, conventional stranding machines with slight modifications can yield the desired result.
2. Due to the smooth surface, the electrical stress developed is contained to minimum values.
3. During extrusion of the semiconducting layer, the thickness can be maintained at a lower level and the consumption of the material will be less and there will be no interstice filling required.
4. So thickness, roundness and ovality can also be contained within a specified value.

5. Overhead ACSR, AAC and AAAC can also be manufactured in the same manner. In such cases, vibration due to high wind is kept within limited values. Nowadays, HTLS (high-temperature, low-sag) conductors with profiled wires offer a good technology solution in place of existing type of ACSR conductors.
6. If desired, the diameter can be reduced by applying two successive layers of stranded flat strip wires.

4.3.2 Messenger Conductor for Aerial Bunch Cables

In recent development, the use of all-aluminium alloy (AAA) is preferred in the manufacture of conductors for overhead transmission lines. The alloy metal is corrosion resistant even when stringed along the sea coast where moist, salty air is always present. Apart from it being corrosion resistant, the metal has a very high tensile value which permits it to be used in place of steel core for ACSR conductors. Particularly, at present, messenger conductors used in aerial bunch cables are made from this alloy, mainly to hold the phase conductor suspended between poles. Such conductors act as both neutral and earth wires. These alloy conductors are best stranded in a skip stranding machine. The conductor made in such units becomes torsion-free as the wires are passed through the rings of high-speed rotating bows. The process is similar to that of a tubular stranding machine (Figure 4.26).

It is to be understood that AAACs are to endure stress and strain being developed under the following conditions and yet should retain its strength giving a long service life:

1. Severe installation conditions in rural areas.
2. Temperature fluctuations from –0°C to 45°C from winter to summer.
3. Sustained wind pressure to be endured along with the weight of cable assembly.
4. To accept momentary short-circuit conditions and overloading as specified.

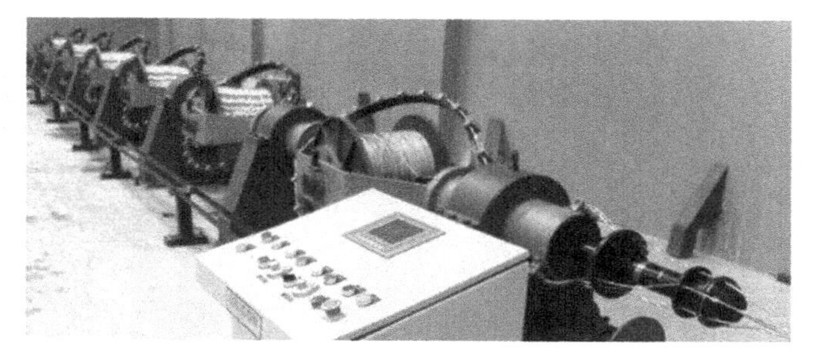

FIGURE 4.26
High-speed skip stranding machine, seven bobbins.

5. Thunderstorm and lightning.
6. In places where snowfall is expected, the weight of ice is also calculated as per suspended length.

Considering all these factors, quality and strength should be maintained without deviation from standards. These conductors are manufactured as per NFC Specification 33–209 and IS 14255. The testing procedure is discussed in Chapter 9.

5

Insulation and Insulated Conductors

5.1 Insulation

The process by which two or more electrically charged elements are kept apart, or isolated from each other and from the surroundings, is called insulation. In an electrical system, insulation keeps two or more conductors separated from each other and does not allow current to pass from one conductor to the other in the radial direction. The insulation should be able to resist the radial pressure of electrical stress when power is transmitted through a conductor. However, a complete isolation is an ideal condition.

It is known that all materials, in whichever form they may exist or synthesise, are conglomerates of atoms and molecules, where atoms along with their electrons vibrate constantly within their lattice structure. In an electrical conductor, electrons are allowed to move freely when pressure (voltage) is applied. But electrons in any insulating material are bound within the lattice structures and are not free to dissociate themselves from their parent atom. However, the frequency of vibration in atoms with their electrons, in an insulating compound, increases whenever heat energy is absorbed by these particles. And at any given time, small increase in the amount of heat generated (due to the flow of current and pressure in the form of voltage) will accelerate the vibration. Part of the energy is absorbed by these particles when power is transmitted through a conductor, in the form of heat. Though these materials act as a barrier between two or more conducting elements, when more heat is absorbed, there is an increase in atomic vibration, which lowers the strength of the insulating material. This absorption of energy is termed dielectric loss and is measured by the loss angle called dielectric power factor, which is, in electrical terms, the power factor angle between voltage and current phasors. An ideal capacitor will be when current leads the voltage by 90°. The power factor for each material is its own characteristic at a given temperature and frequency. This loss increases with an increase in the length and temperature of the insulated conductor because the absorption of energy takes place in and around the inner surface of insulation as well as lengthwise.

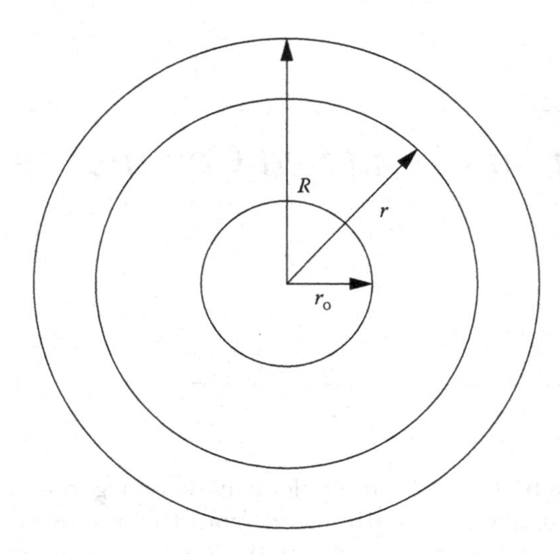

FIGURE 5.1
Cross section of an insulated conductor.

Hence, the resistance value of insulation against the conduction of electrical energy decreases, as the length and temperature of the insulated conductor increase. The reverse is the case of a conductor where resistance increases with length and temperature. In a simple equation, resistance can be described as $R = \delta\ell L/A$, where R is the resistivity of the material, L is the length and A is the area.

It is interesting to note that insulating materials are also bad conductors of heat. Naturally, heat generated in a conductor is not dissipated outwards immediately. It takes certain time to conduct heat from the inner surface to the outer surface, before it stabilises in a condition of equilibrium. A good insulator withstands considerable pressure (voltage) and absorbs as less energy as possible, without deteriorating its characteristics under operating conditions and for a longer period of time.

While absorbing energy in the course of transmitting power, a certain amount of electrical charge, in the form of energy, is retained within the insulating material. The capacity for retaining energy is termed as capacitance and can be computed as shown in Figure 5.1.

When the radius of the inner conductor is 'r' and that of the insulated conductor 'R' over insulation metres and a charge of 'q' coulomb (C) per metre is injected within the conductor, then as per Coulomb's law, an electric flux of a radial nature is produced at an internal radius 'r' ($r_o < r < R$) from the cable axis. The electric flux density is given by

$$f = \frac{q}{2\pi r}\,\mathrm{C\cdot m^{-2}} \tag{5.1}$$

and the electric stress at the radius '*r*' is

$$E = \frac{q}{2\pi r \varepsilon_0 \varepsilon_r} \text{V} \cdot \text{m}^{-1} \tag{5.2}$$

The value of $\varepsilon_o = 8.85 \times 10^{-12}$ the permittivity of free space and ε_r = permittivity of cable dielectric.

Now, the electric field stress at the radius '*r*' is

$$E = \frac{V_o}{r \ln \frac{R}{r_0}} \text{V} \cdot \text{m}^{-1} \tag{5.3}$$

V_o is the potential of the conductor relative to the cable insulation. The equation transforms to

$$V_o = \left\{ \frac{q}{2\pi r \varepsilon_0 \varepsilon_r} \right\} \cdot \ln \frac{R}{r_0} \text{ (V)} \tag{5.4}$$

Thus, the capacitance becomes

$$C = \frac{q}{V_o} = \frac{2\pi r \varepsilon_0 \varepsilon_r}{\ln\left(\frac{R}{r_o}\right)} \text{F} \cdot \text{m}^{-1} \tag{5.5}$$

This gives a simplification for a circular conductor as

$$C = \frac{\varepsilon_r}{18 \ln\left(\frac{R}{r_o}\right)} \mu\text{F/km} \tag{5.6}$$

For shaped conductors, r_o is taken by dividing the periphery of the conductor by 2π.

The permittivity of insulating materials ε_r depends on temperature. With an increase in temperature, permittivity rises and hence the capacitance. Since capacitance is directly proportional to ε_r, and, as such, that of q, the accumulation of charge increases as permittivity increases with a temperature rise and hence the energy loss. Naturally, the best insulation is a material whose permittivity does not change for a wide range of temperature. It also shows that the lower the value of permittivity, the better the quality of insulation. In polythene (PE) and cross-linked polythene

TABLE 5.1

Relative Permittivity of Insulating Materials

Insulating Material	Relative Permittivity, ε_r
Impregnated paper	3.4–4.3
Polythene (PE)	2.4
Cross-linked PE (XLPE)	2.3–2.5
PVC compounds	5.0–8.0

(XLPE), such variations are less, but for polyvinyl chloride (PVC) compounds, it varies from 5.0 to 8.0.

Table 5.1 gives the values of relative permittivity at a frequency of 50 Hz and at a temperature of 20°C.

It is seen that V_o, E, f and q are at root mean square (RMS) values of varying quantities sinusoidal in time, with E and f also varying with space. Capacitive current (charge) with a symmetrical three-phase system is

$$I_c = V_o \omega C \text{ A/m} \quad \text{or} \quad I_c = V_o \omega C \times 10^{-3} \text{ A/km} \tag{5.7}$$

(charging current) where $\omega = 2\pi f$ rad/s, f being frequency in Hz.

It is interesting to note that a charging current increases with an increase in frequency and length. With frequency remaining constant, the charging current increases in higher- and extrahigh-voltage cables and also with an increase in length. It is also found that at a certain length, I_c, it can become equal to the operating current, depending on the size of a conductor, voltage range and length of the conductor. At a particular stage, no current will be available at the end of a cable length. The total energy (MVAr) will be absorbed by the insulation. Here, a shunt reactor is to be used to compensate for the loss. At very high loads, this MVAr loss is not so critical, but at low loads, reactors must function.

Example 5.1

In a 33 kV 120 mm² cable, having an insulation thickness of 8.80 mm, the conductor diameter including the semicon layer is approx 1.5 mm and the capacitance approx 0.17 μF/km. The charging current will be 1.014 A/km, with the current rating of the cable at 285 A. The maximum length at which no current flows is 281 km. With the increase in voltage and conductor size, the charging current will continue to increase.

Electric field in a round single-core cable

$$\text{As per Laplace's equation} \quad \delta^2 V = 0 \tag{5.8}$$

where δ^2 is the operator's symbol, the second derivative. Again, the potential V depends on the radial distance from the cable axis, i.e. 'r'. For a cylindrical form of the cable, it takes the following form:

$$\delta^2 V = \frac{1}{r}\frac{d}{dr}\left(r\frac{dV}{dr}\right) = 0 \tag{5.9}$$

On integration, it gives

$$r\frac{dV}{dr} = p \tag{5.10}$$

where p is the constant of integration. The electric field stress at a distance 'r' is

$$E = \frac{-dV}{dr} \tag{5.11}$$

Integrating the equation with respect to 'r', when 'r' increases to 'R', from the conductor radius to the insulated conductor radius, and where the voltage becomes V_r and the initial voltage being V_o, the result is as follows:

$$V_r - V_o = P\ln\left(\frac{R}{r_o}\right) \tag{5.12}$$

If $r = R$ and $V_r = 0$, where the potential on the outer surface of insulation is zero, the V_o potential on the conductor in relation to surface on insulation becomes

$$V_o = -P\ln\left(\frac{R}{r_o}\right) \quad \text{or} \quad P = -\frac{V_o}{\ln(R/r_o)} \tag{5.13}$$

Differentiating the value of P in Equation 5.13, it comes to be

$$\frac{dV}{dr} = -\frac{V_o}{r\ln(R/r_o)} \tag{5.14}$$

or

$$E = -\frac{dV}{dr} = \frac{V_o}{r \ln(R/r_o)} \quad (5.15)$$

$$E_{\text{max}} = \frac{V_o}{r_o \ln(R/r_o)} \quad (5.16)$$

The maximum value of stress occurs on the surface when $r = r_o$, and thus, the minimum value is found when $r = R$ and on the surface of the insulation:

$$\frac{E_{\text{max}}}{E_{\text{min}}} = \frac{R}{r_o} \quad (5.17)$$

This shows that the stress starts reducing as the distance increases from the conductor surface. This means that the stress decreases on the surface of insulation as the thickness of insulation increases. It will be zero when the diameter over the insulation becomes ∞. This is not a practical solution, and hence, a limiting factor as per acceptable values is taken. The general equation is of logarithmic function

$$E_r = \frac{V_o}{\ln(R/r_o)} \quad (5.18)$$

and is a constant.

Now it becomes

$$\frac{R}{r_o} = e^{\left(\frac{V_o}{r_o E_{\text{max}}}\right)} \quad (5.19)$$

And from here, insulation thickness is found to be

$$t = R - r_o = r_o \left[e^{\left(\frac{V_o}{r_o E_{\text{max}}}\right)} - 1 \right] \quad (5.20)$$

Hence, thickness is determined by taking a maximum permissible working stress. On this basis, the maximum stress on a conductor for a particular insulating material is determined by applying the voltages of various values, until the breakdown occurs. By computing different series of results, the breakdown voltage of an insulating material is determined on an average basis, by applying the Weibull equation. Further, the size of the conductor is also important. For a particular conductor, a minimum thickness of insulation should be worked out to obtain the maximum advantage, rationalising the cost of raw material. But there are other factors that need to be considered. The ratio of the insulated conductor diameter to the conductor diameter in an ideal case is as follows.

When $E_{max} = V_o/r_o$ in (R/r_o), the minimum value of E_{max} rests on the maximum value of r_o, and this being variable, the choice of r_o is an important factor. Generally, the value of r_o needs to be kept higher at elevated voltages so as to reduce the stress on the insulation.

Hence

The value of r_o in (R/r_o) should be maximum to get the minimum value of V_o. Therefore, for the maximum value of r_o the equation becomes $r_o = r_o\ln(R/r_o) = 0$, which can be written as:

$$e^{(R/r_o)} = 1 \quad \text{or} \quad \frac{R}{r_o} = e = 2.718 \tag{5.21}$$

Thus, insulation thickness becomes $t = R - r_o = 1.718r_o$, and the stress on the surface of conductor is obtained as follows:

$$E_o = \frac{V_o}{r_o} \tag{5.22}$$

By choosing this value, an optimum design can be considered.

In general, this does not give practical values. Insulation thickness for low-voltage cables is determined on the basis of mechanical considerations.

Example 5.2

The conductor diameter of 120 mm² is 13.2 mm. The operating voltage is 1100 V. Considering $1.718r_o$, the thickness of insulation '*t*' will be 11.34 mm, which is too thick. Practically, it has no relevance with actual working values. Considering Equation 5.20, we get '*t*' = approx 1.2 E_{max} being approx 1000 V.

In the case of high-voltage and extrahigh-voltage cables, the thickness of insulation is calculated by considering (1) the basic impulse voltage level and (2) the AC breakdown voltage. In this case, '*e*' ratio cannot be made applicable, because it does not yield a practical dimension of thickness. To accommodate an accepted stress level, the thickness of insulation chosen should be such that it should not pose any difficulty in production and installation and should withstand the operating stress level for a longer period of time. Thermal dissipation needs to be considered critically. The excessive accumulation of heat within the system at a weak point can create a hot spot, initiating early failure. To obviate such problems, contamination-free, clean and extra-clean insulating compounds have been developed, raising the basic impulse level and AC breakdown voltage per millimeter basis.

The thickness of insulation, considering the impulse voltage level, is given by the following formula:

$$t = \frac{BIL \times k_1 \times k_2 \times k_3}{E_{L(imp)}} \tag{5.23}$$

where

t is the thickness of insulation (mm)

BIL values corresponding to rated system Voltages are:

Rated System Voltage (r.m.s) kV	BIL V(test) (kV)
33	194
66	342
77	400
110	550
132	650
154	750
220	1050

Further BIL values for higher voltage range can thus be taken from relevant IEC specification

And $k_1 = 1.25$ – thermal coefficient

$k_2 = 1.1$ – ageing coefficient

$k_3 = 1.1$ – coefficient of the unknown factor

$E_{L(imp)} = 50$ kV/mm, the value obtained from the Weibull plot distribution of the impulse breakdown voltage 65 kV for 275 kV and above cables.

These voltages vary from compound to compound, though, for the sake of brevity, one particular value is to be fixed, considering all factors related to the material and processing technique.

The thickness of insulation considering the AC voltage is

$$t = \frac{\left(E_o/\sqrt{3} \times k_1 \times k_2 \times k_3\right)}{E_{l(AC)}} \tag{5.24}$$

where

t is the thickness of insulation

E_o is the maximum circuit voltage, say 69 kV for 66 kV, 80.5 kV for 77 kV and further BIL values for higher voltage range can thus be taken from relevant IEC specification. Here, $E_o = E \times 1.15/1.1$, where E is the nominal voltage of the cable

$k_1 = 1.1$ is the thermal coefficient

$k_2 = 4.0$ is the ageing coefficient

$k_3 = 1.1$ is the coefficient of the unknown factor

$E_{L(AC)} = 20$ kV minimum stress available from the Weibull plot distribution for the AC breakdown voltage (30 kV for 275 kV and above)

While calculating thickness, taking the BIL value

For 66 kV, the thickness becomes 10.6 mm.

And for the AC value, it becomes 9.7 mm.

Of these two values, only a higher value should be considered. Hence, the thickness for a 66 kV cable becomes 11.00 mm (rounding off).

Insulation absorbs a certain amount of energy from the power transmitted through the cable. This should be evaluated by finding capacitance values. Further, insulation exerts a resistance against the energy transmitted through it to approach the adjacent conductor. This resistance is known as insulation resistance. This is computed or measured to determine state of the insulating material applied over the conductor. It does not, however, states the the operational condition of the cable.

Insulation resistance depends on the dimension of the cable, thickness of insulation, type and composition of the insulating material, moisture content and temperature condition. The leakage of current takes place radially around the axis of the conductor to the surface of insulation. Taking a unit length of the cable where 'r_o' is the radius of the conductor and 'R' is the radius of the insulated conductor, for a cable length 'l', it is assumed that the length is a cylindrical section. An elementary cylindrical ring at a distance of 'dr' gives the surface area as $2\pi rl$. The insulation resistance of this elementary cylinder is

$$dR_{ins} = \rho \frac{dr}{2\pi rl}$$

where ρ is the specific resistivity of the material at 20°C. Insulation resistance of the cable for a length of 'l' is obtained by integration (Figure 5.2) and it is found that:

$$R_{ins} = \left(\frac{\rho}{2\pi l}\right)\int_r \frac{dr}{r} = \left(\frac{\rho}{2\pi l}\right)\ln\left(\frac{R}{r_o}\right) \text{M}\Omega/\text{km} \tag{5.25}$$

Insulation resistance is inversely proportional to the length of the cable and reduces with the increase in temperature.

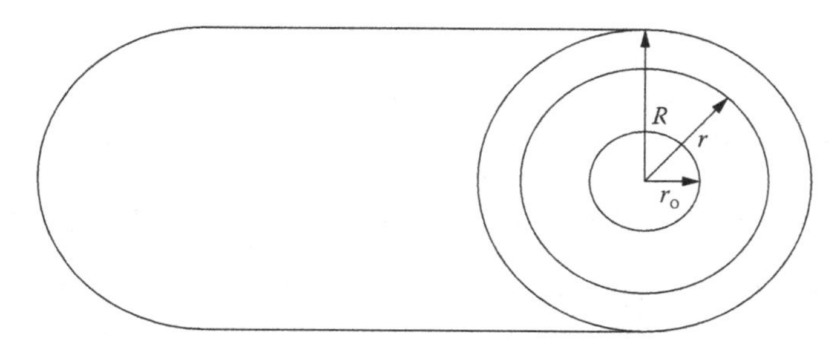

FIGURE 5.2
Section of an insulated conductor.

TABLE 5.2

'p' as a Constant Depending on Number of Wires

Sl. No.	Wires in a Conductor	Value of p
1	3	0.0780
2	7	0.0635
3	19	0.0564
4	37	0.0532
5	61	0.0505

It can be seen that while designing other parameters of the cables as would be required by the customer, planning of distribution lines, keeping the losses minimum and the state of occurrence of fault conditions, it is of prime importance to consider the insulating material and its quality.

Inductance depends on the axial spacing between the conductors 'X'. When 'd' is the diameter of the conductor, then self-inductance and mutual inductance of cores when placed adjacent to each other is given by

$$L = p + 0.2 \ln\left(\frac{2X}{d}\right) \text{mH/km} \tag{5.26}$$

Here, the axial spacing is taken for cables in a trefoil formation. For a single core in a flat formation, it is to be taken 1.2602 times the spacing of each phase, where 'p' is a constant, which depends on the number of wires in a conductor. The acceptable values of 'p' are shown in Table 5.2.

It is clear that capacitance and insulation resistance influence the characteristics of a transmission line. There is a direct relation between insulation resistance and capacitance. To ensure the lowest capacitance values, the dielectric constant must approach zero, which is the ideal case. However, the choice of the insulating material can be based on the lowest dielectric constant value of 'ε' and thermal stability. The mechanical consideration is taken care of during the production and application of the material. Accordingly, processing equipment is designed and constructed. XLPE is one such material where the dielectric constant is lower, and therefore, loss angle is also very low, resulting in lesser losses. With a combination of super clean and super smooth, semiconducting materials and a clean manufacturing process, the best products can be made at a higher voltage.

It has been discussed here before that a part of the current flowing through a conductor is absorbed by the insulating media and, though very small, which varies with a temperature rise and the way insulating media are built. For example, for paper cables, the mode of application of a paper strip during lapping, butt gap, moisture content in the paper, its thickness the and type of impregnating material play an important role in determining the loss within the system. During impregnation, moisture is removed by applying vacuum. But during lapping, if excess moisture remains present, then during drying when contraction occurs, the formation of crease on the paper may become prominent, causing air pockets. At times, it becomes very difficult to completely eliminate air inclusion, which may ionise under high stress and absorb power. This absorption should be accounted for. The butt gap can also contribute to such a factor. A detailed study is needed to understand the behaviour of paper as an insulating material. A thicker paper close to the conductor forms more folds than the thinner ones, which consolidate tighter insulating layers. But this, however, requires sophisticated high-precision lapping machines that have a large number of lapping pads. To make the manufacturing process economical, graded insulation is built up, starting from thinner paper and subsequently increasing the thickness of paper in layers. Paper thicknesses of 0.04, 0.075, 0.100 and 0.150 mm are commonly used for graded insulation. Capacitance as such is to be calculated accepting the thickness of different layers and adding them (capacitance) together.

After the application of voltage at a particular temperature, the loss factor is determined. Voltage is gradually raised. At first, the loss factor dips and rises slowly. This is because at the initial stage, the impregnating compound starts expanding from the normal temperature, and as the temperature rises, all the gaps and voids are filled, attributing a homogeneous character to the system. Thereafter, as it gets more heated, the compound expands more and more, getting thinner when a small amount of gas which remains within the system starts ionising, resulting in changes in the dielectric constant of the total insulating media. The insulating media actually starts absorbing part of the flowing current, increasing the loss factor. This loss factor is called dielectric loss and is measured by the loss angle. This is also true for synthetic plastic, thermoplastic and thermosetting insulating compounds.

Some amount of the flowing current gets absorbed within the dielectric. For this reason, the phase angle becomes slightly lesser than 90°. Charging current can thus be resolved into two components, one, 'I_d', along the voltage, *Vo*, and the other a purely reactive capacitive current 'I_c'. The charging current 'I' thus forms a lead angle of less than 90° with that of the voltage (Figure 5.3a and b). This angle, cos $\varphi = 90° - \delta$, is called the dielectric power factor and the angle δ is termed as the 'loss angle'. For a very good dielectric, δ is very small. In this case,

$$I_c \text{ is almost as } \approx I\text{; hence, } \cos\varphi = \frac{I_d}{I_a} = \frac{I_d}{I_c} = \tan\delta$$

Thus, $I_d = I_c \tan\delta$ A/m.

Now,

$$I_c = V_o \omega C \text{ A/m}$$

where

C is the capacitance in F/m

$\omega = 2\pi f$

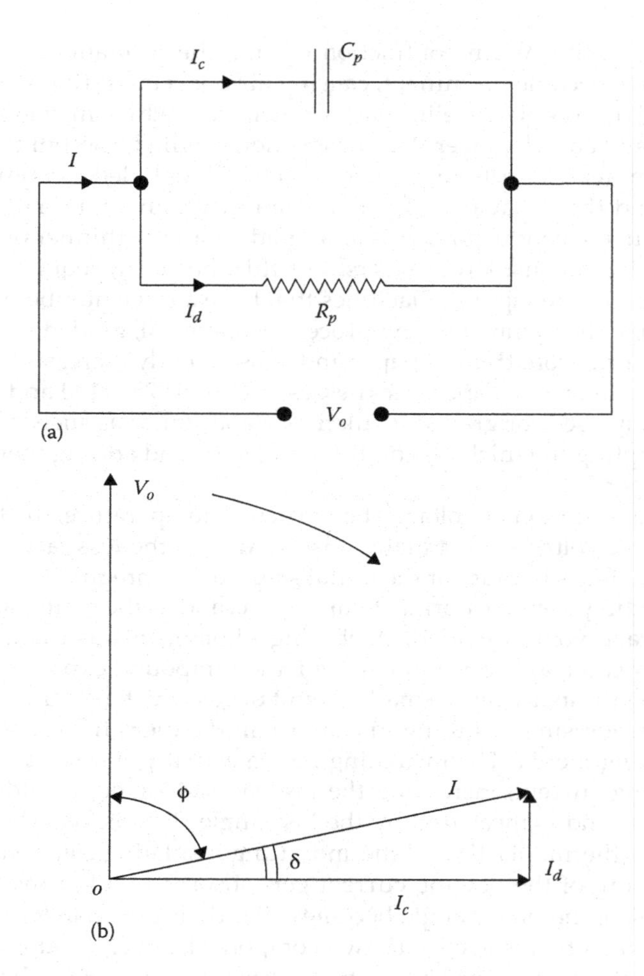

FIGURE 5.3
Showing (a) equivalent circuit and (b) its phasors.

Thus, the dielectric loss (power factor) is

$$W_1 = V_o I_c \tan\delta = V_o^2 \omega C \tan\delta = V_o^2 2\pi f C \tan\delta \tag{5.27}$$

The dielectric power factor determines how good the quality of insulation is.

As indicated, an insulating media shall have a low 'ε' value (as low as can be found). The material should be able to withstand a permissible temperature condition during operation, dissipate heat to the surrounding atmosphere quickly and have the strength to withstand environmental hazards. It should also have the ability to accept handling procedures during installation and withstand the electrical pressure applied in the course of transmitting power. All these qualities should be considered when choosing an insulating material. Naturally, the selection process of an insulating material calls for long-drawn application tests before it is standardised. After ascertaining material characteristics, long-term safe operational parameters are fixed. Conversely, at times there are specific requirements being stipulated, when the material should be selected so as to fulfil all the characteristics as stipulated (such as defence requirements, subsea cables).

The evolution of cable technologies shows that with the change of insulating and other basic raw materials, machinery and equipment should be modified and constructed from time to time, as required, for processing and application. It is interesting to know the history of development of cotton, jute, gum, coal tar, gutta-percha or natural rubber as covering material and the way these materials were processed. Later, paper insulation was introduced, the nature and quality of which had to be established. Paper lapping machine, impregnation systems, lead extruders and aluminium press came into existence. From the days of paper, we have come to the present days of thermoplastic and thermosetting compounds. In the domain of these synthetic materials, a new product is introduced every day, where we have to meet new processing parameters. A new type of extruder with a modified screw profile for a particular compound should be introduced, and processing machinery such as continuous vulcanising (CV), catenary continuous vulcanising (CCV)/vertical continuous vulcanising (VCV) or vertical continuous vulcanising (HCV) lines should be installed, when ever-changing manufacturing conditions are called for. Naturally, in order to design and manufacture electric cables successfully in the present complex technological environment, one must be alert and keep up to date on materials, processes and application technologies.

Research has been going on for several decades to bring in a material that can replace XLPE and operate at an operating temperature higher than 90°C. Only limited success has been achieved by compounding polypropylene as a new insulant. Polypropylene (PP) can operate at a temperature of 160°C. Some attempts have been made up to 11 kV, and we shall soon see that this material will shape and bring revolution in insulation materials.

5.2 Application of Insulation on Conductors

5.2.1 Paper Insulation

Paper as an insulating material is out of application. But the saga of an electric cable insulating material is not complete without understanding the application process of paper tapes and the impregnation system. Paper has to be wrapped in the form of a tape on a conductor. Standard 5 mm or 5.5 mm thick paper tapes were used for manufacturing cables up to 33 kV. Crosswise and lengthwise, the tensile strength of the tape has to be as per specification.

Paper rolls delivered by the manufacturer to cable factories should be cut to the required width. The minimum porosity or soaking capacity, ash content and pH value of the supplied paper should also be tested and measured, before accepting it for insulation. The width of the paper depends on the diameter of the conductor or the core on which wrapping should be done. There exists a definite relationship between wrapping angle, cable diameter and cable width. The width should be calculated considering the permissible butt gap. This gap should be maintained for free sliding of paper, during bending and unbending of an insulated conductor or cable.

In Figure 5.4, let '*d*' be the diameter of the cable or core, '*b*' the required width of the paper strip, '*e*' the butt gap between strips and 'α' the angle of lapping. The angle α is subtended vertically to the right angle of the axis of the conductor with '*h*' being the lay length of the paper strip.

From the similarity of triangles *CBA* and *EDA* (Figure 5.4), it is found that when there is a gap having a dimension of '*e*', the considered equation becomes

$$\frac{b}{(h-e)} = \frac{\pi d}{\sqrt{h^2 + \pi^2 d^2}}$$

Therefore,

$$b = \frac{\pi d(h \pm e)}{\sqrt{h^2 + \pi^2 d^2}} \tag{5.28}$$

Expressing this term considering the lay length as '*h*' where $e = kh$

$$b = \frac{\pi dh(1 \pm k)}{\sqrt{h^2 + \pi^2 d^2}} \tag{5.29}$$

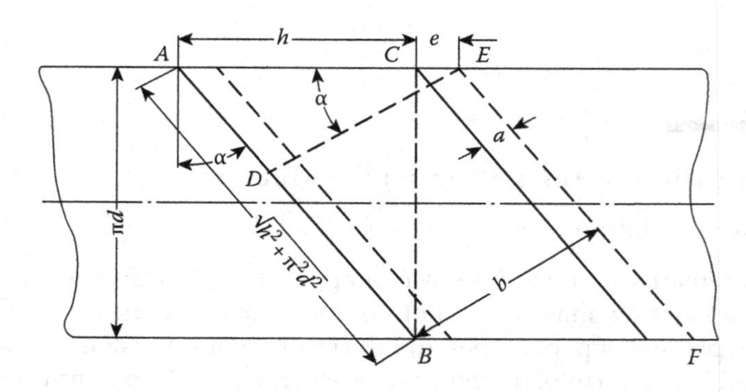

FIGURE 5.4
Lay of paper insulation with gaps.

Now from the triangle *EDA*, it follows that

$$\cos\alpha = \frac{b}{h \pm e} \quad \text{or} \quad b = (h \pm e)\cos\alpha = h(1 \pm k)\cos\alpha \tag{5.30}$$

Further, it can be seen from *CAB* that

$$\tan\alpha = \frac{h}{\pi d} \quad \text{or} \quad h = \pi d \tan\alpha \text{ in that case } b = \pi d \tan\alpha(1 \pm k)\cos\alpha$$

or

$$b = \pi d(1 \pm k)\ \sin\alpha \tag{5.31}$$

Thus, *b*, the width of the paper, is directly proportional to the diameter of the cable '*d*' and also to the angle 'α'. Width of '*b*' increases with the diameter as well as with the increase of the angle.

Example 5.3

If the diameter of a core is 15 mm and a gap of, say, 2 mm should be maintained during lapping, while the lay is 25 mm, then the width of the paper will be as per Equation 5.29, b = approx 23.85 mm, and the lapping angle shall be approx 28° as per Equation 5.32. Here, the apparent difference between lay and width is 1.15 mm, but at an angular position, the actual gap is 1.15/tan 28° = approx 2 mm. Hence, the width of paper, 23.85, mm is correct.

During cutting, the edges of the paper strip must be free from burrs and cut marks. Paper pads should be tight. In a dry atmosphere, paper should be slightly soaked with water. It should retain a moisture of at least 4%; otherwise, even a dry paper will start breaking frequently. During lapping, tension must be made to adjust automatically by a self-adjusting lever arrangement that is connected with the paper pad. The angle should be so adjusted that the face of the paper touches the cable lengthwise, tangentially parallel to the axis of the cable, without any bend, crease or fold. The lapping machine, which allows the tape to be guided as mentioned, is called tangential spinner. The lapping should be tight and uniform. The lapped paper should cover the tape of the previous layer with an overlap of 40%–60% over the butt gap. Here, the butt gap is kept deliberately in order to avoid edge-to-edge rubbing and tearing at the edges, during the bending of the conductor or cable. The paper is also allowed to slide easily without the formation of any crease or fold.

Nowadays, tangential lapping machines with 400–500 rpm are manufactured against 250–300 rpm, which were made earlier. These high-speed units are electronically synchronised and controlled. The lay adjustment is done by the positively infinitely variable (transmission) (PIV) gearbox in combination with a DC drive to facilitate smooth and accurate control of all dimensions. This type of lapping unit is also utilised for wrapping synthetic tapes over the cables.

5.2.2 Process of Impregnation

A dry paper cannot be used as an insulating material for power cables. It has an inherent porosity which absorbs moisture. Further, it remains partially brittle and cannot slide freely during bending and unwinding. But paper with oil or an impregnating compound can resist electrical stress considerably. It can also slide freely within layers during handling. Through the process of impregnation, the porosity of papers is completely filled in. Air and moisture are remove before putting oil. At first, the paper-lapped insulated dry cores are laid together to form a round cable core. These cable cores are wound in perforated trays. A tray should, at least take 5–10 km of cable cores. Two or three such trays, as filled in with cable cores, are placed in a cylindrical vessel, one on top of the other. A vessel is made of cast alloy steel and steam jacketed. Inside the vessel, spiral pipes are kept fixed on the inside wall, to heat the vessel by steam, as necessary. After placing the trays inside, the lid of the vessel is hermetically sealed. The vessel is heated by passing steam through the pipes. Temperature and pressure recorders are attached on the outer surface of the lid. Thermocouples are inserted into the vessel through airtight sealing. Two vacuum pumps, one rough and one fine, are also attached to each vessel. In the case of a very-high-vacuum system, a booster pump is also incorporated for 33 and 66 kV grade cables. For 1,100 V cables, 0.1 bar vacuum is sufficient, whereas for 11,000 V vacuum, it should be raised to 0.02–0.01 bar, and for 33 and 66 kV, it should be 0.001 bar. The evacuation of air and moisture is done simultaneously, raising the temperature to 120°C, while passing steam through the jacket and the inside tubes of the vessel. Temperature and vacuum are monitored constantly. Temperature should not be raised very high, which may deteriorate the quality of the paper by making it brittle. An impregnating compound or oil is introduced after creating a vacuum. Heating the cable, alternatively, can be done by passing current through a conductor, using transformers. As soon as oil or an impregnating compound is put into the vessel, vacuum drops instantaneously. The compound or the oil inside starts frothing, giving out entrapped gas. Vacuum should remain. It should be ensured that even at the very last stage, some adsorbed gas remains adhered to the papers, making it difficult to be torn out of their surface. The compound or the oil must be very pure and free from any contamination. The used compound or oil can be recovered by heating and filtering through a bed of fuller's earth under vacuum. The oil is then tested before pumping in. When frothing subsides, giving a clear surface of liquid, vacuum is broken by introducing an inert oxygen-free nitrogen gas under pressure. This allows the oil to penetrate quickly and properly, within the porous body of paper filling the interstice space of the stranded conductor. During the heating of the vessel, the conductors are also heated by the passage of a high current, hastening the temperature rise within the inner surface of insulation. After a given period, the cables within the vessel are cooled down to a temperature of 40°C–45°C. If required, cold water is circulated through the jacket and pipes to bring down temperature quickly. Cables are not taken out above 50°C because of the following factors:

1. Time has to be allowed for the compound or oil to set within the paper.
2. Bringing out the cable trays in hot condition allows condensation of moisture on the surface of the paper, oil or compound.
3. Improper cooling accelerates the possibility of formation of voids.

Initially, paper was impregnated with mineral oil, having a better dielectric constant and dielectric strength. But it was observed that while laying a cable on a steep gradient, the oil starts draining down, putting undue pressure on the lower part of the cable, at times rupturing the lead sheath and causing failure. Furthermore, the upper part of the cable gets dried up, creating further difficulty. To solve this problem, BICC, in collaboration with M/S Dussek Campbell, developed a compound which does not drain out even if the cable is installed at a steep gradient. The compound contains mineral oil mixed with microcrystalline wax, rosin and antioxidant and is called a mass-impregnated non-draining (MIND) compound. The latest development is to impregnate cables with polyisobutylene compound (PIB) and replace the MIND compound successfully. The cable core, as described, is protected by extruding a lead or an aluminium sheath around it, as soon as it comes out of the impregnating vessel.

5.2.3 Synthetic Polymers as Insulating Materials

5.2.3.1 Their Production Technique and Application

With the introduction of synthetic thermoplastic and thermosetting materials, the use of paper-insulated cables started declining. Gradually, it went out of the market. With the introduction of thermoplastic and thermosetting materials, processing technologies have also changed drastically. The factors related to such a complete switchover to polymeric compounds are as follows:

1. During First World War and Second World War, Central Europe and its allied countries found it difficult to procure insulating paper from the market of unfriendly countries. So they were forced to undertake extensive research and development work and find a substitute to supplement their requirement so as to manufacture electric cables and appliances. As a result, products like PVC, PE and polyester as

thermoplastic materials and SBR, butadiene, polychloroprene (PCP), etc., as thermosetting materials were developed; thermosetting materials came to substitute natural rubber.

2. Realising that natural resources might one day become scarce, Europe and the United States started making efforts to develop synthetic materials and free industries from binding to a few countries who were monopolising the trading of natural products.
3. These countries also had to compete with others in order to retain their economic balance, hence making such a development a must.
4. Realising that the indiscriminate harnessing of natural resources would slowly and gradually create an imbalance within the ecological system, research and development work had to be intensified to be able to generate alternative products.

The production of natural gas and crude oil was intensified. The refining of crude oil yielded by-products, the disposal of which was a problem. Research and development had to be initiated to utilise those by-products. These have paid dividends. New materials were synthesised and new industries came up. This transformed the economy of the countries who were able to keep up to the race. Gradually, the situation has become such that every other day a new compound is pushed into the market for a new application. One of the important areas of development is finding better electrical insulating materials in different areas of application. New processing technologies have been innovated to make all these efforts a success.

These synthetic products are resinous substances, with some having a simple molecular straight-chain structure, such as PE, and some a complex and large molecular conglomeration, such as PVC resin, PCP and ethylene propylene diene monomer (EPDM). These polymers should be blended with various ingredients to make them useful.

PVC was the first synthetic polymer used as an insulating material to manufacture domestic wires, replacing natural rubber. The material became very popular in the market. Unlike rubber, this material can be compounded and stored for a long time. The material being thermoplastic in nature can be reprocessed from scraps, bringing economic advantage to manufacturers. Cut marks and faults can be repaired successfully in situ. Further, the material has inherent fire-resisting properties. The conductor has to be coated with this insulating material by an extrusion process. However, PVC has its limitation in thermal ratings.

Joseph Bramah evolved the concept of extrusion was evolved as early as 1797, who constructed a hand-operated piston press for manufacturing seamless lead pipes. The development of a continuously operating extruder for thermoplastic materials began in the middle of the nineteenth century in the large cable factories of England and Germany. In 1845, the first patent for an extruder for the processing of thermoplastic material was granted to

Bewley and Brooman. It was operated by hand and by 1855 was converted to a mechanical device. The extrusion of rubber compounds has been used from earlier days. The rubber compound was heated to a plastic form and extruded in the form of a pipe, through an orifice, applying pressure by turning a screw, which conveyed the material forward continuously. From then on, the extrusion process has gone a sea change, wherein the design of the extruder is initiated on different concepts, considering semi-fluid viscous quality of the materials at a specified range of temperature.

By the end of the nineteenth and the early twentieth century, with the introduction of current thermoplastic and thermosetting compounds, the coating of wire was established by a continuous extrusion process that had to be operated at a higher temperature. An extruder was designed for a softer plasticised viscous material. The viscosity of the material changes with temperature and pressure as different materials have different characteristics. Naturally, it became necessary to study the properties of materials under various processing conditions to be able to design an extrusion system. Nowadays, one can find many types of PVC compounds and different types of PE materials in the market. Materials like polystyrene, polyurethane and polyamide should also be applied on wires and cables by the extrusion process. To get a proper production result and quality, the design of extruders was studied intensively. Efforts are now concentrated on combining many features within a few extruder performance, though one cannot expect a universal design that extrudes all the polymers by utilising one and the same extruder.

The ratio of the screw length to its diameter is as important as the design and construction of flight and its angle. Considerable work has been done to develop the right type of screw for a particular polymer, depending on its melt-flow index and viscosity. Temperature and pressure also plays a critical role in the extrusion process. To understand the extrusion process, it is necessary to know some basic aspects of extruder construction and the idea behind it (Figure 5.5).

An extruder consists of an assembly of a set of screw and barrel. The screw is fitted within the barrel and have a very close tolerance of 0.2–0.3 mm (modern-day screws and barrel are made with tolerance less than 0.2). The screw should convey the material by pushing it forward in a plasticised form and at a uniform velocity and quantum. The material may be fed in a granule form, sheet form or powder form through a feed hopper. The sheets are fed from the separately designed feed throat in the extruder. The compound entering the extruder is conveyed through the feed zone to a transition zone and then to a metering zone, ultimately coming out as a hot-melt plasticised material to cover the conductor or cable. In the feed zone, the material becomes soft and acquires a plastic form when temperature and pressure are applied. The plasticised material is conveyed forward into the mixing or masticating (transition) zone and last to the metering zone (Figure 5.6). These three zones overlap with each other. At the feeding zone, the flight depths of

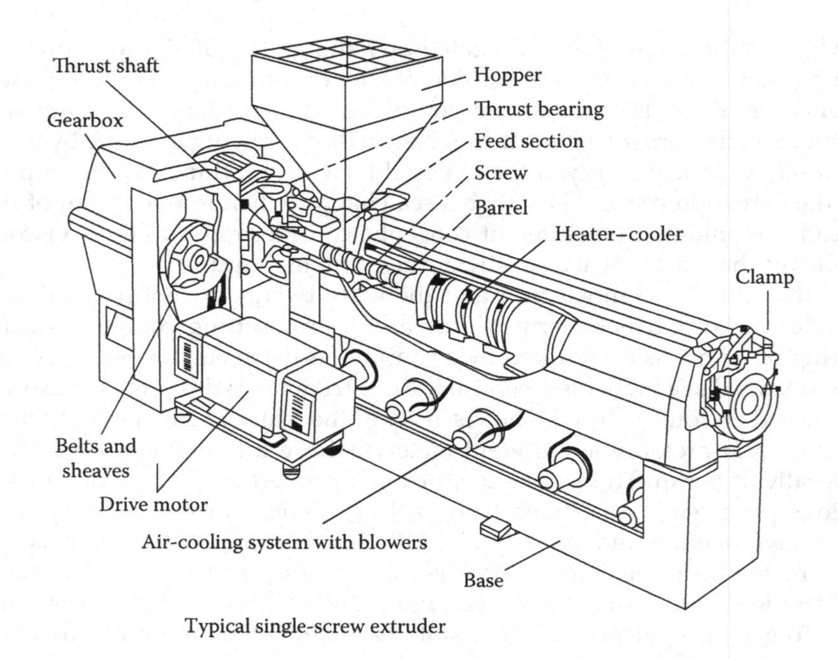

FIGURE 5.5
A typical extruder construction.

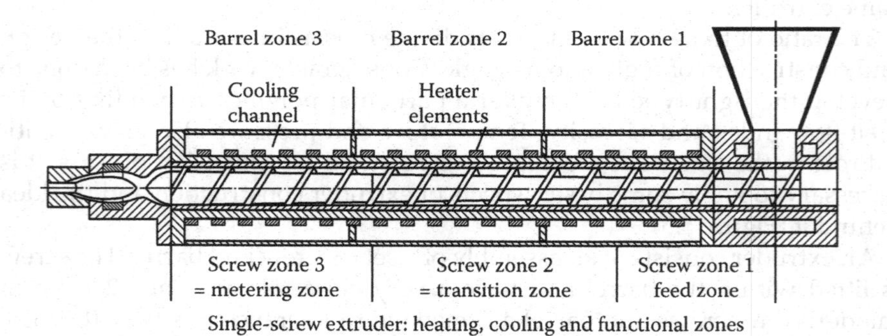

FIGURE 5.6
Screw and barrel.

the screws are deep-cut. The material is heated in this zone through external heating. At this point, the material starts softening and melting partially. From this point, the rotating screw pushes the softer material forward. At this juncture, the top layer of the material sticks to the wall of the barrel and is forced through the clearing between the screw and the barrel. The inner part of the material moves forward, while the top part rubs against the barrel wall and slowly moves with a drag flow. This movement produces a multifold shearing friction within the flow path of the material. The ratio of the area in the metering zone and feed zone is called compression ratio and is lower for materials like high-viscous rubber and high-shear materials like low smoke zero halogen (LSZH). Some of them, experimentally observed, are indicated in the following:

1. Friction develops in the clearance between the screw and the barrel.
2. Friction develops due to the rotational movement of the screw by a dragging effect.
3. Back pressure develops because of restrictions while passing through a metering zone.
4. Different flow speeds in the axial direction.
5. A shearing effect within the molecular structure.

As the material moves forward, a transition from a solid state to a plastic state takes within the mixing zone. In this zone, screw flights are not very deep but are kept at a constant height. The material here is mixed intensively and becomes highly viscous. The next part of the screw is called the metering zone. Here, the depths of flights become shallower as the material moves forward. The distances of successive spiral length of flights increase. The metering zone is the longest of all the zones. Since the flights get shallower and the distance gets longer, the material coming into this zone is put under high pressure. All of the material under high pressure cannot move forward and part of the material is pushed backwards, producing a back pressure within the starting part of the metering zone and the end of the mixing zone. The effect is an intensive homogeneous mixing. The front part of the metering zone pushes the material at a constant speed to allow a smooth flow. These actions develop a frictional heat. Viscosity and flow, however, depend on the nature of the material. At times, the frictional heat of the filled material can become so high that no external heating is required. On the contrary, the screw is needed to be cooled by the flowing water. This type of screw produces an adiabatic condition. When constant heat is required to process the material, it is termed as isothermal extrusion. If frictional heat becomes too high, it can burn the material, degrading its quality. The current screw design is such that most of the time, the machine is allowed to run utilising frictional heat under adiabatic conditions in order to reduce power consumption. Screws are now designed as neutral whereby there is no need of water cooling, and with computerised controls, temperatures are kept within a range. Cooling is maintained by air blowers, which control a temperature rise. Whenever temperature dips below the specified level, heaters are automatically switched on by the control panel's relay systems. Temperature has to be maintained within ±2°C. It is therefore necessary to monitor the temperature profile closely. Current PID system temperature controllers are found to be effective. However, at times, the temperature can rise beyond the limit, even if heaters remain off. With PID controllers, there is a fast response time for the action taken to control the heat, to the extent that heaters are switched off much before they reach the set temperature, with frictional heat taking care of the temperature required and thereby having an accurate control. To ensure a proper process control, a correct choice of channel width and flight depth is worked out and is monitored through a series of mathematical and practical working procedures. With a small channel depth at the discharge end, the pressure flow gets reduced with higher power transformation at a constant speed and where the shearing effect is inversely proportional to the channel depth. Such a screw is ideal for a low-viscosity material such as PE. Screws with higher channel depth are recommended for PVC extrusion, where viscosity is high. Thus, to get a proper output, shallow-cut screws can be used, where a high-pressure range is required, while deep-cut screws give better results where there is low pressure. This is defined through a compression ratio. A compression ratio is the ratio between the heights of the feed zone flight and the metering zone. In the case of low compression systems, the metering zone flight height is relatively higher. Along with this variation of flight, pitch distances are also varied to obtain finer details in the build-up of pressure and mixing. A maddock screw is the combination of such a type. Nowadays, various combinations are available. It is up to the user to choose the system as per the requirement of a particular material. The screw and barrel material must be of a high-quality steel and to be nitrided to get a better polish, smooth flow and resistance against corrosion. The design and fabrication of an extruder is an art by itself. In the early days, a hard chromium plating was used to fabricate polished screws and barrels. In the course of time, this plating chips off, creating problems. With the introduction of nitriding of a metal, the surface life of extruders has increased manifold. Nowadays, the bimetallic screw barrel is made of the Monel metal for better performance and longer life (Figure 5.7).

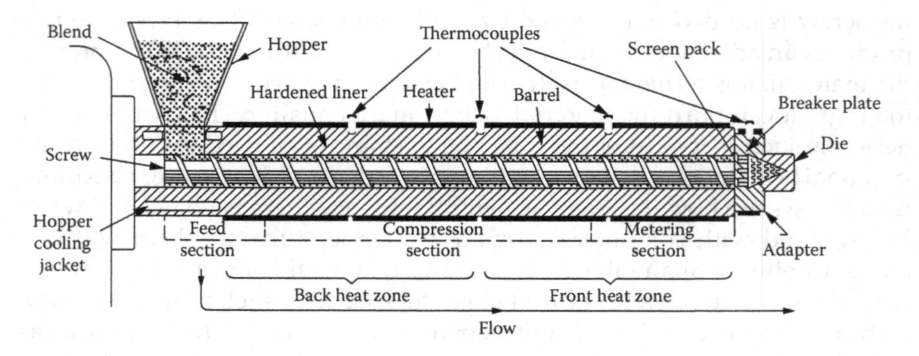

FIGURE 5.7
Constructional features of typical screw and barrel.

To keep the extrudate quality at high standards, it is necessary to observe the following working procedures. These not only allows for the control of the dimensions within the specified limit but also rationalises the consumption of a raw material.

1. The screw and barrel must be nitrided and mirror-polished. Any cut or scratch mark at any point gives rise to accumulated heat and also restrict the flow of materials. Particles get stagnant at the point and start burning.
2. Every time when a particular process is completed, the screw and barrel must be cleaned thoroughly by flushing out the residual materials. It is always advisable to examine equipment to see that not even the smallest particle remains deposited anywhere. The particle, if it remains deposited, sticks firmly on metals and create problems in the subsequent process, when the flow of materials are hindered, burning the compound. It is recommended for the club conductors and cables of the same size to continue the extrusion process for a longer period of time to obtain economic benefit and also to avoid repeating the cleaning process.
3. It is not advisable to keep the extruder stopped for a long period of time, when length or sizes are changed, with the material remaining inside and heaters still on. Sometimes, the material is allowed to purge during intermediate changeover. At this point, purging should be kept at a minimum. It means that the next length must be kept ready with the end prepared. Otherwise, the longer residence time will degrade the quality of the material as well as bring a loss of production generating scrap.

5.2.3.1.1 Selection of the Proper Screw

The design of the screw is the single largest factor for delivering the fully mixed compound at an uniform rate. The selection of the screw is a very critical quality parameter and is also directly related to productivity. By improving the melt characteristics and by selecting a properly designed screw, a consistent cable diameter and a smooth surface can be obtained.

The screws used earlier for XLPE material were of the pin mixing and maddock mixing type. The new barrier, high, medium, and low work maddock screw, is a recent advancement and development in improving the melt quality. It also gives lower pressure and temperature variation at a given output. A maddock barrier screw gives a better melt quality, lesser melt temperature, and lower melt pressure than an ordinary maddock screw. The maddock barrier screw is a combination of a barrier section near the feed zone that does not allow unmelted material to move forward and a maddock section in the metering zone, which helps to give a fully homogenised material.

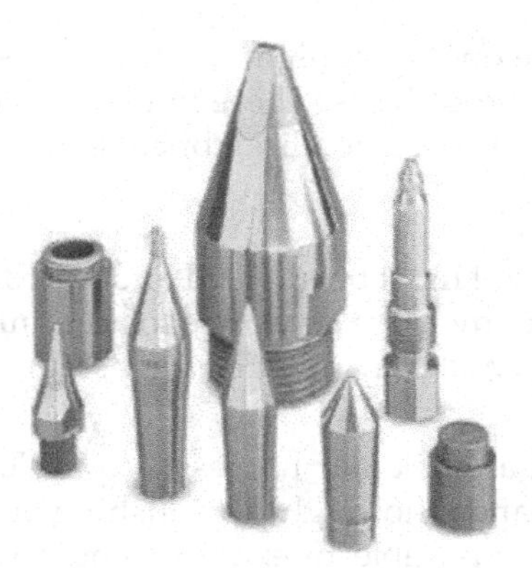

Nipples mirror-polished and hardened

Inside mandrel with flow channel

FIGURE 5.8
Mandrels and nipples of different shapes and dimensions.

The following precautions are to be taken to get quality production:

1. The breaker plate, die and the core must be kept absolutely clean and should be mirror-polished. These accessories should not be cleaned by burning an material sticking on them (Figure 5.8). Rather, immediately after an extrusion is completed, the die and nipple assembly should be dismantled in a hot condition and kept in boiled water and brushed clearn using a soft aluminium or copper rod while the material is still soft. By burning, materials get deposited on the metal, forming a rough surface. During extrusion of PVC the evolution of chlorine will attack the metal during burning, resulting in the formation of numerous tiny pits. Thus, any negligence while cleaning degrades the quality of the material, causing a production loss and increasing scrap formation. It is advisable to keep duplicate tools to cover the time gap.

2. The quality of a material also plays an important role in processing. Highly filled materials may be cheaper but develops high frictional heat. At times, the heat generated becomes uncontrollable. The line speed cannot be increased to the desired level, causing a production loss. The flow of the material becomes erratic. The material starts burning, creating problems. The wear and tear of the machine increases.
3. All synthetic materials must be stored and handled in a clean atmosphere. Many a times the feeding and mixing of materials with master batches are carried out in a casual manner with dirty hands and under the moist and dusty atmosphere. At times, the material is preheated in a conventional manner and left to the choice of an unskilled labour to handle them, leading to a lack of control. It is necessary to consider the surrounding conditions at the time of processing. Accordingly, working parameters need to be adjusted, requiring an understanding of polymers' characteristics. Naturally, operators and supervisors must be trained and given proper knowledge of processing. A clean atmosphere surrounding the extruder is absolutely necessary.
4. After every 6 months or so, the screw should be taken out, examined and cleaned thoroughly. The shine of the barrel should be checked. The clearance between moving parts should be measured to ensure a proper working ability.
5. Granules should not have any dust particles during feeding. Dust particles melts and burns quickly, producing a bad effect and degrading a compound. During preheating in a mixture, rotating fan-type propelling stirrers should not be used, as they may shear material granules during the drying process, producing unwanted fine dust. A vibrating dryer is the best solution.
6. During extrusion, a certain amount of gas is produced, which when trapped in the material generates porosity. The application of partial vacuum helps eliminate this problem. Porosity is also caused by the retention of moisture and water within the material. In this case, the material needs to be dried thoroughly.

   The design of the die and core is very important. The set should be highly polished to allow for an uninterrupted flow. The angle of flow, clearance between assembled equipments, design of contour and a perfect matching on mandrel are all important. One of the most important aspects is the relation between the dimensions of the extruder head and die assembly.
7. The of the material through the hopper should be uniform. To control the temperature within the feed zone, a water cooling system is provided inside. Materials must be dry. Nowadays, a feed hopper is

FIGURE 5.9
Hopper loader.

provided with a dryer and a dehumidifier. When coloured insulation or sheath has to be extruded, a dosing hopper is attached with the main hopper to feed. For a measured quantity of coloured master batch in the main hopper, mixing is done by injecting hot air at a low pressure of 40°C/45°C (Figure 5.9).

In order to avoid manual handling, plastic granules are fed into the hopper using a vacuum suction device.

In the early days, cables were introduced through the backside of an extruder screw. The screw was made hollow. The tip of the screw was made conical. Though the system was simple, it had a lot of disadvantages. The compound used to move inside the clearance of the cable and screw tip. Regulating pressure and temperature was difficult, and screw cooling could not be affected. Later, the system was modified to attach a separate die head. The die head axis was placed at an angle of 30°, 45°, 90° or 120° to the axis of the extruder (Figure 5.10). The recent preference is 90°. This allows for an independent placement of the cable, die assembly and the extruder. The head assembly is termed as cross head. Wire coating dies in this case can be replaced and adjusted independently. It is also distinguished by the nature of the covering process. There are two main basic principles being adopted. In one case, the core tip is kept within the die orifice. In this case, the material flows directly

FIGURE 5.10
Extruder head.

to the wire or conductor under pressure and fills in the interstice gaps, dent or anomalies that may appear on the surface of the wire, or conductor or cable. This process is termed as pressure extrusion. For highly sensitive materials like XLPE and rubber, the selection of insulation shield die should be based on the concept of characteristics of material and the use of drawdown and draw balance ratios. The drawdown ratio is defined as a cross-sectional area through which a compound is extruded to the cross-sectional area of the finished coating. The insulation die is selected based on the linear expansion and temperature of a material. In this process, the compound swells at the die tip. After cooling, though a slight shrinkage occurs, the diameter of the extruded product can remain higher. This is to be adjusted during processing so that pressure and swelling can remain under control. In the other process, the core tip is extended beyond the die orifice. Here, the extruded compound is dragged on to the surface of the wire, conductor or cable, at a distance from the mouth of the die tip, forming a close tube. This process is called the drawdown system. In this case, the extrudate forms a tube and cools down partially before it sits on the conductor. The draw balance ratio is based on the postulate that in a tube of the melted compound from a die if no forces are applied to it other than the tension required for drawing, the ratio of the outside to inside dimensions always remains the same. Naturally, the consumption of the material can be controlled much more precisely. The top surface becomes smooth. In the former case, the consumption of the material increases due to filling of interstices, though the outer dimensions are kept under control, and the shrinkage of insulation is found to be of minimum value. During tube extrusion, the material undergoes a longitudinal stress. This may create problems in case the drawdown ratio is too large.

It is found that the flow of materials between the orifice of die and core is not absolutely linear. The materials tend to adhere to the surface of the metal and a partial shearing effect can be expected. This is obviated to some extent by adjusting temperature but cannot be fully eliminated. The lubricant within the compound produces a thin film on the surface of the die and core, which can facilitate flow pattern. However, to obtain a smooth flow, the surface of the die and core should be highly polished. The existence of lubricant and a smooth surface both help to minimise the drag effect. In such a case, materials do not wet the surface of the metal and can be peeled off easily. If this condition can be maintained, the material flow will be uniform and regular.

A popular choice of die and core is found to have the following dimensional ratio when wire or cables are covered by the tube extrusion process:

$$\frac{d_1}{d_o} = 0.8 \quad \text{and} \quad S_1 = \sqrt{\left(\frac{d_o}{d_1}\right)} S_o \approx 0.9 S_o$$

where

d_o and S_o are the inside diameter and orifice gap, respectively
d_1 and S_1 are the outer diameter and the thickness of covering

Then the drag elongation of covering is

$$d_o S_o = d_1 S_1 = \frac{1}{0.8} \times \frac{1}{0.9} = \frac{1}{0.72} = 1.4$$

For PVC and LDPE, the tube drag is around 1.5; for high density polythene (HDPE), it is about 1.2.

The formation of air pockets and fisheyes on the extruded material is often experienced during processing. The appearance of a crocodile skin is an indication of porosity. A fisheye is formed due to the presence of moisture and foreign particles. Moisture, volatile chemicals like DBP, oxides of metals like antimony trioxide or aluminium trihydrate, excessive chlorine evolution or trapping of gas generated at very high speeds can create problems. An excessive gas formation can be eliminated by extruding materials at relatively lower speeds and drive out moisture and chlorine. These flaws cannot be eliminated if metal oxides and plasticisers used are of cheap quality, and moisture contents within the compound exceed a certain limit. It is therefore essential to get the compound preheated to drive out water or moisture and also to use proper-quality chemicals during the formulation and compounding of plastic and elastomeric materials.

It is found that during cold days (winter time), the material flow gets retarded, creating a hard wavy surface at the bottom part of the conductor or armoured cables. It happens particularly due to the high size of conductors and cables which sag down due to its own weight and touch the tip of the core point. At this juncture, the hot lower part of the core point gives out heat to an incoming cold conductor or armour. This is called chilling effect. During extrusion, compounds coming in contact with this relatively cold tip of core point chill down and become hard momentarily. At the same time hot compound transfers its own heat to the core point, changing the temperature to the desired level. During this brief period of heat transfer compound, the compound starts flowing normally until the core point temperature comes down again touching the cable. The result is an uneven flow of the material, producing an abnormal wavy surface. To eliminate such a phenomenon, the lower part of the conductor or armour must be heated by an electrical heater or by applying an open flame to balance the thermal condition.

During extrusion, stress is built up on the semi-fluid compound, which at a later stage produces an effect of shrinkage in insulation or sheath under a normal working environment, which is not desirable. For a larger-size stranded conductor, this phenomenon may not be perceptible beyond a permissible limit. But on single polished and small compacted conductors, shrinkage is found to be appreciable. Further, the method of tube extrusion causes an appreciable amount of shrinkage. In the case of small conductors, where the surface is smooth and the extrusion is done by the tube method, the drawdown ratio generally tends to be higher. The drawdown ratio for compounds such as rubber is 3. The drawing effect of the compound, from the die tip to the surface of the wire, forms a long cone. In this area, the compound is stretched when the molecular structure elongates, developing an unwanted stress within the compound. On cooling, this stress is retained in the compound (called memory retention). As soon as the insulation is cut at any point, the molecules start retracting to gain their original position and shrink, relieving the stress. This becomes predominant when heat is applied on the cut portion. This phenomenon can be avoided by preheating the wire before extrusion and heating also the insulated wire also it comes out of the cooling trough. With this, process memory is killed and molecules are made stress-free.

An extrusion line consists of a pay-off, preheater, extruder assembly with a vacuum suction device, cooling trough with a telescopic channel, air wiper, spark tester, counter meter, diameter controller, haul-off unit and a take-up unit.

5.3 Pay-Off

The type of pay-off depends on the bobbin dimension, operational requirements and speed.

For wires, pay-off can be of various types. The dimension should be selected as per the requirement (Figure 5.11):

1. The simplest form is a rotating truncated conical frame-type pay-off generally known as 'swift-type pay-off'. The unit is supported by a frame and moves on a bearing. A band brake is kept attached at the bottom which can be adjusted by a spring. The wire is taken out

FIGURE 5.11
Conductor paying-off from a drum.

through a guide ring. In another case, the frame is kept fixed, and the wire is taken out from the top over a guiding pulley.

2. In another type, nonrotating bobbins are put on a shaft at an inclined position. The wire runs over an extended side flange or can be guided by a rotating guide pulley through a conical stationery drum, which restricts flying off of the wire and creates a virtual braking during pay-off, to generate the required tension. These are very-high-speed dispensing units which can pay off wire at the rate of 1000–1500 m/min. Various versions of such units are constructed as per requirements.
3. For a relatively slower speed with an accurate tensioning device, rotating bobbins are deployed. The bobbins are held between pintles. Tension is controlled by a band brake or pneumatic brake, when precision tension control is required. The assembly is mounted on a fabricated structure. In some cases, pay-off is driven by a motor, when very thin and longer lengths of wire are to be paid off. This is to avoid any stretching or sudden snapping of wire.
4. For larger-diameter cables, stands are fabricated from structural steel. The bobbin holding can be of a shaft or pintle type. In the case of a pintle type, columns are made electrically moveable. Up and down movements are also actuated by motors. The tension control is actuated by a brake band system. In a pay-off stand, normally a traversing stand is not required, unless it is in the case of HV and EHV cables,

where centring of the cable is necessary during feeding. For very heavy cables, pay-off is driven electrically for smooth running.

5. There are various versions of pay-offs being manufactured and wanted by users as per their requirements.

In the case of a heavy cable, pay-off can be combined with a caterpillar for smooth and uniform running.

5.3.1 Pre-heater

During extrusion, when a conductor has a surface temperature much lower than that of the extruded compound, the hot compound coming in contact with a cold metal experiences a quenching effect on its inner surface and suddenly becomes hard, getting stuck on the point. It has a dragging effect on the compound. The inner layer of the compound remains at a static position, whereas the flow of the outer layer continues uninterrupted. This creates an uneven rough surface on the extrudate. Further, while running forward, the relatively cooler conductor touches the tip of the core point. The thickness at the end of the core point being very less, the core tip transfers its heat to the conductor and becomes cold. An increase in both these effects create problems during extrusion. This phenomenon is known as chilling effect and can create serious problems during the winter. It is therefore advisable to preheat the conductor at least 80°C–100°C, even during normal atmospheric conditions. The armoured cable need to be preheated before outer sheathing to avoid the chilling effect. It is advisable that the applied wire, steel tape or any other metallic surface on the cable needs to be warmed up before extrusion to get a smooth finish. Electrical induction heating or a band-type electrical heater enclosed in a tube can be employed. A gas burner is also used to heat the conductor. An open flame has the advantage of burning oil and other impurities on the conductor.

5.3.2 Cooling System

It is a long uncovered channel made of a galvanised sheet or stainless steel sheet, open on both ends. In the front-end side, a sliding telescopic short channel is provided. Water is held by putting stoppers at the ends along with slots to allow the cable to run smoothly. Guide rollers are provided at an interval to support the wire or cable. It is advisable to have graded cooling for all the polymers, though it may seem costly at times. Generally, cold water is employed. But for small conductors/wires and particularly for HDPE, linear low density polythene (LLDPE) or medium density polythene (MDPE), Sioplas (XLPE) and PVC, graded cooling is a must. HDPE tends to crack if cooled with cold water immediately after extrusion because of the stress which develops due to the chilling effect. HDPE is also very sensitive to environmental stresses, and therefore, it is important that at the time of processing, no stress is entrapped in the material. For other materials, the cooling system is modified as required. The cooling length should be sufficient enough to cool down hot polymers to such an extent that during passing through a haul-off caterpullar or capstan, no deformation occurs. Though the material become sufficiently hard after cooling, it takes time to dissipate the internal heat of insulation or sheath, and hence, a certain amount of time has to be given for conditioning before testing the characteristics of the material.

There exists a definite relationship between the thickness of the material, water temperature and the cooling length. The heat absorbed by a compound is retained within for some time. This depends on the thermal dissipation characteristic, thermal conductivity and flow of water, in relation to the line speed. This relation can be calculated by applying a modified Nusselt factor, which expresses the heat transfer coefficient as

$$Nu = \frac{\alpha D}{\lambda_k} \tag{5.32}$$

where

D is the diameter of the pipe length
α is the heat transfer coefficient
λ_k is the thermal conductivity of the cooling medium

Empirical Nusselt equation of the form is

$$Nu = f(Re, Pr) \text{ for forced convection}$$

enabling the relationship between the heat transfer with the cooling media flow velocity and the physical properties of the medium to be represented. In this equation, $Re = \omega D/\nu = \omega D\rho/\mu$, are the Reynolds numbers.

where

$Re = \omega D/\nu = \omega D\rho/\mu$, here $\nu = \mu/\rho$, ω is the flow velocity, μ is the kinematic viscosity, ρ is the density of the cooling medium
Pr is the Prandtl number $= c\mu/\lambda_k = c\nu\rho/\lambda_k$, c is the specific heat of the cooling medium

From these equations, cooling lengths and temperature gradients can be determined for different types of flow pattern.

These equations are further modified and worked out for a particular application. The cooling tube length for the CCV line is determined by such equations and it is not done arbitrarily. Similar is the case of normal cooling trough for extrusion lines. Naturally, one needs to take cooling systems seriously for polymer cable extrusion lines. Further, immediately after extrusion, the material remains slightly soft and in an expanded form. Internally, numerous voids remain in existence. By forced and gradual cooling, these voids are allowed to contract or get eliminated. A sudden cooling though squeezes the voids quickly but will also produce a quenching effect on the polymer making it brittle. The crystalline structure of PE gets particularly distorted. Hence, gradual cooling (from hot to cold water) for a longer period of time is preferred, whereas a cooling system which takes longer time will be costly and time consuming. It is seen that the cooling trough is measurably long so as to cool the material to make it hard so that it does not get distorted during winding.

The cable or wire should be wiped dry after it comes out of the cooling trough. If any water droplets remain on the surface, it will affect the measurement of the diameter over the coating. Spark testing is done to find the weakness, if any, present within the insulation during processing.

5.3.3 Haul-Off System

For smaller-diameter insulated wires, a high-speed dual capstan or belt-type wrap capstan is used. A dual take-up with an automatic bobbin changing system is employed. These are electronically controlled. For larger-size cables, pneumatically controlled constant-speed caterpullers are used. In most of these cases, a DC drive is preferred.

5.3.4 Take-Up Unit

A take-up unit is made of a fabricated steel having two columns attached to reel holding pintles. The columns can be moved sideways. Up and down movements along with sideway movements are motorised. The movement of the reel is actuated by a DC motor. Traversing can be done by an AC impulse motor or by a rolling-type traversing system. Modern take-up winding is done accurately layer by layer through electronic control. There are several varieties of take-up units being manufactured as per the requirements of users.

5.3.5 Thermosetting Cross-Linked Polymers

A thermoplastic material becomes soft and changes its physical form when heat is applied but regains its original form when the thermal source is removed. The thermosetting material, however, acquires a permanent set even after the heat is removed. It never regains its original form once its physical state is changed.

The thermosetting process was accidentally invented by Mr. Goodyear when he was experimenting to set natural rubber by mixing sulphur. It was observed that by heating a compound of natural rubber mixed with sulphur, a spongy but soft flexible material could be obtained.

Working with a thermosetting material involves three steps:

1. The compounding of base polymers with different chemicals, along with a catalyst and accelerators – such as curing agents and fillers – to obtain desired electrical and physical properties. This should be done selectively by choosing a proper base material and chemicals for a particular application.
2. Applying a material to cover conductors and cable assembly by the extrusion process in a normal extrusion line or in a CV, CCV, or VCV line.
3. Cross-linking (also termed as curing) the material under inert gas pressure and temperature to obtain the desired properties. Rubber and elastomeric materials are cross-linked in a steam-heated autoclave under pressure.

The extrusion of thermosetting materials like natural rubber and various other elastomeric compounds is similar in nature as described in a thermoplastic material. In these cases, the extruder screw diameter-to-length ratio can vary from 1:11 to 1:20. Nowadays, to rationalise capital expenditure and also to take a maximum advantage of machine capacity, thermoplastic and rubber compounds are extruded in the same machine at times. A rubber extrusion is best done at an L/D ratio of 15:1, but attempts have been made at a 20:1 L/D ratio too, and therefore, the design of the screw should be different to get proper quality and output. A deep-cut screw should be employed for elastomeric compounds. The masticating of a material inside the screw must be intensive and done at a relatively slower speed and lower temperature. The viscosity needs to be under a controlled condition. In the earlier days, steam-heated extruders were used. This was abolished and replaced by electrical heating with water/air cooling to control temperature. The controlling of temperature is extremely critical considering that materials should not get scorched (partially vulcanised) within the extruder. For sensitive materials, the extruder head temperature is controlled by oil circulation through a heat exchanger system. Even then, resting time within the extruder must be controlled to avoid an undue scorching phenomenon. Materials can be fed as a sheet or in a granular form. Feeding must be uniform and uninterrupted. It is very important to note that a compound processing system, storing and feeding must be contamination-free and handling must be done with proper care. For longer lengths of rubber cables, some companies in Europe have mastered the art of running the sheathing of even bigger sizes, also on CV lines, thereby making a cost-effective as well as a good-quality product.

For shorter length elastomeric cables, processing in a CV or CCV line is expensive. In these cases, cables are vulcanised in an autoclave as batch process under steam pressure.

All thermosetting compounds as soon as they are heated start swelling, acquiring porosity within their mass and rendering it useless. To contain swelling and void formation, rubber-/elastomeric-coated core/cable lengths were kept under pressure during heating. This was done by taping the lengths with proofed cotton tapes before applying a pressured steam.

In the early days, when CV or CCV lines were not in existence, vulcanisation was done in autoclaves called batch process. Here, the length of the cable remained restricted because of the following factors:

1. Take-up bobbin dimensions were kept within a given limit, considering the dimension of the autoclave.
2. Steam and pressure were kept within a limit.
3. an unvacanised cable after extrusion can get deformed within the haul-off system when pressure is applied to pull the cable, as well as on the take-up bobbin due to its own weight, if the length is long.
4. Rewinding of longer lengths in trays and on bobbins can be very difficult, and chances of the cable length getting damaged could not be eliminated.
5. At times, the outer surface of the elastomeric insulated core or elastomeric cable remains rough, retaining the impression of tape marking (if the proofing of tape is not of good quality) after the tape is removed from the surface of the cured lengths. This also affects the abrasive quality of surface and resilience. At times, covering starts cracking from the depression marks during repeated bending and unbending processes. On a smooth surface, such deficiencies are not predominant.

To eliminate all the aforementioned shortcomings and to produce longer lengths without an interruption, a CV process has been developed. Though a smooth surface can be obtained by extruding a lead sheath on the covering, the limitation of producing continuous lengths still remains unresolved. With the introduction of a lead extruder, longer lengths can be covered with a lead sheathing, but the problems faced are enumerated as follows:

1. The weight of the cable would become heavy.
2. Larger-diameter cables could still get deformed due to its own weight.
3. Cable could get cut marks and get damaged during stripping of lead sheath.
4. Occasional bulging of lead covering observed at the places where cable diameter would be found to be slightly oversized. The material under sheath would be squeezed and a dragging effect could be observed.

A CV process would eliminate all the problems stated above.

The advantages of a CV process are as follows:

1. Theoretically, a cable can be produced at any length. Scrap generation is minimised to a large extent. Further, multiple delivery lengths for the same size of a cable/core within the same voltage range can be extruded on a one-time basis and vulcanised in one go.
2. Production planning also becomes rationalised and different orders of the same size and voltage can be clubbed together, which eliminates frequent loading and unloading of the machine, improving efficiency and reducing downtime.
3. Repeated checking and winding and unwinding processes are eliminated, reducing the use of auxiliary equipment vis-à-vis capital expenditure and running costs, such as manpower and electricity.
4. The surface of a cable becomes very smooth, and dimensions can be controlled precisely. Naturally, abrasive surfaces do not affect the sheath easily. An installation becomes less cumbersome. As for the manufacturer, the raw material consumption can be kept within a given limit while maintaining all the parameters.
5. Vulcanisation becomes uniform all over the length.

Extrusion process is an art by itself. It requires thorough knowledge of the product being used, if not of the detailed formulation but at least the basic characteristics of the generic product. Naturally, the production engineer and the supervision staff must acquire sufficient information through theoretical studies and practical knowledge. Every day and every time, a new phenomenon is observed, and a new problem may crop up. All those are to be tackled and solved by analysing the root cause of the problem. Well-equipped laboratories are needed to help in resolving such problems. Further standardisation of process parameters and strict in-process quality control measures can bring fruitful results.

Amongst these polymers, PE was introduced in 1943 as an insulating material. It has a low melting point and a low dielectric constant, 2.3. The operating temperature could not be raised above 70°C. In France, PE was used at the initial stage to manufacture 45 kV HV cables. Thereafter, 70 and 132 kV cables were produced and successfully installed. But the operating temperature could not be raised above 70°C and short-circuit rating was restricted to 130°C. In 1960, ethylene propylene monomer (EPM) found recognition as a good insulating material with a dielectric constant being 2.5–2.8. This compound was rubbery in nature but could not be cross-linked, as it was a saturated polymer. After modifying EPM as EPDM, the compound could be cross-linked successfully as a flexible product. EPDM found much use in Spain and Italy. By impregnating with glycol, this compound was used to manufacture cables up to 110 kV. But it did not gain popularity as an insulating material for regular EHV cables. The compound was opaque and needed to be blended with various ingredients

TABLE 5.3

Decomposition Temperatures of Cross-linking Agents

Name of Cross-Linking Agent	Decomposition Temp.
Benzoyl peroxide	133°C
Di-*t*-butyl peroxide	193°C
Di-cumyl-peroxide (DCP)	171°C
2,5-Dimethyl-2.5 di(*t*-butyl peroxy)	179°C
Hexane (DMDBH)	
1,3-bis (*t*-Butyl peroxy isopropyl)	182°C
t-Butyl hydroperoxide	>200°C
Cumene hydroperoxide	>200°C

to improve its characteristics. Naturally, the probability of retaining contaminants became a concern. Further, cost factor (density 1.5 against 0.95–0.98 for PE) came into play. However, this compound is extensively used to make flexible mining cables up to a 33 kV range without any problem.

In the course of time, the cross-linking of PE became successful. Cross-linked PE, now termed as XLPE, was found to be an excellent insulating material.

It is an established fact that polymers can be cross-linked by irradiation or by a chemically initiated action. In the case of irradiation curing, high-energy electron beams (EBs) are impinged on the extruded layer of insulation. Chemically, cross-linking is actuated by organic peroxides. One of the most popular peroxides used is di-α-cumyl-peroxide (DCP). DCP decomposes at a certain elevated temperature into two radicals which react with PE, thereby causing cross-linking. Some of the cross-linking agents are shown in Table 5.3.

Of all the agents mentioned earlier, DCP is most commonly used.

A CV system consists essentially of the following:

1. A pay-off system to accept a bare conductor, stranded or flexible, and/or a cable to be sheathed. A dual-type pay-off is preferred to run the unit continuously for several days. When the conductor of one pay-off is at the end, the conductor from the second drum is kept ready to be welded to form continuous lengths. During the time of welding, the conductor is held stationery at the welding point, whereas on the extruder side, the line is kept running uninterrupted at a constant speed.
2. An accumulator is placed between the pay-off and the extruder to hold a sufficient length for the purpose and to facilitate such an action and it also keeps a constant tension on the line.
3. A metering capstan is placed between the extruder and the accumulator. It allows wire to be fed into the extruder at a constant speed, even though the line tension may fluctuate between the pay-off and the accumulator, which is normal, as the winding on drum goes on decreasing. Here, the tension is to be kept constant and under control. Thus, the function of the metering capstan is of extreme importance.

4. A conductor which should be heated for better adherence and low shrinkage of insulation and to avoid a sudden quenching effect of the inner surface of insulation touching the cold surface of the conductor (chilling effect). Naturally, a preheater should be placed between the metering capstan and the extruder.
5. For insulating a low-voltage conductor, one single-screw extruder is employed. For a high-tension conductor, a triple-head extrusion is preferred. In this system, a three-layer extrusion is carried out in one go. The first layer is of a semiconducting compound to be extruded over a bare conductor. This is done to smooth out electrical stress and distribute it radially around the conductor periphery. The second layer is of an insulating compound which is vulcanised by the cross-linking process followed by the third extruded layer of the semiconducting compound. All these three layers are extruded simultaneously, employing a die head connected to three extruders, through special connecting neck pieces. By this process, the adhesion of all the three layers is perfectly ensured, eliminating the inclusion of any air pocket and outside contamination. This allows the distribution of undistorted radial electric field in a uniform manner, perpendicular to the longitudinal axis of the conductor. Further, the longitudinal stress developed over the surface of insulation is minimised within a predictable limit. This unique production process technology has made it possible to extrude insulation up to 400 kV, with surety, and attempts are made up to the 750 kV range.
6. The splice box and vulcanising tube. The curing of the insulating layer is allowed to be carried out under an inert gas pressure. In this case, oxygen-free nitrogen gas is preferred for its stable performance. The tube is heated electrically by an induction heating system, where the tube is used as a low resistance path and heated directly by supplying DC through low-voltage transformers. In the conventional system, separate heaters were used to heat the tube. The heated tube supplies the necessary heat to raise the temperature of the gas by a convection process as well as by radiation. The running insulated conductor coming out of the extruder is put under the pressure of the heated gas for curing. The die tip of the extruder is connected tightly to a moving section of the tube part, which can slide over polished railings called splice box.
7. The calbe which moves through the tube in the form of a catenary curve. In this case, there is a definite connection between the cable weight, pulling force and the angle of catenary. Accordingly, the line speed is determined kept fixed. The cable is pulled within the tube in such a manner that it does not touch the wall of the tube and is controlled precisely by the pulling tension. The sag is controlled by a magnetic sensor specially designed and built for the purpose.

This sensor controls the tension constantly to maintain the position of the cable within the annular space of the tube, by transmitting electrical signals to both the metering capstans. The sensor is kept in a special section of the tube installed at a fixed measured position.

8. At the lower part, cooling tubes which are attached to the heating tube separated by a neutral zone (called buffer tube). A water–gas interface is installed where pressurised gas is held back by the cooling water introduced by pumping under equal pressure. At this point, the cable passes from the curing zone to the water cooling zone, where water pressure is kept at the same level to that of the gas. Between the two zones, a bleeding pipe is kept to drain out the sludge material which formed by the partial reaction of peroxide coming in contact with nitrogen. The cable is cooled gradually in this zone and is allowed to come out through a water seal box. At that end, another metering capstan is employed to pull the cable with a constant tension coupled with a caterpullar. Finally, the cable is wound on a take-up bobbin. Two take-up units are placed side by side to keep the process in a continuous running system. The dimensions of the cable are checked and kept ready for the next process.
9. The line speed of the insulated conductor, which depends on the time of curing. Curing is again dependent on the duration of heat transfer through insulation vis-à-vis rise in temperature. Heat absorption or heat transfer within the insulation depends on the specific heat of the insulating material, specific heat of the conductor material and initial and final temperatures of the cable while moving within the tube. Through a series of complex calculations, the rise in temperature and speed can be determined.

Thus, there exists a connection between the temperature and the line speed. The line speed is determined by the time the insulated the conductor is cured while passing through the heating zone. Cure time depends on the conductor material, insulating material, thickness of insulation and the quantum of heat applied to raise the temperature to the required level. The rise in temperature is dependent on the specific heat of the material, conductivity and coefficient of heat transfer through the insulating material. If nitrogen gas is used, the conductivity of the gas, the flow of the gas per unit time, etc., are also considered. The equation is a complex one. Empirically, it can be taken that rise in temperature T is proportional to the line speed 'S', with thickness remaining constant: Thus,

$$T \propto S$$

where

T is in °C

S is the line speed (cm/s)

If the thickness 't' is changed, heat quantum also changes proportionately. With the increase in thickness, the line speed will decrease, with the temperature remaining constant.

Then $t \propto 1/S$. Thus

$$t \propto \frac{T}{S} \quad \text{or} \quad t = \frac{kT}{S} \tag{5.33}$$

where 'k' is a constant, depending on the nature of the material, and is called material constant. This remains constant for a particular diameter (thickness) of a particular compound whose thermal characteristics are known. By knowing 'k', a graph for line speed can be plotted against temperature.

The line speed depends on the amount of material to be cured and is predetermined using heat transfer equations like those before, and currently, online curing software is used extensively as well. It should be understood that the line speed must be computed for both curing and cooling. Generally, V cooling is considered over V curing to get the best of the product.

5.4 Horizontal Continuous Vulcanisation System

5.4.1 Concept of Vulcanising Line

It is known that any suspended string between two points tends to form a curvature, however small it may be. This curvature is very small when the weight and diameter of the string are less. As the weight and diameter of the string (cable) increase, sag formation becomes deeper and deeper, taking the shape of a catenary. Normally, when diameter and weight are uniform, the catenary takes the form of a parabola. Naturally, to process heavy cables, this natural sag formation is utilised to build a CCV line.

For a smaller-size conductor having less weight per unit length, an HCV line is used. This line can be kept in a horizontal position without touching the inside wall of the tube when tension is applied at both ends. A pulling force is exerted at the exit end. Let the tension be denoted by 'T'. With the increasing weight and diameter, the curvature increases and produces a natural sag at the midpoint, forming a parabolic curvature. Considering weight to be 'W' per unit length of the cable, the depth of the sag is 'd'.

The tension applied is given by

$$T = \frac{WL^2}{8d}$$

with L being the horizontal length between two points holding the string. If the tension applied is large compared to the weight of the cable, then 'd'

tends to become 'zero', that is, very small ($d = WL^2/8T$). In this case, the line becomes almost straight. For a conductor area of 25 mm², having a diameter (stranded compacted) of 8.7 mm, with an average thickness of rubber 1.2 mm (sp. wt. 1.5), an average weight of 0.3 kg/m, requires a pulling tension of 375 kg for a span of 100 m. In this case, the line is to be inclined at 1.5° from the horizontal position. This is a safe pulling tension that can be applied on a copper conductor of 25 mm² (tensile being 11 kg/mm² as taken, and if we allow a sag of 2.5 m, the tension reduces to 150 kg). Therefore, all the small conductors are extruded and vulcanised in an HCV line having an inclination around 1.5°–2°. For conductors up to 10 mm², the CV line can be made practically horizontal (with an inclination <2°). A slight inclination is always advisable for safe performance, as weight can never become zero.

5.4.2 CCV Line

As the weight and diameter of a cable increase, the cable cannot be kept in a straight line between two points, and unless the span is very small, neither will a small amount of slope be able to hold it in position. During vulcanisation, the cable has to be suspended freely between two points. Further, as the thickness of insulation for higher-voltage cables increases, the temperature profile should be changed at a higher setting point but cannot be increased beyond a limit so as not to overheat or overcure the compound. Naturally, the suspended length should be increased as also the length of cooling. This, however, forms a natural sag in the form of a catenary. If the depth of the sag is too low, the compound will flow downwards, slipping off the conductor surface and causing a drooping effect. Whereas higher depths allow a smoother flow of the compound, forming a concentric layer over the conductor, the angle of suspension IS large and the point of suspension higher. This increases the cost of building and infrastructure. To control this depth of sag at a particular level and restrict the height of the suspended point, a predetermined tension should be applied. The level sag (depth of catenary) allowed determines the parameters of a catenary. The design of a CCV line depends primarily on the tension applied as per the weight of the cable. This also pre-fixes the horizontal span. All these parameters are interrelated. Those are to be worked out based on the catenary equation, as defined as the catenary constant.

The catenary equation is given by

$$T = \frac{WL^2}{8d}$$

where

L is the length of the catenary
d is the maximum sag point (lowest point) of the catenary, known as '0' point
W is the weight per unit length of the insulated core
T is the applied tension

This equation can be written as

$$L^2 = \frac{8Td}{W} = 4\left(\frac{2T}{W}\right)d$$

where

$$\frac{2T}{W} = a$$

$$d = x$$

If $W = 1$, then $a = 2T$, which becomes an equation of parabola $Y^2 = 4ax$.

When $W = 1$ (unity), then $T_1 = L^2/8d = C$, a constant for a particular CCV line whose horizontal span length is L. 'C' is called a catenary constant. Therefore, for any weight of a cable, tension $T = CW$, that is $T/W = C$.

Considering the previous picture of parabola (Figure 5.12)

$$AP = BA = a; \quad \text{then } Y^2 = 4ax \text{ or } a = \frac{Y^2}{4x}$$

Knowing x and Y, the value of 'a' is determined.

To determine the angle 'α' of the catenary at the point of suspension (entry point of the wire), the angle θ is determined first, where $\alpha = 90° - \theta$.

Now, $\tan\theta = m$, and the equation of a tangent at the point is given by

$$Y = mx + \frac{a}{m} \quad \text{or} \quad m^2x - mY + a = 0$$

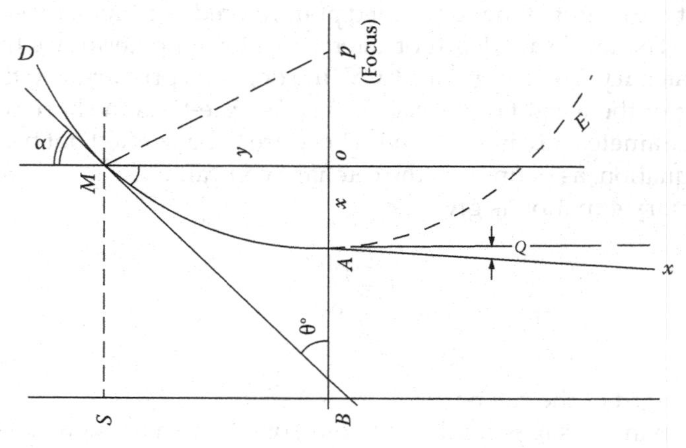

FIGURE 5.12
Schematic diagram of a CCV line design features.

hence,

$$m = \frac{Y \pm \sqrt{Y^2 - 4ax}}{2x}$$

Since $Y^2 - 4ax = 0$, $m = Y/2x = \tan\theta$.

Knowing θ, α can be found.

Normally, a line is designed as a half-catenary. This is done to cool down the cable under water or gas pressure. From the deep point, a straight tube is drawn at a 1°–2° angle, through which cold water or gas is circulated under pressure to cool down insulation to 70°C–80°C; thereafter, it is run through the atmospheric environment. Cooling down insulation is as important as heating and curing to minimise void formation and deformation. The length of the cooling pipe is determined by the inlet water temperature and pressure. The cooling length could also be determined by applying Nusselt constant, which depends on the pipe diameter, water pressure, the temperature of the hot body, the specific heat of materials and the pattern of water flow.

In Table 5.4 , an example is given of how a catenary design can be worked out. Knowing the values of *X* and *Y*, a curve can be plotted for a given angle of a catenary. The higher the angle, the higher the touchdown point. On this basis, the height of a building should be determined. The best angle for a 220 kV insulated cable is 20°. This estimates the height of a building as approx. 17 m.

At the initial stage, following the concept of rubber cables, a CV was carried out under steam pressure. In this case, steam could be injected at a maximum pressure of 12 (kg/cm²) or 11.76 bar at a temperature of 190.7°C. The decomposition temperature of di-cumyl peroxide is 175°C and is needed to initiate the cross-linking process within the extruded material. Naturally, the line speed must be restricted to allow time for the curing of insulation. A higher temperature could have accelerated the curing process, allowing the production speed to be increased. To achieve this condition, steam pressure needs to be raised, which means that the plant should have an arrangement to produce superheated steam. Typically at 18 bar of steam pressure, 210°C temperature is achieved in the tube of a CV line, and in order to raise the temperature to 300°C, the steam pressure of 90 bar may be required; it is not only uneconomical but also difficult to make such a large system. This involves additional cost for a steam-generating plant and higher safety norms. This leads technologists to develop a process where both temperature and pressure are independently controlled. Further, during curing under steam pressure, the insulation material absorbs more moisture. This results in the initiation of a number of water tree formations and early failure of insulation. The curing process must be initiated in an oxygen-free atmosphere, under pressure and temperature. At first, dry oxygen-free carbon dioxide was used. However, it was found that oxygen-free nitrogen could be generated easily and used to pressurise the tube independent of temperature. This led to raising the temperature level and increasing the line speed

TABLE 5.4

Catenary Design in Relation to Catenary Angle

Catenary Calculation Considering the Catenary Angle (Inverse Angle Can Be Calculated, Also Taking the Catenary Constant)					
1	$Y^2 = 4aX$ (equation of a parabola)	$X = Y^2/4a$	$X = Y/2m$		
2	$m = 2a/Y$ (angle at which tangent [$\tan\theta$] touches the curve of the parabola)	$\tan\theta$			
3	Here $a = X = d$				
4	α = catenary angle	$\alpha°$	To be selected		18
5	$\theta = (90° - \alpha)$	$\theta°$			72
6	Y = half-catenary length in metres	$L/2$ (metres)			61
7	T = tension on the catenary				
8	X = touchdown point (axis of parabola)	$X = d$ (metres)	Is $Y/2a = Y/2m$	$Y/2 \tan\theta$	9.91
9	a = length of focus from vertex on x-axis	$2T/W$			
10	A	$2T$	When $W = 1$		
11	C	T/W			
12	C = catenary constant	T (kg)	When $W = 1$	$L^2/8d$	187.74
13	a {focus of parabola = $Y^2/(4d)$}	1/2 C (metres)	When $W = 1$	$Y^2/(4d)$	93.87

P.S.: Accepting a particular value, a catenary curve of the tube can be plotted exactly.
Knowing the catenary constant, multiply the cable weight per metre to get the required maximum and minimum tension.
To get a catenary constant of 110 m/kg, the angle should be a minimum of 29°.
And the height of the extruder from the touchdown point shall be approx. 16.91 m = 55.5 ft.
Taking all allowances, the height of the building shall be max. (17 + 10 = 27 m i.e. 88.6 ft).

vis-à-vis production. Water absorption within insulation/sheath was lowered down considerably. This increased the life span of the cable containing failure rates. The dry cure process using inert nitrogen got stabilised very fast. Pressurised nitrogen gas heating and cooling, thereafter, enhanced the quality of the product considerably. The water content during the steam curing stage was restricted to <3000 ppm, whereas after the introduction of nitrogen gas, it came down to <300 ppm. Naturally, the failure rate due to water tree formation was practically eliminated.

The CCV line consists of the following:

1. Dual pay-off stand
2. Accumulator
3. Metering capstan
4. Preheater

5. Triple extrusion system with a diameter thickness and a diameter controller
6. Control panel
7. Oil cooling system with a heat exchanger
8. Catenary line consisting of
   a. Splice box
   b. Heating zone tubes with transformers (normally 6–8, or sometimes 9 as tubes)
   c. Tension monitoring sensor device within the tube
   d. Buffer zone with an affluent purging nozzle
   e. Pressurised water cooling tube
   f. Water seal and wiper
   g. Additional cooling trough
   h. Pulling caterpullar (main and auxiliary)
   i. Dual take-up unit

Apart from these, the following auxiliary equipment is to be made available:

1. Nitrogen-generating unit to produce dry gas, having a purity of 99.5% minimum at a minimum of 20 bar with a dual-type gas reservoir. Nitrogen is to be fed into the tube constantly at a pressure of 12 bar.
2. Conductor surface wiping and cleaning devices.
3. Conductor crimp joint tools with ferrules.
4. Multistage water turbo pump to pump water in the tube at a rate of 18–20 bar for dual-cooling type.
5. Contamination-free water reservoir.
6. Water recovering and recirculation filter and pump (if required).
7. Conditioning chamber.
8. Uninterrupted power supply system for continuous running of the CCV line for days together.
9. Clean contamination-free compound storage room with vacuum suction devices to feed three extruder hoppers.

For MV cables up to 33 kV, the normal clean XLPE compound can be used, where the minimum contamination level is specified. It is also experienced that water and chemical tree formation have been considerably minimised by a triple extrusion process, where three-layer extruders are employed to extrude inner semiconducting, insulating and outer semiconducting layers in a closed-circuit system employing a single triple extrusion head.

The compound in the hopper should be heated by circulating dry hot air. In turn, hoppers are loaded by a vacuum suction device directly from the bins. Master batch and a catalyst are fed into the hopper in measured quantity through a doser. Manual handling is thus completely eliminated, ensuring a contamination-free operation. Nowadays, grafted polymers with anti-oxygen and catalysts are supplied by the compound manufacturer. The material can be fed directly into the extruder hopper. The extrusion system – the metering capstan, triple-extruder system, operator control panels, extruder heating and cooling device (coolant devices), up to the splice box assembly with a diameter control x-ray system and the material feeding chambers – is to be kept in cleaned pressurised chambers to ensure a contamination-free operation. Control panels and electronic monitoring systems should also kept in a separate conditioned chamber for smooth functioning, adjacent to the operating panel (Figure 5.13).

For HV and EHV cables, the compound is made super cleaned.

They are tested and examined after their receipt in laboratories before use, as per specified norms.

In order to ensure the proper dimension apart from the x-ray system, samples should be checked intermittently for the thickness of semiconductive and insulating layers. Eccentricity and ovality measurements are to be ascertained. A hot set test should be done after conditioning the samples taken from the running length. Once the parameters are found satisfactory, the machine can run days together. It has been found that a longest run can be for 20 days without interruption. For this to happen, maintenance schedules, operational skills and conditions should be followed and monitored strictly.

To view the contamination level within insulation and protrusion, a hot silicon oil bath should be used during extrusion. A cut portion of the sample

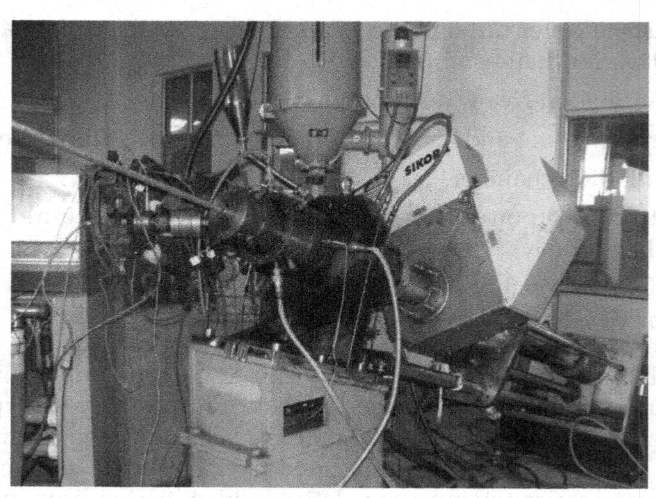

FIGURE 5.13
Triple extrusion head.

of about 25 mm long should be taken. Conductor wires are taken out, forming a hollow cylindrical tube of a three-layer extruded material, which should be dipped in a silicon oil bath heated to a temperature of about 120°C.

After a certain time, the insulated layer becomes transparent, showing inner faults such as voids, contaminant particles, protrusions and wavy surfaces clearly, along with the surfaces of semiconductive layers. Faults occur within insulation and the semiconducting layers during processing. These are generated either from faulty extrusion conditions or from using improper materials. A contaminated insulation material particularly generates a large number of voids. Protrusions and an uneven surface along the boundary of insulation and semiconducting compounds are the result of having coarse carbon black particles within the semiconducting compound, or having higher imperfections during the production (Figure 5.14).

Due to an inconsistent temperature profile, it is likely that insulation can start degrading, showing discoloration called amber formation. This is very serious. Considering the nature and level of faults, corrective measures should to be taken. These observations are essential during the ongoing production, particularly for EHV cables (Figure 5.15).

Apart from the CCV line, CV is also being done in the Mitsubishi Dainichi continuous vulcanising (MDCV) lines. Initially, the concept was developed by Anaconda. In this system after triple extrusion, the insulated core is passed through a closely fitted tube containing hot pressurised molten liquid salt. Salt remains in a molten condition at high temperatures and can withstand more than 400°C without any deterioration. The insulated core

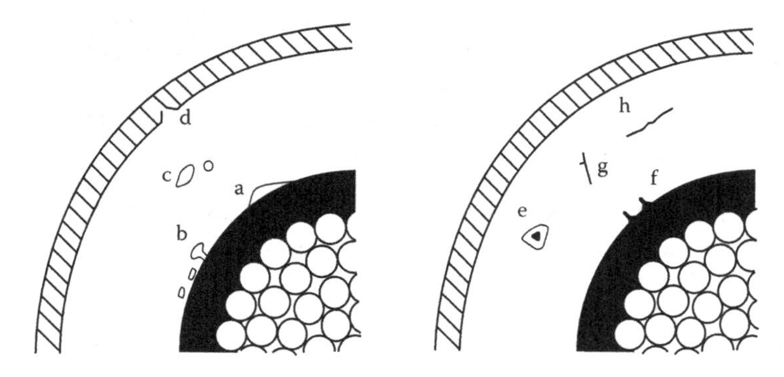

a. Loose semiconductive screen
b. Bubbles caused by gas evolution in the conductive screen
c. Cavities due to shrinkage or gas formation in insulation
d. Defects in the core screen
e. Inclusion of foreign particles that separate gases, often due to moisture in the particles
f. Projections or points on the semiconductive screen
g. Splinters
h. Fibres

FIGURE 5.14
Void formation within extruded XLPE insulation layer.

FIGURE 5.15
Vulcanising and cooling tubes of a continuous catenary line.

remains enclosed by the molten salt and floats under buoyancy and is prevented from touching the wall of the tube called long land die. XLPE core is thus vulcanised under temperature and pressure while passing through the die tube. The core as vulcanised is cooled by passing it through water. The advantage of the system is that the equipment can be installed in almost a horizontal position, where the building cost is minimal. The biggest disadvantage is that for almost every core size, one long land die is to be kept in ready stock, which is very costly. Further, if any of the die gets damaged, it becomes unusable if handled casually. Replacement is very costly. A particular salt should be procured specifically for the operation and should be absolutely contamination-free (Figure 5.16).

It has been experienced that during the extrusion of heavy cross-sectional conductors above 1000 mm², such as 1200 mm² and above, at higher-voltage systems, such as 132 and 154–400 kV, where the conductor is of Milliken construction and insulation thickness is very high,

1. The conductor may deform while passing through the CCV bent tube.
2. Further, due to heavy weight, the touchdown point of the CCV line needs to be deeper, calling for a steep angle. This calls for a high-rise building.
3. A thick insulation layer tends to sag towards the bottom part of the conductor in a CCV line, called drooping effect. It occurs even when the conductor is of a high cross-sectional area. A conductor insulated in an MDCV line (where 400 kV at higher cross section can be taken for processing) tends to float down within the insulation, causing

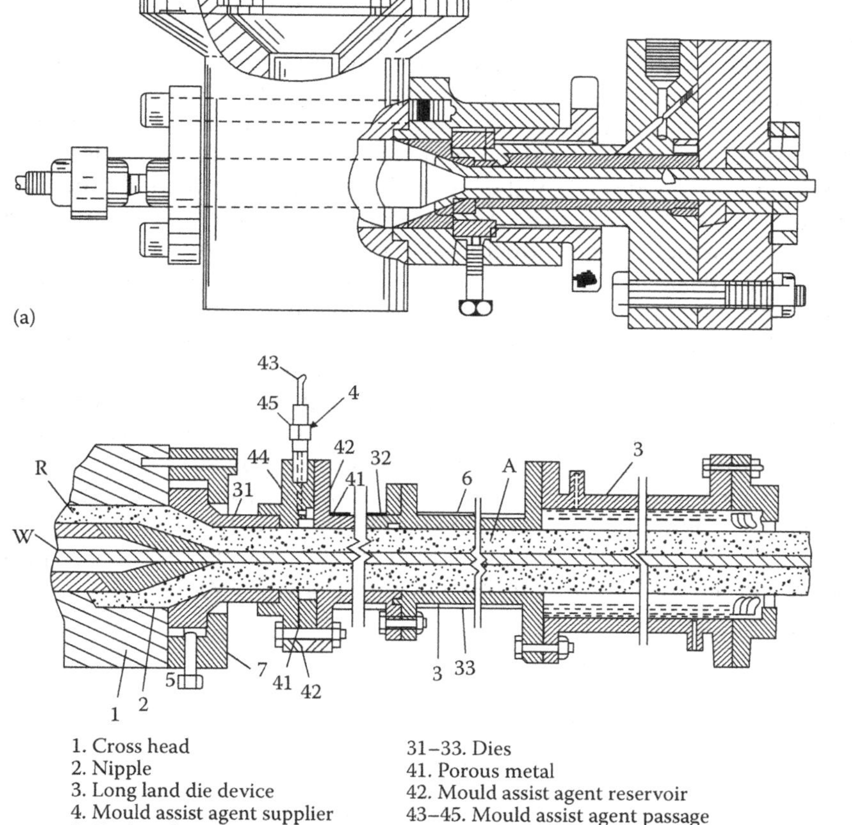

FIGURE 5.16
(a) MDCV triple extrusion head (b) MDCV long land die.

> high eccentricity. This, however, was compensated by deploying a rotating caterpullar at the take-up end of the line. It is found that this has a very limited effect in keeping the insulation in the centre position. The only remedial measure is offsetting the position of the conductor in relation to the die and core. It depends on how accurately the operating personnel can position the conductor in relation to the die and core point within the cross head.

In order to get a more consistent and better result, a VCV line is used. In fact in the initial stage, the use of a VCV line was recommended before the CCV line was brought forward. To install a VCV line, a tall tower should be constructed. The tower may become approx 100 m high. Triple extruders should be kept fixed at the highest position, with the vulcanising tube running down vertically along with the cooling tube. In this case, the extruded insulation remains concentric. While starting, a certain amount of insulation rundown is experienced. This is compensated by a higher insulation thickness, which drags down slowly as the conductor comes down towards the cooling zone, as also hardening of the outer layer by controlled heating. Taking into account the higher cost of building along with the required infrastructure, VCV lines are not preferred for producing cables of up to the 220 kV range, which have conductor sizes of up to 1000 mm^2.

Elastomeric flexible cable cores can be effectively produced in a CCV line, but for longer lengths of elastomeric cables having a smooth finish and a controlled diameter, the pressurized liquid continuous vulcanising (PLCV) line is the best choice. A modified version of an MDCV line is the PLCV line. The PLCV line is mostly used to produce higher-diameter heavy elastomeric cables in longer lengths, which cannot be accommodated within the batch system. In this case, the insulated cable core or cable sheathing can be done without using any special long land die. The cable or core is allowed to float in the molten salt in a buoyant position under pressure. The machine can run continuously for longer periods, provided proper take-up units are kept in position with all the handling infrastructures (Figure 5.17).

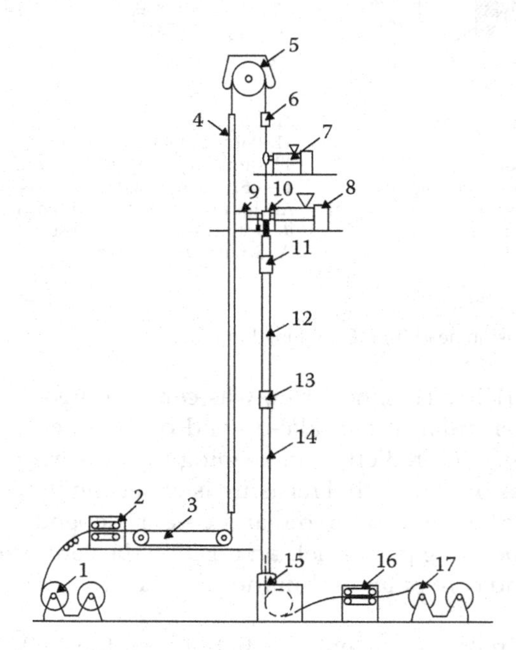

FIGURE 5.17
Vertical continuous vulcanising line. 1, supply stand; 2, capstan; 3, loop car; 4, conductor preheater; 5, pull-up capstan; 6, conductor preheater; 7–9, extruder; 10, two-layer common head; 11, x-ray TV set; 12, heater; 13, liquid height control; 14, cooling tube; 15, double seal; 16, capstan; 17, reel.

5.5 Degassing: For HV and EHV Cables

During curing, peroxide compounds react with PE and catalyst, giving out by-products like, methane, acetophenone and cumylalcohol on decomposition, along with some amount of water. As the curing process continues at a slower rate even after coming out of the CV, CCV or VCV line, the reaction to the formation of these by-products continues within the insulation. If the cable ends are kept sealed, these gaseous products accumulate within and can become hazardous to life and the surrounding environment. These products can explode when mixed with atmospheric oxygen and create a problem if allowed to be retained within the insulation for a long time. It has been found that at an elevated temperature, the formation of these products accelerates and exudes out of the cable ends. On observing this phenomenon, methods have been worked out to keep the cable lengths within an enclosed but ventilated chamber, at an elevated temperature of 80°C–90°C, for precalculated hours, and to let all these gaseous products out of the cable lengths until the safest limit is achieved. If these products are retained within the cable length, the following types of problems may appear in the long run:

1. Pressure of these gaseous products can explode, displacing the jointing accessories. Methane being inflammable should be promptly removed.
2. It can increase dielectric loss in the system.
3. It can rupture the metallic sheath, creating faults within the cable.
4. It may show abnormal test results; particularly, values of partial discharge may be shown as much lower than the actual ones, as these gases fills in the voids under pressure and allows the actual ionisation value to come out. In the long run, this affects the cable life.

The more the thickness of the insulation, the more the formation of these by-products. Degassing needs to be done before any metallic sheath encloses the cable core.

5.6 Sioplas System of Curing and Manufacturing XLPE Cables

It has been observed that PE can be grafted with silane. Mixing with a catalyst, the extruded material can be cross-linked by the absorption of water or moisture at an elevated temperature. This process has the advantage that it can be processed in a normal PE extruder for the production of low-voltage cables. For processing medium-voltage cables, a triple extrusion system is applied to keep materials free of contamination. Here, a grafted polymer is extruded and mixed with a catalyst and master batch. Extruders are kept on the floor in a horizontal position. In order to eliminate chilling effect, the extruded polymer coming out of the extruder is cooled down in a graded manner, passing through a hot water trough in the beginning and then passing through water at normal room temperature so that the polymer cools down. The insulated cable core thus produced is conditioned at room temperature for a few hours to stabilise its condition. It has been established that any void generated during extrusion remains in an expanded form as long as an insulated material is in hot condition. The gas entrapped within the void also remains under low pressure. Naturally, any electric pressure applied ionises the entrapped gas easily and may show a high discharge rate and even initiate a premature fault. As the insulation cools down, the dimension of the voids becomes smaller, raising the gas pressure inside. According to Paschen's law, the breakdown strength is a function of the distance of the electrode and gas pressure. As the gas pressure increases, the voltage required to ionise the gas increases. After conditioning, the cable core is cured under a low-pressure steam (sauna), or in a hot water bath at about 80°C–90°C, for a predetermined hour. The time required to complete a curing cycle depends on the thickness of the insulation, say 5–6 h for a thickness of 2.5 mm. During this time, the insulation absorbs a certain amount of moisture and water. After bringing out the cable from the curing zone, it is kept on the floor for certain hours to cool down at room temperature, before the material is sent for further processing. During this time, the insulation exudes out excess water and moisture into the atmosphere, reducing the chances of forming water trees. If these steps are not properly followed, chances of failure in the long run become inevitable.

It is advisable to cure the cable core under a low-pressure steam at 80°C–90°C. This has the advantage that the steam could spread over the total area of the cable uniformly with the moisture, while temperature at all points can be maintained at a constant level by a proper circulating system. The curing chamber should be double walled and thermally insulated to minimise heat dissipation. The door should also be double walled and thermally insulated and should be sealed hermetically. In the case of a water tank, it is difficult to maintain a constant temperature throughout, unless the water is circulated uniformly all over by an agitator. Second, an underground tank is difficult to insulate, unless silica/magnesia bricks are used all around to keep the tank thermally stable. Further, a considerable amount of steam is to be wasted in order to bring the temperature up to a specified level. It is also seen that insulated cores while winding on a take-up bobbin are kept slightly loose to be able to allow moisture to penetrate around the insulated core. In a tight winding, contact points between cores act as a rubber or plastic washer and penetration of moisture cannot be fully ensured.

In the case of a steam chamber, as soon as the curing operation is over, the flow of steam can be stopped immediately, saving considerable steam and power. In the case of a curing tank, a large amount of steam is required initially to raise the temperature for curing, which is time consuming. Further, maintaining a constant temperature throughout is a problem. This can lead to an uneven curing of insulation over the total surface of longer length cables.

Curing does not get completed within this given span. It continues for days, weeks and even months to complete the cycle. As days progress, the quality of insulation gets better and better.

In the Sioplas system, the biggest advantage is that cables can be produced in desired short lengths, reducing the scrap content, unlike CCV lines where a minimum length needs to be discarded during starting and stopping of the production process. Further, the machine can be installed on the shop floor without raising the building cost and heavy infrastructure. In a triple extrusion system, nowadays, the outer semiconducting layer can be bonded or stripped as required. With special precautions, cable cores of up to 33 kV can be produced with the desired quality standards.

5.7 Irradiation System of Curing Polythene Insulation

While a constant effort was exerted to produce a better quality of insulated conductors, it was found that by impinging an EB on PE and on a selected plastic material, a higher-quality product could be obtained by cross-linking them by an irradiation process. An EB has thus been found to have a high degree of potentiality in modifying the characteristics of various compounds as well as other materials.

Radiation processing can be defined as the treatment of materials and products with radiation or ionising energy, to change their physical, chemical or biological characteristics, increase their usefulness and value and reduce their impact on the environment. Accelerated electrons, x-rays (bremsstrahlung) emitted by energetic electrons and gamma rays emitted by radioactive nuclides are suitable energy sources. They are all capable of ejecting atomic electrons which can then ionise other atoms in a cascade of collisions. Thus, they can produce similar molecular effects. The choice of energy source is usually based on practical considerations, such as absorbed dose, dose uniformity (max/min) ratio, material thickness, density and configuration, processing rate, capital and operating costs. In the case of EB processing, the incident electron energy determines the maximum material thickness, and the EB current and power determine the maximum processing rate. In the case of x-ray processing, the emitted power increases with the electron energy and beam power.

For high-throughput industrial processes, capital costs and operating costs of an irradiation facility are more competitive than those of more conventional treatment methods.

Successful irradiation processes provide significant advantages in comparison to typical thermal and chemical processes, such as higher throughput rates, reduced energy consumption, less environmental pollution, more precise control over the process and products with superior qualities. In some applications, radiation processing can produce unique effects that cannot be duplicated by other means. Radiation processing was introduced more than 60 years ago, and many useful applications have since been developed. The most important commercial applications involve modifying a variety of plastic and rubber products.

5.8 Basic Concepts of Radiation Processing

5.8.1 Absorbed Dose Definition

The most important specification for any irradiation process is the absorbed dose. The quantitative effects of the process are related to this factor. An absorbed dose is proportional to the ionising energy delivered per unit mass of a material. The international unit of the dose is gray (Gy), which is defined as the absorption of 1 joule per kilogram (J/kg). A more convenient unit for most radiation processing applications is kilogray (kJ/kg or J/g). An older unit is rad which is defined as the absorption of 100 ergs/g or 10^{-5} J/g. So, 100 rad is equivalent to 10^{-3} J/g or 1 J/kg or 1 Gy. Absorbed dose requirements for various industrial processes cover a wide range of 0.1 kGy to more than 1000 kGy.

5.8.2 Temperature Rise versus Absorbed Dose

If energy transfers from chemical reactions are negligible, then the adiabatic temperature rise (ΔT) from the absorption of thermal energy per unit mass (H) is given by the following equation:

$$\Delta T = \frac{H}{c}$$

where
- ΔT is in °C
- H is in J/g
- c is the thermal capacity in J/g °C

Similarly, the adiabatic temperature rise from the absorption of ionising energy is given by

$$\Delta T = \frac{D(\text{ave})}{c}$$

where

D(ave) is the average dose in kGy (kJ/kg or J/g)

ΔT and c are the same as in the equation

The thermal capacity of water is 4.19 J/g °C, so the adiabatic temperature rise is 0.24°C with an average absorbed dose of 1.0 kGy. Most other materials have lower thermal capacities and higher rises in temperature with the same dose. For example, the thermal capacity of polyethylene is 2.3, polytetrafluoroethylene is 1.05, aluminium is 0.90, copper is 0.38 and tantalum is 0.15 J/g °C. On the other hand, when an electrical wire receives a typical dose of 100 kGy to cross-link the insulation, the temperature rise of the copper conductor could be as high as 260°C. This excessive temperature rise can be reduced by passing the wire many times back and forth through the EB to allow most of the heat to dissipate in the air and in the underbeam wire handling fixture.

The latest development is to cross-link PE insulated cores by EB radiation. Even PVC can be cross-linked in the same process. A beam having varying power is applied to cure materials like plastics, adhesives, diamonds and certain types of metals. An e-beam cross-linked insulation has all the advantages, such as better tensile strength and cold and hot impact tests, wherein cables can withstand very low and high temperatures such as −40°C to 125°C (for automobile cables and cords); for defence applications, it has higher creep resistance; durability; solvent, lubricant and chemical resistance; abrasion resistance and environmental stress crack resistance (Figure 5.18). The material gets a smooth finish, increases impact resistance and gives a higher electrical strength. Nowadays, high-power e-beam to cure higher insulation thickness, even for MV cables, has been developed. Generally, 1.5–2 MeV beams were used. A 3 MeV system can be used to cure larger thickness of an insulating material. The system generally consists of an electron gun, a accelerating tube, an RF transformer, a scan magnet chamber and a scan horn. Through the scan horn, beams are impinged on the product to be irradiated. In the case of wire and cable, the same should be rotated several times, passing through special rotating pulleys with pay-off and take-up systems to get the radiation from all sides. A double scan horn system has been developed to irradiate the object from all sides at a time.

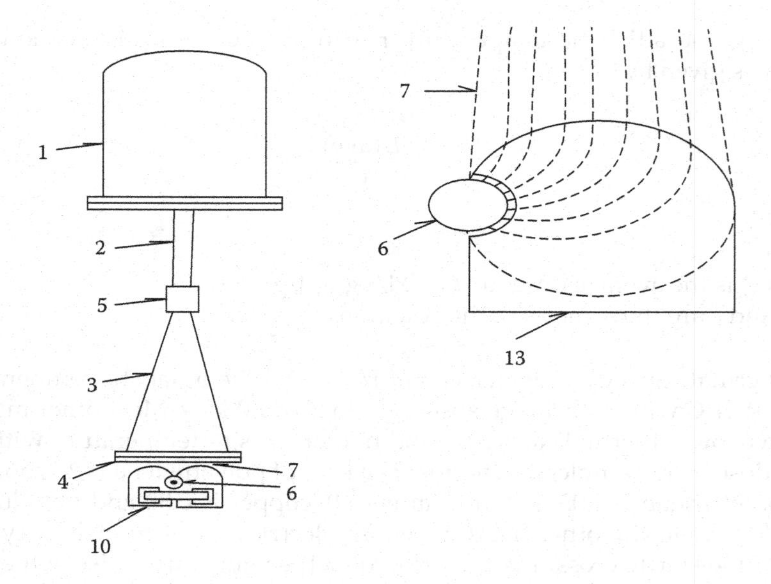

1. Electron accelerator
2. Evacuated extension tube
3. Spreading portion
4. Electron aperture
5. Beam scanning device
6. Object to be irradiated
7. Electron current
10. Magnet pole face
13. Vacuum pump

FIGURE 5.18
Schematic diagram of electron beam accelerator.

5.9 Processing Methods

Small-diameter wires or cables are usually EB cross-linked by passing them many times through a wide scanning beam. This method has several benefits: (1) A narrow, high-current EB must be scanned to increase its width and reduce the average current density in order to avoid overheating the thin metallic beam window of the accelerator; (2) The scanning beam will be much wider than the wire or cable diameter so that such products can pass through the beam many times and intercept most of the beam current; and (3) Multiple passes also avoid the possibility of overheating the insulated wire or cable by allowing some of the heat from EB processing to dissipate between passes (Figures 5.19 and 5.20).

FIGURE 5.19
A small compact type accelerator (5.19 and 5.20 are similar types hence one can be omitted).

Small wires can be EB processed by passing them several times under the beam. These wires are placed on sheaves, winding in several turns. The wires are slowly turned on 360° rotation to allow the beam to get projected on the wires in all directions over the diameter. These are a special type of undercarriages which are installed under the accelerator for the purpose.

Large wires and cables require a larger bending radius and can be processed by keeping on a special type of drum fixtures.

Cables with larger drums in the forward pass are closer to the beam window and receive a higher dose than the more distant reverse pass. There are various types of fixtures used to pass cables under the beam and allow full penetration of EB energy. The divergence of the scanning beam causes the sides of the wire or cable to be treated, as the wire progresses from one end to the other, of the multiple-pass underbeam fixture.

The outer sheaths on very large cables can be treated by rotating the cables as they pass through the EB along the scanning direction. With this method, both the pay-off and take-up reels are rotated outside of the shielded treatment room. The electron energy needs to be just enough to penetrate the radial thickness of the sheath. In this way, a near-uniform dose distribution around the cable will result. Large cables may have to be cooled inside and outside of the treatment room when a relatively high dose is delivered in

FIGURE 5.20
Compact accelerator.

a single pass through the beam. However, proper formulation of the cable jacketing can reduce the dose needed to impart cross-linking and thereby reduce the temperature rise.

Only the choice of insulating material, their processing technique, a strict adherence to parameters and quality system can ensure the reliability of the cable to be supplied to a user. This is the core of cable designing and production system. It remains up to the manufacturer to ensure a reliable performance of their product on which depends the quality of today's life.

6

Assembling and Laying Up of Multicore Cables and Protective Metal Sheathing

In a three-phase AC system, all the phases are loaded to distribute power in a balanced manner; otherwise, the system remains unbalanced. In such a case, a neutral is incorporated to divert the excess load of any phase and balance the system. At times, a neutral is used for supplying loads for lighting purposes as well. Generally, three insulated conductors have to run in parallel with a neutral (earthed condition). There are two types of systems used predominantly in any power generating system: a star type with three phases and a neutral. The delta connected system is generally without a neutral, where the phase voltage is equal to the line voltage. In a star connected system, the line voltage is equal to the root of three times of the phase voltage. This is true for low voltage (LV) distribution lines. For high voltage (HV) and extra high voltage (EHV) cables, the three cores have to be laid in parallel. Normally, all the phases are loaded in a balanced condition.

During installation, if the phases are single core, then three single cables for medium voltage (MV) and three single cores with a single neutral for LV system are to be laid adjacent to each other. As the current flows through the conductors, an induced voltage develops due to a unidirectional magnetic field. This voltage can disturb the adjacent cable(s) by superimposing an induced voltage and current on the existing phase voltage and current. This can add or oppose the already transmitted power, creating more unbalance initiating fault conditions and unwarranted accidents. This voltage needs to be restricted by covering the individual phase (cable) with a nonmagnetic metallic lead/aluminium sheath or an armouring of nonmagnetic materials like aluminium/copper wire/strip or stainless steel having a low impedance path. This sheath or armour is grounded to divert the induced current restricting the rise in voltage on adjacent phase(s). A considerable power loss, thus, could be experienced on the individual cable length through sheath and armour (termed as sheath or armour loss). As a matter of fact, for low-voltage cables, the production of individual lengths on the basis of the earlier constructional features, and laying and installation of each phase singularly, becomes uneconomical and time consuming. This is because the quantum of route lengths for LV distribution cables is enormous.

This is also true for medium-voltage cables of up to 33 kV, where route lengths become quite considerable at times.

It has been experienced that by transposing phase conductors, a large amount of the magnetic field, vis-à-vis the induced voltage, could be neutralised. In this case, only a very limited amount of loss would be generated by a stray magnetic field, which would be taken care of by a normal steel wire armouring. This led to the concept of assembling low-voltage, individually insulated three-phase cores, with a neutral, by twisting them spirally with a regular lay ratio, called the laying-up process. The same procedure is also followed for medium-voltage cables up to sizes of 33 kV, where three individual screened cores are assembled together.

The laying-up process has the following advantages:

1. The cable becomes compact, saving considerable raw material consumption during subsequent processes.
2. Twisting during assembling ensures automatic transposition of cores, which would neutralise a large part of the induced magnetic field acting on an individual core.
3. The axial spacing between conductors has the least possible distance, and hence the influence of magnetic flux reduces considerably. This is more so for sector-shaped conductors.
4. The assembled cable remains flexible during repeated winding and unwinding, compensating for the unequal movement of core lengths during processing in the factory and also during laying and installation.
5. When all the cores are laid together with a twist giving a natural state of transposition, induced magnetic forces get neutralised to a large extent. In such a case when a common armour with steel wire or strips (magnetic material) are brought on the assembled cable core, armour loss gets reduced to a minimum level. Further, steel wire or strip armour ensures better mechanical as well as earth protection. This also reduces the cost of the armour material and the multiple production process.
6. Common jointing and termination kits can be used for all the phases.
7. The number of drums required will be less, eliminating considerable transport and handling costs.
8. Overall, the laying and installation cost is reduced considerably.
9. Generally, it is observed that the longer the lay length, the more the circularity and length of the cable and the production output.

Cables above a 33 kV range are made of an individual core. All the three phases are laid and installed individually, in parallel or in a trefoil formation, while transposition is done by the cross-bonding method, which will be discussed in Chapter 8.

Though the primary principle for the twisting of cable cores is somewhat similar to the stranding process, the production procedure remains entirely different. In this case, apart from power cores as stated earlier, various sizes of control cables, instrumentation cables, highly flexible cables, etc. are also twisted together to form multicore cables. Generally, the operational method needs to be based on a clear and sensitive perception. The machines used differ in construction, considering the type of cable that has to be assembled.

In one case, multicore insulated conductors such as control cables, rigid or flexible, are assembled together. In another case, larger sizes of insulated conductors are laid up to assemble power cables, either rigid or flexible. Apart from these, a special type of machine is used for making screened instrumentation cables which simultaneously undergo combined twisting, lapping and assembly.

In principle, it is desirable to assemble round or sector-shaped insulated cores, having solid, stranded or flexible conductors in such a manner that during laying, the operation spools or drums holding the insulated cores do not experience a 360° twist with each revolution of the machine carriage. If the bobbins are kept fixed in position while the bobbin-holding carriage rotates, then the bobbin along with the insulated conductor will also rotate around their axes by 360°. With each turn, the conductors get twisted in the space, as it moves forwards. This develops a stress on the conductor and on the insulation. As the length progresses, the twisting stress moves forwards, ultimately leading to a failure of insulation and/or the conductor. This effect is aggravated when the layer of insulation is thick, and the three-layer components exist along with a metallic screen, as in the case of medium-voltage cables. Generally, during the laying operation, the conductor axis should not get twisted. It should remain the same on the horizontal plane. This can be achieved by rotating the bobbins opposite to the rotational movement of the carriage. This opposite movement is actuated by a set of gears connected to the shaft of the bobbin-holding yoke, which in turn is engaged with a gear attached to the centre shaft of the carriage. The gears are thus arranged with a calculated ratio so that the opposing movements remain at 360° with each full turn of the carriage, keeping the bobbins virtually in a horizontal position. This device to arrange opposing rotational movements is called an anti-twist device, and the movement is called 'anti-twist movement'. One full turn of the carriage with an opposite full turn of the bobbin is called a 100% anti-twist action. Though the gears are said to be matched properly, a percentage of error does occur during manufacturing and installation. This error slowly gets multiplied with many rotations and brings a slow twist on the core length as it moves forwards. If the twisting length is very long, it does not create much of an adverse effect. If the error is more, it needs to be corrected promptly by an electrically operated sensor correction device or manually by rotating the bobbins in the opposite direction from time to time. The machine can be stopped manually. The lay length of the laid-up cores is kept in the same ratio as that of the lay provided while making the strand.

For straight sector-shaped conductors, the pay-off bobbins should remain fixed (no pre-spiral twist is involved). The apex of the shaped core must always remain directed towards the centre. This can be accomplished with a 360° turn of the bobbin-holding carriage for (1 + 3) laying-up machines (Figure 6.1) or in a drum twister keeping the pay-off bobbins locked in position. In this case, though the core gets a twist of 360° because the lay length is kept considerably long (apx. 35–40 times the laid-up diameter), chances of getting the cores damaged are very rare. Further, these cores have a conductor cross-sectional area of not more than 300 mm². Larger area conductors should be made with pre-spiral twists, from 300 mm² and above. It is observed that above 300 mm², there is a tendency of a back twist of the cores, and therefore straight lay conductors are restricted up to 300 mm², and in some laying-up machines, not above 185 mm².

In order to minimise machine cost operation time, some manufacturers have tried to run 400 mm² with straight lay, and because they did not adhere to the basic technical norms, ended up with misshapen cables. This also led to a misshapen laid-up core, damaged insulation and an increased diameter. With the pre-spiral conductor, the lay length has to match rotational movement of the carriage, along with the drawing speed of the capstan/caterpillar. In this case, the bobbin-holding yoke should remain on the same plane. Here, an anti-twist device is needed to be actuated in order to keep the apex of the conductor towards the centre of the cable during the laying-up process. In practice, some variation of lay length between the conductors and a rotational movement of the machine remained. To compensate for this anomaly, the pay-off bobbins have to be rotated individually, either left or right as required, to correct the apex

FIGURE 6.1
1 + 3-core laying-up machine.

position without disturbing the anti-twist movement. This is normally done by operating an electric motor through a sensing device (sector correction device). The sensing device senses the apex position of the core. When it senses any of the apex of the conductor being not in line with that of other conductors, it transmits a signal to operate the sector correction motor of the corresponding bobbin and brings it to a correct position by rotating the bobbin-holding yoke. At times, this type of correction is done manually as well (because of sensor failure). When a cable is laid up properly, the contour of the cable core becomes perfectly round. This ensures a control on the laid-up cable core diameter, which in turn guarantees the consumption of raw material in subsequent applications as per design parameters.

These cores coming out of the bobbins are guided through a front guide plate. Here, the angle of inclination of the pay-off cradles is very important. Figure 6.2 clearly shows that when no guide plates are used, the angle of each core converging on the closing die remains unequal.

This leads to an uneven length entering through the closing die during the assembling operation. Generally, the tension on each core differs. This results in deformation and twisting of length at the entry point. The shape of a laid-up cable will be non-uniform with a wavy contour. In a laying-up machine, the angle of inclination for the pay-off unit should be around 20° to avoid a sharp bending at the entry point of the guide plate. This is critical for MV cable cores with thicker insulation and metallic screen. The inclination should be towards the front side of the machine. It can be easily understood that if the cores are drawn straight, then a sharp bending will result while leading to the guide plate. This will damage the cores. On the other hand, if the inclination is too small, then the length of the machine would have to be made longer. Through the front guide plate, the cores are guided into the closing die with an equal inclination. The cores should also be led through the guide plate with an equal spacing. For instance, three cores should have a spacing of 120° and four cores 90°, and 3½ core cables should have three cores of 100° and a half core of 60° and so on. (Divide 360°

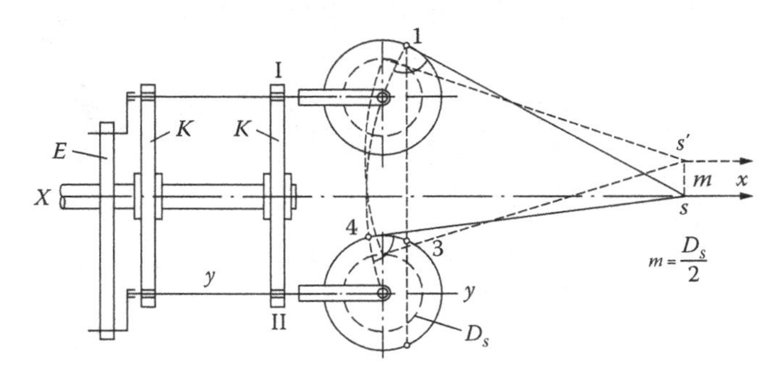

FIGURE 6.2
Stranding machine with back twist device.

with the number of cores. This is also true for multicore cables. The number of entry/exit holes required should be drilled on the guide plates.) If angles and/or spacing on guide plates are not equal, then the laid-up cable core will come out with a snaking effect. This will create variations in diameter and thickness of the extruded sheath (inner or outer), misshaping the cable and increasing the material consumption. The construction and design of the machine is done after giving a full thought to the working principle and precautions that are to be taken to maintain production stability. Though all the laying-up machines look robust and big, the operational features are very sensitive. An efficient technician can understand the delicate operational procedure of assembling the cores during laying up and acts accordingly. The assembled cores coming out of the die are held together as required by applying a binder tape or thread.

Apart from the classical rotating carriage-type laying-up machine, where pay-off bobbins are held between two discs mounted on yokes and rotated along with the carriage (1 + 3-core laying-up machine), a machine has been developed and employed to lay a cable core, where the individual take-up bobbin is rotated independently at a moderately higher speed. Here, the pay-off bobbins are placed on yokes mounted on fixed stands. During operation, the pay-off bobbins are either kept fixed or rotated as required by rotating bobbin-holding yokes. A rotating caterpillar is employed to pull the cable. The rotation of the caterpillar is synchronised with the rotational movement of the take-up bobbin. These machines are called drum twisters (Figure 6.3). The basic advantage here is that the centrifugal and centripetal forces acting on the pay-off and take-up bobbins of a drum twister are lesser than conventional large-diameter rotating layer of 1 + 3 laying-up

FIGURE 6.3
Drum twister for core assembly.

units, which allows the total assembly to rotate at a higher speed, giving an enhanced production rate. Furthermore, the height of the machine gets reduced. The rotational movement of a pay-off system can be monitored independently or synchronised as per the machines' construction. Maintenance of the machine is also easier. If required, the number of pay-off stands can be increased as per requirement to produce multicore cables as well. In the case of a drum twister for straight sectors, the pay-off bobbin remains fixed, whereas the caterpillar rotates in synchronisation along with the take-up unit. For pre-spiral conductors and round-shaped conductors, the pay-off bobbins rotate in synchronisation with the caterpillar and take-up unit. In this case, the conductor axes always remain on the same plane.

This has led to the fact that by modifying certain features, such machines can been used for armouring and for the production of stranded conductors. These types of machines are also employed to lay multi-unit communication cables. With the introduction of these machines, called drum twisters, a longer length at a higher production speed is achieved by deploying less manpower.

A special feature of these machines which hitherto had not been conceived and utilised is now being exploited. This is to lay up low- and medium-voltage aerial bunch cables. A special working principle of laying up these cables is the fact that the messenger conductor (either stranded all-aluminium alloy insulated or bare) must remain straight at the centre (under maximum tension), around which all other phases, neutral and lighting conductors having a definite lay ratio, are twisted in a spiral form. The messenger conductor has to support the total weight of the cable while it is laid by a stringing process from pole to pole. This type of lay design is conceived to achieve the following benefits:

1. The messenger conductor can be hooked to support the cable at any given point, without damaging the assembly.
2. Phase conductors can be tapped at any point for service connections.
3. During temperature fluctuations, the cores can slide over the messenger without putting undue stress on the phase lengths, and at the pole end, while expansion and contraction take place along the axial lengths of the cores.
4. When remaining exposed in the air, the current-carrying capacity becomes higher than the underground versions and heat dissipation becomes quicker.
5. Installation is easy, but for medium-voltage cables having a layer of extruded screen layer over the conductor, splicing and installation work should be done carefully. The screen and the messenger should be grounded solidly at a regular interval; otherwise, induced voltage and current can create fault conditions.

While assembling AB cables, the twisting of insulated conductors around the messenger has to be done carefully. These cables cannot be assembled in a conventional manner like three-core, four-core or multicore cables.

Coming back to the conventional laying-up process, it is imperative to understand that while assembling, the cores have to be guided through a front guide plate. This plate, as described earlier, allows all the conductors to converge on the closing die at an equal angle. This facilitates proper tensioning and equal length paying off to give a smooth and round contour to the cable core as laid up. The laid-up cable core as it comes out of the die opens up slightly and releases its internal tension, which is called the feathering effect. As a result, the actual cable diameter increases slightly, which is not desirable. Every 0.1 mm increase in diameter will consume more material in the proportion of πR^2 at every subsequent stage. Hence, a mylar or polypropylene tape of 0.015 mm thick could be applied as a binder to hold the cable diameter, after which an extruded inner sheathing is applied. In the case where a thermoplastic tape (polyvinyl chloride [PVC] tape) is applied as bedding, mylar tape binding has to be avoided. In order to keep the consumption of tapes within the given limit, it is necessary to consider the mechanism of tape application. A clear relation exists between the width of tape, diameter of cable, angle of application and thickness of the tape. Normally, at this stage, while calculations are carried out to determine the mean diameter of the cable, the thickness of the tape, being very less, can be ignored. A minimum overlap of around 5%–7% is more than enough to meet the purpose of holding the cores. In earlier days, when paper cables were in vogue, the technology for the application of tape on conductors, shaped or round, on the laid-up cable core was studied extensively. In those cases, every bit of overlap and gap was important. Even a fraction of a millimetre made a huge difference in consumption and quality variation. At present, when large quantum cables are processed, every fraction of a millimetre can show a large variation in tape consumption. It is therefore necessary to standardise tape width, thickness and angle of application in relation to the cable diameter (refer to Figure 5.4). In any case, the diameter should not be allowed to exceed the tolerance limit, as per the production design sheet. It needs extensive work to establish the parameters, depending on operational skill, machine parameters, material quality and environmental conditions such as temperature, humidity and elongation.

Dimensional stability has to be established from the very beginning, during the time of conductor formation, either sector or round. In the case of a round conductor, the compactness and diameter have to be controlled precisely. For the sector-shaped ones, the apex, sector angle and width of the sector have to be checked as per factory design. It should be seen that an allowance is kept while the sector angle is fixed – say 119.5° in place of 120° – allowing a recess of 1.5° for three-core cables. It is expected that a slight feathering effect will increase the dimension of the sector height by 0.1–0.2 mm. Further more, during insulation, a small unavoidable variation will creep up, which needs to be kept strictly under control. This is where the skill of the operation lies. Even the best electronic controlling units cannot stop a creep. However, efforts should be made to minimise such problems as much as possible. But all gadgets must perform satisfactorily according to the set parameters, minimise human error and limit the variations within the given data. Notwithstanding, human supervision cannot be avoided. One has to remain watchful and ensure that no anomaly arises during production. These small unavoidable deviations are adjusted during laying up. The recess of 1.5° takes care of the variations, allowing the cores to sit easily in their places without causing undue stress and distortion. The diameter of the cable will be compact and round. Apart from the shape of a conductor, the form of guide plate, the angle of cores coming out of the guide plate, the distance of closing die from the guide plate, and the finishing and landing length of closing dies are all of great importance. The guide plate in front of the laying-up machine is made slightly conical, with an angle as per the distance of the pay-off bobbins. If the distance is short, the guide plate angle should be bent towards the front side. A short distance between the guide plate and the die entry point will form sharp bending points on the cores. This may damage insulation and break wires. At this point, the die head has to be shifted away from the guide plate to allow a comfortable entry angle of 15°. But this may not be possible for larger-size conductors. Hence, the yokes of the bobbins have to be placed far backwards and at an angle that will allow the conductor to move without a sharp bending. If the angle is wider, the die should be made with a wider bell mouth. A small amount of angle adjustment can be made by shifting the die block longitudinally forwards or backwards. While closing through the die, insulated cores experience a rubbing effect and generate frictional heat. This can injure the insulation. Too tight a die will generate more friction, that is, more heat. This will soften the plastic material. Insulation will start bulging and stick to the die mouth, damaging the cable length. Too loose a die will give a free passage resulting in overriding and increase in the laid-up diameter. A rough surface inside the die will result in high frictional force, which will cause heating up of the insulation. It is therefore essential that the internal surface of the die be very smooth (mirror polished) and accurately dimensioned. Die faces must match perfectly without having any edge or gap. The bell mouth should be smooth and regular in contour without any sharp angular ridges. These machines are well designed and fabricated with proper calculation and understanding. The machine is fabricated with a scientific approach, and no shortcuts are allowed. Operators are fully trained and therefore have a full knowledge and understanding of the laying-up procedure. It should be noted that a nylon or Teflon die should not be used while laying-up polythene, XLPE or PVC, or any other plastic insulated core. This will generate more heat, and the materials would tightly grip each other and create problems. For all production purposes, therefore, only polished and hardened cast iron die should be used.

Starting from 1880 to 1970, insulated and shaped 11 kV cable and elliptical conductors and 33 kV cables conductors have to be manufactured as per the design parameters. Here, the conductor can be solid or stranded as per requirement. Currently, 11–33 kV shaped and elliptical cores have to be extruded and cured in a CCV line. In these cases, pre-spiralling of the conductor is not considered, because of the limitations in the extrusion system. It is experienced that XLPE can absorb some amount of twisting force during laying up where the lay length is large enough. Nowadays, the force of torsion can be minimised by selecting a longer lay length with minimum lay angle during the laying-up procedure.

1. Pre-spiralling was essential in earlier days when insulating papers had to be wrapped around the shaped conductor. During laying up of insulated cable cores, the paper had a crease-like surface leading to stress which could damage the layers of paper tapes. Hence, pre-spiralling was necessary.
2. With the introduction of polymeric insulation, it was found that the twist in the conductor during insulation was absorbed by the force of resilience and flexing property inherent within the compound. The material could adjust itself in time and absorb the twisting force, as it could stretch and retain its form by elongating a limited amount as required.
3. When the angle of twist 'α' is made larger (angle subtended by the normal to the horizontal plane of laying), then torsion can be kept to a minimum. Torsion per one complete turn is given by $T = \sin 2\alpha/D$.

Example 6.1

When $\alpha = 30°$, then torsion per unit length is (when the laid-up diameter is 79 mm approx.) $T = 0.011$ rads per revolution of carriage. If lay length is $40 \times 79 = 3160$ mm, then for a length of 1000 m laid-up cable, the torsion will be 3.5 rad approx. Cables with such dimensions have to be made in lengths of 350–500 m. Hence, the torsion within will be much lesser and will not have any effect on insulation and cable cores.

Thus, keeping a longer lay and a smaller angle, the torsion can be controlled considerably. In such cases, to contain the feathering effect, one layer of a strong binder tape has to be applied. The result is a near circular cable.

In practice, laying operation is done to keep three cores together in a transposed condition to minimise the induced sheath voltage and reduce power loss during transmission.

6.1 Taping

From 1880 to 1970, paper taping was done extensively on conductors. Considerable studies were carried out to obtain a crease-free taping, either with gaps or overlapping. To economise consumption and reduce wastage,

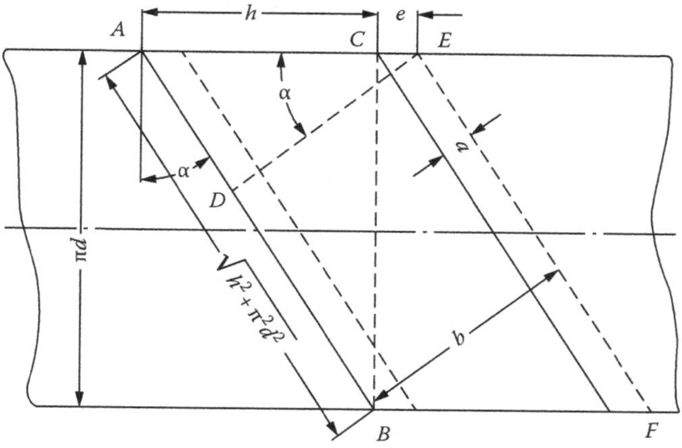

FIGURE 6.4
Insulating paper lapping angle with lay length.

exact mathematical relations were established to standardise paper width and sizes. The same can be accepted for the taping of thermoplastic and other types of tapes. The tape has to be applied in a helical manner as shown in the figure. If the paper is unwinded about the diameter of a cable 'd' and the lapping length is 'h', then the width of the tape is given by (Figure 6.4)

$$\frac{b}{h} = \frac{\pi d}{\sqrt{h^2 + \pi^2 d^2}} \tag{6.1}$$

In cases where the tape is applied with an overlap or with a gap having the measure as 'e', then the equation becomes

$$\frac{b}{h \pm e} = \frac{\pi d}{\sqrt{h^2 + \pi^2 d^2}}$$

or

$$b = \frac{\pi d(h \pm e)}{\sqrt{h^2 + \pi^2 d^2}} \tag{6.2}$$

If 'e' is expressed in terms of lay length 'h' and 'k' times 'h' then $e = kh$, and the expression becomes

$$b = \frac{\pi d h(1 \pm k)}{\sqrt{h^2 + \pi^2 d^2}}$$

Again, if thickness of the tape is considered in the equation and is denoted by 'Δ', then the mean diameter is $(d + \Delta)$, and the equation finally takes the following form:

$$b = \frac{\pi h(d+\Delta)(1 \pm k)}{\sqrt{[h^2 + \pi^2(d+\Delta)^2]}} \tag{6.3}$$

Example 6.2

Let
The cable diameter $d = 25$ mm
Thickness of tape $\Delta = 0.15$ mm
Overlap of tape is 15% $k = 0.15$
Lay length of the tape $h = 30$ mm (as desired)
Then

$$b = \frac{\pi \times 30 \times (25+0.15) \times (1+0.15)}{\sqrt{[30^2 + \pi^2(25+0.15)^2]}} = 32.25 \text{ mm}$$

For brevity of production, the width can be selected as either 32.00 mm or 32.50 mm.

In case the thickness of the tape is negligible in comparison to the diameter of the cable, then the value of Δ can be neglected during calculation.

The angle of lapping of tape has to be determined as follows:

$$b = (h \pm e) \cos\alpha = h\,(1 \pm k) \cos\alpha$$

Further, considering the triangle $\tan\alpha = h/\pi d$ or $h = \pi d \tan\alpha$

Thus by combining the previous equation, we get

$$b = \pi d \tan\alpha (1 \pm k) \cos\alpha = \pi d (1 \pm k) \sin\alpha$$

Thus,

$$\sin\alpha = \frac{b}{[\pi d(1 \pm k)]}$$

In this case, $\alpha = 23°$ approx.

To make 'h' larger, the angle 'α' will also be greater.

During lapping, the tension of the tape must be maintained correctly. The strength of the taping must be ascertained so that while wrapping the tape does not rupture, stretch or elongate abnormally and develop folds and creases. Normally, for thermoplastic tapes, the breaking load along the length should not be less than 15 N/mm^2 and elongation not more than 200% to obtain a trouble-free lapping. The tapes are applied with 10%–15% overlap.

Nowadays, the taping head has been designed with a high revolution of about 400–500 rpm, with a constant tension regulating device and smooth pay-off adjustable angle, so as to get the perfect tangential touch on the cable surface parallel to the cable axis. Therefore, the better the rounding of the cable during laying up, the better will be the lapping, and any formation of crease and folds, if not totally eliminated, would be reduced almost to zero. The taping will be tight and uniform. This will reduce the consumption of material during subsequent productions and give a high-quality performance.

6.2 Protective Sheath for Laid-Up Cable Core (Mechanical and Electrical Protection)

At the initial stage, metal sheaths were used to protect the insulated cable core, power and communication from mechanical damage and ingress of moisture. Apart from mechanical protection, the metal sheath provides a path for circulating current that may come into play due to magnetic fields which emanate from adjacent power cables or conductors with a high current-carrying capacity. Otherwise, conductors of communication cables would be influenced by a magnetic field emanating from power lines, inducing a voltage within the circuit, creating a very high noise level and rendering the transmission system defunct. In the case of power cables, the metal sheath provides a path for diverting short-circuit current to the earth and protecting the cable against damage. Normally, metals having a low impedance path, such as lead and aluminium, are used as sheathing material. Developments have also been made to use steel as a protecting material.

6.3 Metal Covering (Lead)

At the initial stage when no other materials are available, a metal covering is found to suit the purpose. The metal covering has to be such that it can be bent easily and has a considerable mechanical stability. It also should not get easily corroded. The metal should also be formed easily to extrude as a pipe. This was a difficult proposition as all the previous characteristics could not be found easily in many known metals except lead. It was then accepted that lead could be the metal that could fulfil most of the requirements and could be used for the purpose. The metal is soft and can be formed to a desired shape. It also does not get corroded easily, although its mechanical strength is very poor. The specific weight of the metal is very high but can be wound and unwound several times without deforming. During installation and jointing, it did not pose any problems. Electrical conductivity allowed the metal to be used at earth potential. A circulating current on any exposed length of a lead sheath touching the ground, or any grounded conductive material, can generate a localised eddy current, while acting as an anode, the metal causes heavy corrosion at the exposed part. This causes failure during service. The metal is also attacked by microorganisms, causing gradual deterioration, and hence needs to be protected. A vibration on the installed cable can cause fatigue due to recrystallisation in the metal.

At the initial stage, lead covering was done by two tapes cut in size. One tape was placed at the bottom of the cable to be covered and the other one at the top longitudinally. Both tapes were pressed around the cable guiding through a set of grooved rollers. The seams of the tape at edges got welded under pressure, forming a tube. The rough edges were smoothed by passing through a die. This process was very slow and costly. At the next stage, an oversized lead pipe was drawn over the cable core, reducing the diameter and successively drawing through a series of dies. This process generally has a joint at every charge change from the lead press and is described in detail later. This was also a very tedious job. Lengths obtained by this process were limited. Later, an extrusion technique was developed to form a seamless pipe over the cable core continually to get longer lengths and larger diameters. Pipe was formed by extruding under a given temperature (300°C) and pressure. Lead used for the process should have the characteristics listed in Table 6.1.

Considering lower melting temperature and soft plastic nature of the metal, a press was developed to extrude lead at 400 kgf pressure in plasticised form. In earlier days, the press was of a vertical type and hydraulically operated and consisted of the following:

1. Lead melting pot with a charging trolley. The pot had a nozzle for pouring molten metal into a holding pot. An additional heating arrangement was incorporated, when required.
2. A holding pot to accept molten metal from 300 to 500 kg covered with a top lid.
3. A hydraulically operated ram which presses the plasticised metal. The diameter of the ram closely fits the diameter of the holding pot.

TABLE 6.1

Characteristic of Primary Lead Used for Cable Sheathing

Purity	99.9%
Melting temperature (°C)	327.4°C
Specific weight at 20°C (g/cm³)	11.34
Tensile strength (kg/mm²)	1.3–1.8
Elongation at break (%)	30–45
Specific resistance at 20°C (Ω mm²/m)	0.22
Temperature coefficient 1/°C	0.00421
Specific heat (0°C–100°C) (cal/g °C)	0.031
Melting heat (cal/g)	5.5

4. Die holding box to hold mandrel and die with an adjustable screw.
5. Pressure pump for circulating water for cooling the die box and the extruded sheath.
6. Hydraulic pump and oil container.
7. Control system and operating console.

During operation, the molten metal is poured into the ladle after skimming off the oxide layer from the surface of the liquid metal (Figure 6.5). The metal is then poured into the vertical hollow barrel from a ladle. During pouring, precaution should be taken to see that unwanted particles of dross or any other foreign material do not get mixed. After pouring, a time gap is allowed to bring the air and oxygen bubbles to the surface. The top layer is skimmed off again. The ram is then lowered slowly pressing the metal downwards. At the bottom of the press, the cable is led through the centre of the mandrel from the rear. In the beginning, the extruded metal tube coming out of the annular space of the die and the mandrel is checked for concentricity and thickness. The surface of the extruded pipe is examined and adjusted to gain the correct dimensional parameters. The cable is then led through an extruded pipe and moved forwards to a winding system. Simultaneously, the temperature of the hot lead pipe

FIGURE 6.5
Vertical hydraulic lead extrusion press sheathing on cables.

is brought down by spraying water onto its surface as soon as it comes out of the die face. Cooling should not be very intensive as it may cause a chilling effect, affecting the grain structure of the metal which ultimately may develop into cracks. Hence, post cooling is allowed to progress slowly under atmospheric conditions. As the metal is soft, a cable with a lead sheathing can be wound and re-wound on the drums easily for several times. While winding, layers of hot metal surface touch each other and are likely to stick together. To prevent this, a coat of bitumen is applied on the hot surface of the lead sheath followed by a spraying of water. To prevent microbiological attack and from surrounding affluent, the metal sheath is protected by applying a layer of bitumen-impregnated fibrous material. Nowadays, an extruded thermoplastic coating has become very popular. As the metal is soft, further mechanical protection is given by applying steel tape, steel wire or a strip armouring over the protective coating. Pure lead cannot be used for sheathing as the metal recrystallises under constant vibration and, being very soft, develops fatigue. It is used after mixing with 0.4% tin and 0.2% antimony (lead alloy E). Where the vibration is severe, even alloy 'E' is not strong enough to sustain the stress-when laid alongside a railway track or bridge where high-speed traffic passes frequently. Cables transported by ship are also subjected to severe vibrations. The vibration reorients the crystalline structure of the metal, causing fatigue and early fracture. To prevent this, pure metal is mixed with a small percentage of antimony (0.85% – lead alloy 'B'), or at times with tellurium/arsenic/bismuth (0.05%–0.15%), to increase the strength of the metal further to withstand severe stress and strain. These lead alloys, 'E' and 'B', are recommended by the BIS standard in the United Kingdom and subsequently used in Asian countries and the Middle East, whereas European countries like Germany and Central Europe prefer alloys containing copper (0.03%–0.05%) in combination with tellurium (0.04%). At times, alloys containing antimony are also used in proportions from 0.03% to 1%. The extrusion parameters of such alloys are difficult and hence need careful monitoring. The thickness of the sheath depends on the cable diameter and its installation condition. For submarine cables, the thickness is much higher and an alloy consisting of tellurium and copper in small percentages is used to withstand severe marine turbulence (current, water pressure, hurricane, etc.).

The content of the metal in the internal pot of the vertical press is limited by its dimension. Generally, after each fill, the extruded pipe remains limited by its length. The press has to be stopped to recharge fresh metal and to continue the extrusion process. During this time, the cable remains static at its position. The heat sustained at that point can damage the cable insulation. Hence, special precaution has to be taken every time the press is stopped for recharging. Further, when a new charge is being extruded, a distinct ring mark at the meeting point of the freshly extruded metal pipe is observed. This mark is popularly known as a bamboo ring. This is a very critical point of the extruded sheath. If the joint is weak and does not amalgamate perfectly, it can get ruptured. An expansion test is done after ascertaining the extrusion conditions and the time to be taken for restarting. At the holding point, high temperature affects the crystalline structure of metal. A recess of 3–4 min is allowed between the stoppage and the restarting time. Efforts are made to reduce the stoppage time and also to extrude metal continuously without stoppage. To that effect, double-stroke ram-type vertical and horizontal presses were developed. A better solution offered was the development of a continuous screw-type lead extrusion press after the Hansson–Robertson system manufactured by H. Folke Sandeklin AB, Sweden. In this process, the molten lead is kept in the holding furnace from where liquid metal is transferred through a pipe into the barrel. The rotating screw inside the barrel of the continuous press the plasticised metal forward, under pressure through annular the space of mandrel and die in the form of a pipe covering the cable and coming from the rear. In this case, a continuous pipe can be extruded in kilometres faultlessly without stoppage. In one case, 'M/S Pirelli General' of the United Kingdom could continue sheathing 50 km of submarine cable without stoppage. Initially, the press was developed for a pure lead extrusion. Subsequently, after modification, the lead extruder could take the load of the lead alloys as well. During extrusion, heating, and cooling, the temperature must be kept at constant parameters. Dross or any foreign particles liable to contaminate metal should not be allowed to move through. Mixing of the alloy must be uniform and alloying metals are accurately proportioned. It is better to prepare the master batch with virgin metal kept ready before extrusion commences. Any deviation and/or contamination hastens the development of an undue stress within the metal, causing an early failure (Figure 6.6).

The thickness of lead and aluminium sheaths for communication cables and power cables of up to 33 kV depends mainly on their mechanical strength and manufacturing process. It should also accept the force of deformation during bending. The nominal thickness of the sheath 't' depends on the diameter under the sheath 'd_k'.

In the majority of cases, this relation is given by

$$t = m + fd_k \tag{6.4}$$

As for practical value, it can be accepted that 'm = 0.6–1.7' and f = 0.02–0.04.

In the case of paper- and oil-filled cables, the sheath experiences an internal pressure 'p' kgf/cm^2, and then the stress σ_1 on any cross section of the sheath may approximately be determined by the following equation:

$$\sigma_1 = \frac{pd_k}{2t} \text{ kgf/cm}^2 \tag{6.5}$$

FIGURE 6.6
Hansson and Robertson–type continuous lead extrusion press.

The force acting on the sheath in the longitudinal direction is

$$N \approx \frac{p\pi d_k^2}{4}$$

and the stress in this direction is

$$\sigma_2 = \frac{N}{F} = \frac{p\pi d_k^2}{\left[4\pi(d_k + t)t\right]} \approx \frac{pd_k}{4t} \tag{6.6}$$

This shows that pressure can be contained by increasing the value of 't'. To retain an economical value of 't', reinforcement has to be brought in the form of an armour to balance the force of pressure. If this is not done, the sheath will rupture along the cable axis. This is found to be the case in actual practice.

Besides internal pressure, the metal sheaths are also subject to stresses due to bending of the cable. During bending, the outer part will stretch while the inner side will get compressed. Stress due to bending is given by

$$\sigma_b = \pm\frac{Ed_k}{D_b} \tag{6.7}$$

where

- d_k is the diameter of the cable under the metal sheath
- D_b is the diameter over the metal sheath
- E is the modulus of elasticity

For lead it is 0.17×10^6 kgf/cm², and for aluminium it is 0.67×10^6 kgf/cm². Aluminium has a higher strength than lead to sustain internal pressure; hence, the thickness could be reduced, but pressure required for bending would be four times higher than lead and hence too much reduction in thickness could cause a sheath rupture.

A compromise thus has to be established in order to arrive at a practical value.

For EHV XLPE cables with a lead sheath as a protection layer, care has to be taken to keep a gap between the sheath and the core and allow for a thermal expansion of the core during load. Position of the die in the die block and its size has to be selected in such a way that a gap of 0.1–0.2 mm is created.

6.4 Aluminium

During Second World War, Germany suffered from an acute shortage of lead. At that time, lead was extensively used to make cartridges for weapons. The basic metal was mostly under the control of the Allied forces. Germany could not use lead for sheathing of cables. Under compulsion, they started developing aluminium as an alternative metal for sheathing of electrical cables. Before this, aluminium had been tried as a sheathing material: The cable core was inserted in an oversized aluminium pipe and then drawn down by a subsequent operation. It was a very tedious job. The metal corroded easily when left exposed to the atmosphere. This exercise was then discontinued.

Under compulsion to find a better way, Hydraulik Duisburg developed a vertical aluminium press (Figure 6.7) resembling a lead press. The initial press had a capacity to develop pressure of up to 3800 M ton. Pressing of aluminium faced considerable difficulties: the temperature of the metal during pressing had to be raised to 420°C–470°C against the 300°C of lead. To start with, the purity of the metal had to be 99.98%–98.99%. Later with the improvement of press design, a purity up to 99.5% was established.

The working surface of the ram has to accept higher temperatures, and a force of 86–92 kgf/mm² against that of 40–45 kgf/mm² for a lead press is to be applied. To build up such a high pressure, the ratio of the hydraulic piston to the ram piston was increased.

For lead press, the ratio was $\eta = F/f = 13.8$–14.7, whereas the first design of an aluminium press had a ratio of $\eta = F/f = 22.7$–23.0.

The fusion temperature of aluminium being >600°C, it was difficult to pour molten metal into a holding cylinder. Further, the molten metal formed an oxide layer very quickly on the surface during pouring, and at the top layer within the cylinder, making it very difficult to use the metal

FIGURE 6.7
Four thousand ton hydraulic vertical aluminium press.

in a molten state. The metal was then extruded by bringing it to a plasticised state under high pressure and allowing it to be heated somewhat lower than the melting temperature. The press was therefore modified to accept a machined round billet of metal at a preheated stage. The billet was preheated in an induction oven at 500°C–550°C and then fed into the press by an automatic holding and feeding device. The initial press had a single-stroke system. The press had to be stopped for recharging. The time of recharging was not more than 30 s. The temperature of the metal being high, the cable could not be held at stopping point for long. This needed a special cooling process. Like the lead pipe, a bamboo ring would develop at the point, and recrystallisation of metal had to be controlled precisely to avoid cracking. Further developments were made to extrude the metal sheath continuously. This was accomplished by a double-stroke ram press. In this case, a secondary ram was kept ready for charging a smaller-size billet, just before the main ram pressure is stopped. As soon as the main ram pressure ends, the secondary ram starts pressing the smaller billet, continuing the extrusion process.

Simultaneously, a horizontal double-stroke press was developed to extrude the aluminium sheath.

The press is named Schlömann press (Figure 6.8), after its designer, Schlömann. A vertical press needs a higher ceiling, but a horizontal press

FIGURE 6.8
A 1600 ton horizontal-type Schlömann aluminium press.

can be accommodated at a lower ceiling height. This continuous press with an automatic high-frequency induction heating and feeding system has found enormous popularity in the market. The press is of 2×1600 ton capacity.

Aluminium being stiff becomes flattened at the top during bending and produces a crease at the bottom. To avoid such a phenomenon, the winding of the cable has to be done on a very-large-diameter barrel. Further handling becomes a problem. This problem is solved by corrugating the metal sheath immediately after extrusion. In this case, the bending diameter could be reduced considerably. After various trials, corrugated aluminium sheaths were made with unsymmetrical bellows, as shown in Figure 6.9.

With such a profile, it is possible to decrease the thickness by 50%, whereas a sinusoidal profile allows a decrease of only 25%–30%. An optimal parameter could approximately be obtained as

$$h = 20t \quad \text{and} \quad \tau = \frac{b}{h} \approx 0.38$$

The corrugation profile of an aluminium sheath depends on a minimum permissible bending radius and is given by the ratio

$$n = \frac{D}{d}$$

where

D is the diameter of bending of the cable
d is the diameter of the cable over the sheath

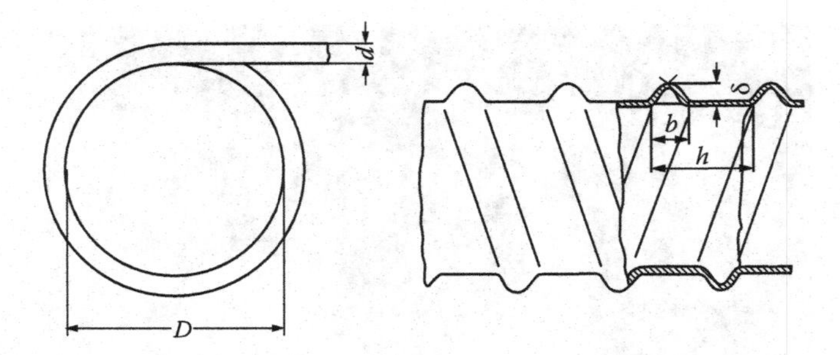

d, cable diameter; *D*, minimum bending diameter

$$k = \frac{D}{d}$$

h, corrugation pitch (depending on the cable core diameter); *b*, width of corrugation

$$\tau = \frac{b}{h}\tau$$

$$\text{Corrugation height } \delta = h\sqrt{\frac{1}{4(k+1)^2} + \frac{\tau}{2(k+1)}}$$

For example, when $k = 10$, then $\delta = 0.14h$ and $\tau = 0.38$.

FIGURE 6.9
Pitch of corrugated aluminium sheath.

Elongation for one turn of bending, thus, is given by $\Delta l = \pi(D + 2d) - \pi(D + d) = \pi d$, and by approximation taking the protrusion as a triangle and flattening it, $\Delta l_{el} = 2x - b$.

To counteract this elongation, the additional number of turns required will be

$$\frac{\Delta l}{\Delta l_{el}} = k = \frac{\pi d}{(2X - b)} \quad \text{again } k = \frac{\pi(D+d)}{h}$$

Hence,

$$\frac{(D+d)}{d} = \frac{(2X-b)}{h}$$

Now from Figure 6.9 $X = \sqrt{\delta^2 + \frac{b^2}{4}}$.

FIGURE 6.10
Corrugated aluminium sheathed cable.

Thus,

$$\delta = \frac{h\sqrt{1+2h(n+1)}}{2(n+1)} \tag{6.8}$$

The dependency of the height of corrugation on its lay length has to be considered as an approximation (Figure 6.10).

A more careful study is being carried out to evaluate the proper height of corrugation and get more stable results. A bending diameter of $n < 30$ gives away the sheath after several bending.

The sheath is corrugated gradually after it comes out of the press by means of a calibrated spring which runs with gradual turns.

Aluminium gets easily corroded. The metal sheath, either of lead or aluminium, is protected by applying a layer of bitumen compound and polyester tape, on which a final protective covering of extruded PVC or polythene (PE) is applied. Apart from its rigid structure and being impervious to moisture, the sheath protects the cable during short-circuit conditions. The sheath also protects communication cables from outside magnetic influence generated by adjacent power cables or from overhead traction lines in railways. The induced magnetic field from the power line generates a circulating current and voltage along the length of the cable sheath. The circulating current has to be earthed to protect the inner communication conductors from generating high noise levels. The sheath thickness needs to be computed considering all the earlier factors where sheath resistance plays an important role (see calculations in Section 8.3).

An entirely different approach was taken to bring aluminium sheathing on the cable core by a seam welding process (Figure 6.11).

In this case, the aluminium sheet is taken in the form of a large coiled pad. It is then straightened and trimmed at the edges. As the strip reaches the first pair of horizontal forming rolls, the cable core is introduced. The strip is now allowed to close over the core progressively. Edges are prepared for welding by scratch brushing at a high speed. After welding, the cable passes through horizontal positioning rolls, which correct any tendency of the sheath to rotate and displace the welded seam, in relation to the position of the arc. In order to avoid oxidation, a jet of argon gas is injected at the point of welding. To protect the cable core from the heat generated from the arc welding, an oversized metal shoe is extended beyond the welding point. After welding, the oversized

FIGURE 6.11
UNIWEMA machine: Showing welding of metal tape on cable core.

FIGURE 6.12
Corrugation unit of metal sheath: "UNIWEMA" machine.

sheath is rolled down to the proper size or closed down by corrugating the sheath (Figure 6.12). The method has shortened the length of the process.

Aluminium as a metal sheath has the following advantages:

1. With the development of the seam welding process, the need for a costly hydraulic powered high-pressure press is eliminated, reducing the building height and initial equipment cost. Processing and maintenance costs are also reduced.
2. Flexibility of the cable is gained by corrugating the sheath.

3. Thickness is reduced, enabling reduction in cable weight.
4. All types of cables can be protected by the sheath.
5. The mechanical strength of metal being stronger than lead, at times, armouring with steel tape/wire is not required.
6. As the resistance value is low, a higher short-circuit rating can be accepted by the sheath.
7. The sheath also acts as a non-magnetic screen and can protect the cable from outside magnetic influences.
8. The metal is available in plenty on the earth's crust as bauxite ore.
9. The jointing process has been developed to assure safety.
10. The sheath complies with RoHS norm.
11. The cable can be installed in a conventional manner.
12. The sheath can sustain higher-temperature conditions.
13. Snaking of cable during temperature fluctuation is taken care of by corrugation.
14. The current rating is not affected in any way.
15. Finally, the cost of the cable comes down drastically. In order to meet the conductivity parameters between the sheath of lead and aluminium, the thickness of the aluminium sheath is increased. In spite of this increase in the cost, aluminium sheath cables offer a very cost-effective solution and are to be beneficial to customers and to the environment.

6.5 Corrugated Steel Tape

With continuous effort and development, one can find a substitute material and economise product cost. The availability of aluminium in Germany is limited. Steel was a major product in the country. During subsequent years, efforts were made to use steel and replace lead and aluminium as protecting sheath material in the manufacture of cables. Extrusion of steel as a pipe was out of the question. It has been found that the metal in strip form can be bent over the cable core and welded continuously to form a continuous tube and corrugated to make it pliable for repeated bending and unbending like the aluminium strip. This also led to the use of universal and continuous corrugating sheathing machine called Universal Wellmantel Maschinen (Corrugating machine of metal sheath Universal Type) (UNIWEMA). In it, a polished clean steel tape is brought over the cable core by folding a set of bending rollers around it. The cable core is led through a metal pipe acting as an inner electrode. The folded metal sheet is then closed over the cable with an overlap and allowed to move over the inner electrode. A circular rotating metal electrode is pressed over the metal folding to act as a second electrode which continuously goes on welding the tape at its folded seam. As the cable with the seam weld tube is drawn forward by a gripping caterpillar, an eccentric rotating die forms a corrugation pattern over the sheath. To prevent the oxidation of the welding point, a jet of inert gas such as nitrogen or argon is injected at the point. The welded corrugated sheath is tested by injecting nitrogen under pressure to ensure the integrity of the sheath. The sheath is protected by applying a coat of bitumen followed by a polyester tape wrapping. A final protection is given by extruding a sheath of PVC or PE. The corrugated steel sheath found extensive use in Germany, initially for communication cables. Later, it was used to sheath a power cable. To accept the required short-circuit current, copper or aluminium wires were used underneath. Sheath is more flexible and can be bent easily to a shorter bending radius. It also possesses a higher mechanical strength. The thickness of the tape used is in the order of 0.3–0.5 mm, reducing the weight and diameter of the cable considerably. Copper-coated steel tape is used to manufacture the high frequency (HF) cable, keeping a copper plating as an inside layer. Corrugation in this case is sinusoidal in nature. In Europe and America and now in China, this technique is being adopted to manufacture cables with corrugated aluminium, copper or a steel sheath.

6.6 Calculation of Short-Circuit Rating Capacity for Metal Sheath

Sometimes, it is required to divert a part or the full quantum of short-circuit current through the metal sheath. It is particularly important for mining, HV and EHV cables. In such cases, short-circuit current is large and has a duration of almost a second, the heat developed is considerable. If the conductor is to accept the total current singularly, the heat developed will be large enough to fuse the conductor metal, causing a severe disruption in the supply system. At the same time, the expansion of the conductor and sheath takes place longitudinally throughout, generating high stress. In such situations, quick protective devices have to be installed to reduce the probability of completely damaging the system. By diverting a part or the whole of the short-circuit current through the sheath, heat can be dissipated quickly over a large surface area, reducing the danger of disruption of power. With the aid of the following formula, short-circuit rating can be calculated. On the other hand, knowing the value of short-circuit current requires the sheath area of a particular metal.

The formula of short-circuit rating is given by

$$I_{sc} = \frac{KA}{\sqrt{t}} \tag{6.9}$$

where K is the constant, depending on the characteristics of the metal at an initial and final temperature, before and after commencement of the short circuit.

$$K = \sqrt{\frac{\left(S_h \delta \theta_{sc} 10^{-6}\right)}{\rho_0 \left\{1 + \alpha\left(\theta_{\max} - \theta_a + \frac{\theta_{sc}}{2}\right)\right\}}} \tag{6.10}$$

where

S_h is the specific heat of the metal (J/kg/°C)
δ is the density of the metal (kg m³)
θ_{sc} is the temperature allowed to rise at the end of the short circuit (°C)
ρ_0 is the resistivity of the metal at 20°C (Ω mm²/m)
α is the temperature coefficient of the metal (°C⁻¹)
θ_{max} is the allowable maximum temperature above the conductor operating temperature (°C)
θ_a is the ambient temperature (°C)
A is the area of metal in question (mm²)
t is the time (s)
I_{sc} is the short-circuit current

Example 6.3

Let the inner diameter of the cable core be 50 mm and the metal sheath as lead thickness be 2.00 mm. The area of lead $A = 326.73$ mm².

Let the insulating material be XLPE, $\theta_{sc} = 200°C$, with an operating temperature of the conductor 90°C

where $S_h = 127$ J/kg/°C
$\delta = 11{,}370$ kg m³
$\rho_0 = 0.214$ at 20°C (Ω mm²/m)
$\alpha = 0.004°C^{-1}$
$\theta_{max} = 70°C$ (temperature on the surface)
$\theta_a = 30°C$
$\theta_{sc} = 200°C - 70°C = 130°C$
$t = 1$ s
Therefore $K = 28$

Thus, a short-circuit current which can be accepted safely by the lead sheath for 1 s is

$$I_{sc} = \frac{28 \times 326.73}{\sqrt{1}} = 9148 \text{ A}$$

From these results, one can understand how important it is to provide a metal sheath over the cable core. The rating is directly proportional to the area of the metallic covering. A part of the short-circuit current can also be diverted through the armour of the cable, either of steel strip or of steel wire.

When single-core cables are laid side by side with a protective metal sheath (as in the case of MV, HV and EHV cables), some losses occur within the sheath due to magnetic flux generated by the flowing current within the cable, which induces a localised voltage in the sheath, called 'eddy current loss'. Further more, lines of force influence adjacent cables, inducing a voltage within its sheath. It is imperative that in a three-phase circuit, single-core cables be laid either in a trefoil or parallel formation. In both the cases, induced voltage comes to play. If the sheath is not grounded, no current will flow, and the voltage generated, at times, may become very high. If the sheath is earthed at one end, only a limited amount of current will flow. It will be more near the grounding point. If both ends are grounded, a circulating current will flow through the sheath. In any case, energy is lost in the form of heat which influences the flow of current within the conductor, due to which the current rating will be reduced to some extent. This reduction or loss is termed 'sheath loss'.

These losses appear as heat within the cable and need to be dissipated quickly. Various losses occur such as the sheath loss due to sheath resistance, armour loss due to armour resistance and conductor loss due to metal resistance, and the heat needs to be dissipated through insulation, bedding and outer sheath. This can only be possible if the thermal resistivity of these materials is low. Even surrounding materials, such as soil and water, should be considered, whose resistivity also affects the rating of the cable.

6.7 Earth Bonding of HV and EHV Cable Sheath

From voltages of 66 kV and above, cables are made as a single-phase system as they cannot be laid as three-core cables. These cables are manufactured with a metallic sheath as a protective layer to prevent the ingress of moisture and other affluent from surrounding areas. Metal sheaths, whether of lead or aluminium, have to be protected against corrosive elements by applying a thermoplastic sheath of PVC or high-density polythene. During installation, these cables are laid either in a trefoil formation or in parallel. In all such cases, the current flowing through the conductor produces an electromagnetic field, inducing a voltage within its own sheath (self-inductance), and within the sheath of the adjacent cable (mutual inductance), and is known as the sheath voltage. In case the metallic sheath is not grounded, this voltage can rise to a great extent, creating fault conditions and posing a danger for the working personnel. As per international norm, this voltage has to be restricted to 30–60 V for overhead cables and a maximum of 100 V for underground cables. In a short-circuit condition, the voltage has to be restricted to 300 V. It is therefore necessary that the sheath be grounded. When the sheath is grounded, the current flows through the sheath reducing the current-carrying capacity of the conductor (as this energy cannot be created from the outside). The amount of power which is lost is known as the sheath loss. Grounding at one end produces a current near the grounding point. This is tolerable when the cable length is short. For longer lengths, the voltage will rise at a remote distance from the grounding point. The grounding of the sheath at both ends will generate a high circulating current, again at the ends, but for a longer length the voltage will show a significant value at the midpoint. To obviate such a situation, the cable needs to be transposed during installation to balance inductive forces, which is not possible physically. The solution is cross bonding the phases after every three lengths.

The nature and intensity of induced voltage and current under different conditions are dealt with in detail in Chapter 8.

7

Armouring and Protective Covering

7.1 Armouring

7.1.1 Steel Tape Armouring

Today, unlike paper cables, most low-tension polymer-insulated cables need not be protected by a metal sheath, unless specifically required. It is very risky to bring a lead or aluminium sheath by the extrusion process on a polymeric cable, as the high heat would damage the cable. It is found that polymer-insulated cables having an extruded inner and outer sheath can sufficiently be relied upon against ingress of moisture and the surrounding affluent since they are not chemically very reactive. Also, polymers do not get corroded under normal conditions. In some cases, as in Germany, the United States and Japan, a corrugated steel sheath was applied by a seam welding process. Notwithstanding, it is essential that the cable have a mechanical protection to withstand severe laying and installation conditions, when buried underground or laid in any other manner. The environmental conditions, impact of a pick axe during excavation, abrasive forces during laying and soil subsidence could all inflict mechanical damage. Furthermore, to protect the cable electrically, particularly for mining cables, a part of the short-circuit current has to be diverted through armouring. It is mandatory for mining cables to have a minimum armour resistance of 75% of a phase conductor.

The assembled insulated cable core is protected inherently by an inner covering, either by a tape wrapping or by an extruded PVC or PE sheath. This assembled cable core is protected by mechanically applying a layer of galvanised flat steel strip or by a galvanised steel wire. The use of a galvanised strip is more popular in India from a cost dynamics perspective. The use of a flat steel tape was discontinued, as were PILC cables. Its use remained confined to some special applications, such as railway signalling cables to be laid alongside the traction line, where the steel tape lowers the impedance value of the screen. Communication cables are also armoured with steel tape for the same purpose. It is also used as a binder for wire-armoured shaft cables for mines, where at times the cables need to be installed at an inclined or a vertical position. In such cases, the wire armour keeps the cable from sliding down and acts as a strengthening member.

It is of interest to study the application of a steel tape on the cable. During application, it is necessary to determine the exact width of the tape considering the gap to be allowed when applied helically. There exists a relation between the cable diameter, the angle of application and the lay length.

The width of the tape is given by

$$b_B = \frac{100 \pm k}{100} \times \frac{\pi d_B}{\sqrt{\{1 + \pi(d_B/h_B)^2\}}} \tag{7.1}$$

where

b_B is the width of the steel tape (mm)
d_B is the diameter of the cable under armouring (mm)
h_B is the lay length of lapping (mm)
k is the overlapping or gap in each lapping in percent ((+) for overlap and (−) for gap)

Example 7.1

$d_B = 30$ mm
$h_B = 40$ mm
$k = 25\%$ gap

Then

$$b_B = \frac{100 - 25}{100} \times \frac{\pi \times 30}{\sqrt{[1 + \pi(30/40)^2]}} = 27.60 \text{ mm}$$

say 28 mm and the angle of lapping is 23° approx., while the lay length is given by

$$h_B = \pi d_B \tan \alpha$$

The best angle for lapping lies between 20° and 35°.

It is clearly shown that in order to determine the tape width, the relation between the lapping angle, lay length and diameter of cable needs to be considered.

Where heavy corrosion is experienced, a galvanised steel tape is recommended. The magnetic influence of the steel tape is used to lower the reduction factors of the cables under the influence of power cables (see Chapter 8).

A steel tape armouring line (Figure 7.1) consists of a pay-off stand to hold the bobbin within the inner sheathed cable. This can be a vertical column type with a self-traversing system. Up and down and sideways movements are motorised in order to accept different sizes of bobbins. Bobbin holdings used today are of a pintle type. Brake adjustment has to be pneumatic for larger-size drums or of a band type. Safety devices are incorporated within the system. A cable guide roller and a counter metre are provided after a pay-off stand. A second metre can be provided before a take-up unit. The steel taping head

FIGURE 7.1
A complete steel tape armouring line as installed.

FIGURE 7.2
Steel tape armouring head unit.

is of a tangential type (Figure 7.2). The steel tapes used must be smooth. The edges must be free of burrs and kinks.

Two steel tape pad holding hubs are provided on both sides of the taping head arms. The weight of the pads varies from 500 to 800 kg. The width of the pad depends on the cable diameter and varies generally from 25 to 60 mm. The pads are wound on a steel core, as sent by the supplier. The inner diameter of the core is predetermined. After placing in position, the pads are securely locked between two round steel locking frames. A tension control device is provided to keep the tension on the tapes constant while running throughout the length by an automatic tension control adjustment. Indications are also given when the tape is at the end. The tape angle has to be adjusted as per the cable diameter. An angle adjustment device is provided by shifting positions of the guide rollers with a threaded screw. This type of a machine can run at a speed of 200–250 rpm, having a line speed of 40 m/min. Fine adjustment is done by a PIV gearbox. By incorporating a direction change gearbox unit, the machine can run in either left or right direction as required. A pneumatic operated caterpillar is provided to pull the cable at a constant speed. The take-up unit can be of a self-traversing type or of a vertical column type. The bobbin holding is of a pintle type. The columns are adjustable in order to hold different sizes of bobbins. The sideways movement of the columns and the up and down movement of the pintles are electrically manoeuvred. The bobbin is driven by a geared motor synchronised with the line speed. For traversing of a vertical column type, the take-up unit is motor driven. Proper safety devices are incorporated within the system.

7.1.2 Galvanised Steel Wire or Flat Steel Strip Armouring

There are various conditions under which the cables are to be laid and installed. The steel tape/strip armoured cables are buried directly under the ground. The laying conditions here are not very stringent. When the cable has to cross a road or a culvert, or when it has to be laid in a shaft of an inclined mine or to be hung vertically, then a considerable pulling strength is required to install the cable. Also, the weight of the cable has to be borne by an armour. In such cases, wire armouring is a must. The wire armour is also grounded to accept short-circuit current of the cable, particularly in the case of mining cables, and 75% of the short-circuit current is diverted through the armour (Figure 7.3). Hence, the resistance of the armour must be 75% of that of the phase conductor. To achieve such a condition, sometimes a double-wire armouring has to be considered. Double-wire armouring is also needed to protect the cable from falling debris within the mines and hold it in a vertical or an inclined position. At times, the cables have to be shifted from place to place and laid under severe rough terrains. The armour has to accept such abuses, intermittently protecting the cable and ensuring an uninterrupted power supply. In mining application, sometimes in order to meet the armour resistance, a tinned copper wire combination is adopted with the steel.

FIGURE 7.3
A typical GI steel strip/wire armour machine with guide rollers on lay plates for wires/strip die head with an adjustable die holder box.

The wire armour is laid spirally and is similar to the stranding of a conductor. The cable is protected by an inner sheath of plastic material to accept the pressure of the armour wire or tape.

There are special-purpose cables made for the navy and for other defence purposes, such as float cables for mine detecting and sweeping, where the armour is made of stainless steel. In such cases, the armour must be made completely torsion-free to keep the cable floating and operating without twists and turns. Stainless steel being hard, the armouring machine is provided with preformer rollers to allow torsion-free wires to be wrapped on the cable's inner sheath (Figure 7.4).

Since wires are laid helically around the cable's inner sheath, the cross section of the wire when cut at right angles to the cable axis will be elliptical. Taking the major axis of the wire to be 'd_a', the number of wires (n) that are placed over the inner sheath is given by

$$n = \frac{\pi(D + d_w)}{d_a}$$

where

D is the diameter under the armour wire

d_w is the diameter of the armour wire

Now

$$d_a = \frac{d_w}{\sin \alpha} \tag{7.2}$$

FIGURE 7.4
Armouring head with preformer for torsion-free stainless steel and hard steel armour to make special cables for defence requirements.

Hence,

$$n = \frac{\pi(D + d_w)\sin\alpha}{d_w}$$

The angle 'α' is determined as $h/D = m$, where h is the lay length of the wire and is given by the number of times of cable diameter under the armour, m.

When

$$h = mD$$

then

$$\tan\alpha = \frac{h}{\pi D} = \frac{m}{\pi}$$

Therefore,

$$\sin\alpha = \frac{\tan\alpha}{\sqrt{1 + \tan^2\alpha}} \tag{7.3}$$

or

$$\sin\alpha = \frac{m}{\sqrt{1 + m^2}} \tag{7.4}$$

Therefore,

$$n = \left\{\frac{\pi(D + d_w)}{d_w}\right\} \times \frac{m}{\sqrt{1 + m^2}} \tag{7.5}$$

By selecting the value of 'm', the number of wires to be wrapped on the given cable core for armouring is determined.

Example 7.2

Let the diameter of a cable be 40.56 mm and the wire size required 2.50 mm. Selecting $m = 12 \times 40.56$, $h = 486.72$ mm and $n = 52.923$.
Since 'n' cannot be a fraction, the number of wires shall be 52 and the angle of the armour wire shall be $\tan\alpha = 12/\pi$, as such $\alpha = 75°20'$; in this case, the value of $\sin\alpha = 0.967$ approx.

The weight of the wire can be calculated by the following formula:

$$\text{Weight of wires:} \quad n \times \left\{\frac{\left(\pi d_w^2\right)\rho}{4}\right\} \times \frac{1}{\sin\alpha} \text{ kg/km} \tag{7.6}$$

where ρ is the density of the wire.

In the case of a flat strip, the dimension of the width of the strip is accepted in place of 'd_w' and the mean diameter is $(D + t)$ where t is the thickness of the strip. And the formula becomes

$$n = \left\{ \frac{\pi(D+t)}{d_w} \right\} \times \frac{m}{\sqrt{1+m^2}} \tag{7.7}$$

and the weight is given by

$$Wt = n \times (d_w \times t)\rho \times \frac{1}{\sin\alpha} \text{ kg/km} \tag{7.8}$$

Normally, the lay length for the wire is taken as 10–12 times the diameter of the cable under the armour for round wires and 8 times for flat strips. In these cases, sin α becomes 0.967 for round wires and 0.93 for flat strips.

While selecting and ordering wire or strip, it is always advisable to keep the dimensions towards the lower limit of tolerance. This will ensure meeting the standard and saving material without jeopardising quality.

For vertical shaft cables to be installed in mines, the safety factor of the strength of the armour wire should be considered as five times the weight of the cable. This has to be selected by a proper calculation. Therefore, either the dimension of the wire or the number of wires has to be increased. This is necessary to hold the total weight of the cable by armour wires and to prevent the cable from slipping downwards.

To achieve this strength, the wire sizes selected may be higher in dimension, keeping the number of wires minimum but ensuring full coverage. In this case, the number of wires is calculated as follows:

$$d_w = \sqrt{\left\{ \frac{D^2}{4} + \frac{4F_b}{\pi^2 \sin\alpha} \right\}} - \frac{D}{2} \text{(mm)} \tag{7.9}$$

The cross-sectional area of wires is given by

$$F_b = \frac{lg_0}{\sigma - lg_b} (\text{mm}^2)$$

where

l is the length of the cable when hung vertically straight (m)

g_0 is the weight of 1 m cable when hung vertically straight without armouring but with an inner sheath and a protective layer under the armour (kg/m)

σ is the specific tensile strength of the armour wire (kg/mm²)

g_b is the weight of 1 m armour wire having 1 mm² area (kg/m mm²)

D is the diameter of the cable under the armour

Example 7.3

Let l = 1000 m
g_0 = 5 kg/m (taking five times safety factor = 25 kg/m)
σ = 45 kg/mm^2
g_b = 0.00787 (kg/m mm^2)
Then F_b = 673.31 (mm^2) as such
D = 50 mm, d_w = 5.03 mm diameter

In this case 34 single-layer wires of 5.03 mm diameter are used. However, this diameter would be too stiff for winding over the cable. During operation, the pressure of the wire may damage the cable. The torsional force during twisting of wires has to be considered. Hence, the application of 3.55 mm wires in two layers would be advisable.

To prevent bird-caging of armour wires, a single layer of steel tape can be applied helically with an open spiral form.

For all types of mining, the cable conductivity of the armour should be 75% that of a phase conductor.

In the case of horizontally suspended cables and aerial cables, the weight and suspension point, along with the tension, have to be calculated to ascertain the diameter of the wire that is required for armouring. In these cases, a tape armour cable cannot be used. Special consideration is also needed to design submarine cables. Here, the depth of the water column on the cable, the weight of the cable between suspension points, the stress which develops due to the speed of ocean current and the maximum stress which develops during a cyclone or tornado are all to be considered. Precaution is needed to protect the cable from saltwater, whales, and sharks.

For thermoplastic cables, a metallic sheathing is not required. The material itself is moisture proof. It seldom reacts with affluent environment. These cables are protected by a galvanised steel wire or a flat strip armouring, depending on the short-circuit condition and fuse rating.

A steel wire/strip armouring machine consists of a pay-off stand to hold the bobbin containing an inner sheathed cable. The stands can be of vertical column type with a self-traversing system. In the past, pay-off drums were loaded in the pit, which enabled a better handling of the product. The up, down and sideways are motorised movements to accommodate different sizes of bobbins. The brake adjustment is a pneumatic type for larger-size bobbins or of a band type. Safety devices are incorporated within the system. A cable guide roller and counter metre are provided after a pay-off stand. A second metre can be provided before the take-up stand. Wire/strip holding bobbins are placed on cradles. Tension adjustment and locking devices are connected with the pintle holding the bobbins. Cradles are placed between flayers, which are mounted on a hollow centre shaft. The design of the machine varies as per customer requirements. The machine can be made of single, two or three cages running in tandem, or operated individually. Normally, these machines are of a rigid type and are constructed with a back-twist system, when higher-diameter wires such as for extrahigh-voltage cables or submarine cables are used. The machine can be used for wire and for strip armouring. In the case of a strip armour, a cone and ring–type guiding attachment is used in front of the carriage(s). For wire armouring, a guide plate with a requisite number of holes is provided. An adjustable die-holding block is provided before the carriage. The cable can be pulled, preferably by a capstan, to give a smooth operational movement at a constant speed. In this case, caterpillars are not recommended. The lay length of the armour wire/strip is adjusted by a lever-operated lay change gearbox. The rotational direction of the carriage(s) can be altered by a direction change gearbox. The take-up unit can be of a self-traversing type or of a vertical column type. The bobbin holding is of a pintle type. The columns are adjustable in order to hold different sizes of bobbins. Sideways movement of the columns and upward and downward movements of the pintle are motorised. The take-up bobbin is driven by a geared motor synchronised with the line speed. Traversing for the vertical column take-up is motor driven. Proper safety devices are incorporated within the system.

A new S–Z-type armouring machine has been developed, where the wires are laid in a back-and-forth movement from a 180° to 270° rotation, to allow an 'S' and 'Z' formation. In this case, the bobbins holding the wire/strip remain stationary on the stands. The cable is drawn by a heavy-duty caterpillar. Wire/strips passing through the guide plates of the S–Z oscillating device are allowed to cover the cable surface uniformly. The device is easier to handle. The loading and unloading time of the bobbins is less. In this case, large-size carriages are eliminated. The speed of the machine is also higher than the carriage-type armouring machine. During installation, the armour wires/strips can be handled in a much easier way. Wire and strips can be fed directly from the coils, eliminating the bobbin winding process. During installation and jointing, the armour wires can be taken out easily for earthing and jointing.

For single-core cables, armour wires should be of a non-magnetic type, like aluminium. In this case, the current flowing through the conductor emanates a magnetic field, which induces a voltage within the armour. If the armour is of a magnetic material, hysteresis losses become significant. Further, adjacent cables will influence the armour, while their magnetic field will induce an additional voltage, raising the total value which is too high. Naturally, a non-magnetic armour with a low impedance value becomes preferable for diverting the circulating current quickly and for earthing at end points.

Armour loss is sustained due to a magnetic field induced by absorbing a part of power from the current flowing through the conductor.

Calculation for armour loss for different types of armours can be taken from IEC 287-1-1 – for 'Calculation of the Current Rating'.

7.2 Protective Covering

7.2.1 Extruded Thermoplastic (Inner and Outer Sheath)

Cables, whether armoured or unarmoured, should have a protective covering irrespective of the installation area and surroundings. Cables installed underground may have to withstand aggressive soil with moisture, chemicals and various corrosive affluents. Microbiological elements, termites and rodents also attack the outer covering, allowing passage to corrosive elements and moisture and creating a fault condition. When exposed to air, polluted surrounding, chemical fumes, UV radiation from sunlight and rainwater, the moist air slowly acts on the protective covering bringing down the operational safety of the cable.

Assembled, bedded or inner sheathed cable cores, along with a metal sheath (where applicable) on cables or armouring as applicable, have to be protected from the corrosive surroundings and underground soil. The mechanism of corrosion is an electrochemical process. The current within the conductor and also of the conductors from adjacent cables emanates a magnetic field which induces current and voltage within the metal sheath and armouring. The sheath current of the metal sheath, or armour, if unprotected by an insulating media, will form a local cell with the moisture contents of the surrounding elements, such as air and underground soil or affluent around it. The formed electrolytic local cell gradually erodes the surface of the metal creating a fault condition, by allowing the moisture to penetrate inside the layers of the cable. Initially, a protective covering is formed by the combination of bitumen, bitumen-impregnated hessian cloth or jute yarn. Later, with the coming of thermoplastic and thermosetting materials such as PVC and polythene, elastomeric compounds replaced the conventional serving materials. Nowadays, a thermoplastic sheath of PVC or polythene is applied on the metal covering or on the armour.

An outer protective covering is applied below the armouring, which consists of paper-insulated mass-impregnated lead-covered and double steel tape or wire armoured cables consisting of one layer of bitumen compound, two numbers of bitumen impregnated paper and one layer of impregnated cotton tapes, along with two layers of bitumen-impregnated hessian tape or jute roving. Over the armour, two layers of bitumen-impregnated hessian cloth or two layers of impregnated jute roving are used. For unarmoured cables, a similar type of serving is applied on the metal sheath. This type of protective covering was found to last more than 40 years. Initially, PVC-insulated cables were also protected in a similar manner. Soon they were replaced by an extruded PVC coating. Gradually, all types of armoured and unarmoured cables were protected with an extruded PVC or PE compound. PVC compounds have the following characteristics in general.

The polymeric compounds used as the sheathing material should have the following characteristics and should be confirmed by tests on samples, as per the given specification.

The density of compounds can be higher than those of insulating-type compounds to make it cost-effective.

1. The thickness should be determined as per the diameter of the cable (as per the given standard specification).
2. It should withstand mechanical stress during transportation and, when installed, in an abrasive condition.
3. It should be sufficiently flexible to accept strains of pulling, bending and unbending during manufacturing, transportation and installation.
4. Water absorption should be very less, since the sheath may remain in moist air and water intermittently or permanently.
5. It should withstand all weather conditions, such as UV radiation. To protect from UV radiation, the sheath should be coloured black by mixing with carbon black.
6. It should not deteriorate easily under hot and humid conditions; the ageing process should be slow or minimum.
7. Normally, it should be fire retardant and not emit toxic fumes.
8. It should not crack under cold and hot, humid conditions.
9. The sheath material should not be affected by affluent and chemicals present in soil or in the surrounding atmosphere.

In addition, compounds are expected to have the following properties:

10. Fire extinguishing property
11. Low smoke and low halogen emission
12. Heat-resistant property
13. UV-resistant property
14. High abrasion resistance and flexing properties

The compound should pass all the tests to determine its stability under different operational conditions.

The sheath is provided by an extrusion process. The process of extrusion has been explained in Chapter 5 and Section 5.2 on thermoplastic insulation. The sheath material may be of PVC or polythene and should be black in colour to protect the polymer from prolonged exposure to ultraviolet radiation. Two to three percent of carbon black should be added to gain the desired result. At times, conductive carbon black is added to the PE sheath to protect the cable from lightning failures and to detect pinholes which may appear during the extrusion process. This is particularly applied in the outer covering of HV and EHV cables. At times, a conductive graphite coating is also suggested to be applied on the outer covering. After the application of a conductive layer, the sheath should be tested by applying a suitable voltage to ensure the integrity of the covering. The choice of any other colour normally is not advisable for the outer protective layer. For EHV cables, with MDPE or HDPE as sheathing materials, care should be taken of extrusion temperatures, as a very high temperature in the feed zone of the extruder with a rising profile leads to stresses in the extruded material. The MDPE with high stress levels are suspect to environmental stresses and produce pinhole-type breaks in the sheath and fail a DC test prior to the charging of the cable. Therefore, care should be taken in MDPE extrusion with a properly selected temperature profile.

During the extrusion, care has to be taken to keep the concentricity and sheath thickness within the specified limits. Normally, a tube extrusion is preferred to give a smooth finish. In tube extrusion, one of the advantages is that once the cable is centred, it does not change, and the gap between the core tip and the laid-up cable is 3–4 mm, depending upon the size. This gives a lateral play for the snaky finish of larger cables and therefore extrudes safely. It also helps keep material consumption under control. Sheathing materials are made cheaper by incorporating filling compounds like whiting, China clay and calcium carbonate. The mixing of these materials would reduce the cost, but the resulting specific gravity becomes higher and production slows down. Hence, the cost of production and weight of the material per unit volume need to be checked against an apparent calculated saving.

Example 7.4

A 120 mm extruder as per the calculated production criteria should give an output of 220 L of PVC by volume per hour and 308 kg/h, with a specific gravity of 1.4.

For a specific gravity of 1.45, the weight of the PVC will be 319 kg/h. To extrude 319 kg, the extruder time required will be 1 h 2 min 8 s.

Now suppose a cable with a diameter of 50 mm is covered with a coating of PVC with a thickness of 2.50 mm. The volume of PVC required per km will be 412.33 L. For a specific gravity of 1.4, the weight of the PVC compound will be 577.26 kg/km. But the extruder will give out only 308 kg/h, which covers 533.55 m of cable approx. per hour.

In a compound having a specific gravity of 1.45, the requirement will be 597.88 kg/km, giving 515.15 m/h.

The cost of the first compound per kilometre is 577.26 × 38.00 = 21935.88 at the rate of 38.00/kg.

The cost of the second compound per kilometre at the rate of 36.00/kg is 597.88 × 36.00 = 21523.68, bringing a gain of 412.20/km. The time required for extruding 1 km of the first compound is 1.8742 h and for the second compound is 1.9412 h.

In this time difference of 0.067 h, the first compound will produce 35.74 m more cable.

If extruder cost is 3000.00/h (electricity, wages, water, etc.), then the total cost will be, for the time 1 km is extruded, 5823.60 for the second compound, against a cost of 5622.60/h for the first compound. Therefore, the total gain is 211.20/km when a higher-density compound is used.

This saving can be obtained if the operational procedure, maintenance of die and core, cleanliness of extruder and the use of contamination-free compound are strictly adhered to. One of the innovative ways of saving the sheath material is to save the head waste from becoming scrap. As soon as the hot plasticized compound comes out of the extruder head, it is allowed to enter in an online granulator. Otherwise, a casual approach may offset the operational cost and the quality of the extrudate, bringing a loss to the company. Highly filled compounds can increase the wear and tear of the extruder. Hence, a balance has to be worked out in order to get the maximum benefit. A close study of compound behaviour, standardisation and a consistency in quality will yield the desired benefits.

7.2.2 Grade of Material

It is an established fact that PVC is a material that is extensively used as the outer sheathing for electric cables. The required characteristics of the outer sheath have been discussed here earlier. Initially, PVC was extruded over a bitumen-coated metal sheath. It has been observed that over time, the plasticiser from the PVC compound starts migrating slowly, making the outer sheath stiff. In turn, bitumen becomes softer. With the increase in temperature, bitumen starts flowing towards a lower gradient, exerting pressure on the outer sheath. Subsequently, the application of the bitumen compound is withdrawn. It has been found that PVC alone was sufficient to protect the cable from outer corrosive elements. However, PVC gets softened by aromatic solvents like MEK, MIBK and furfuran and in a prolonged atmosphere of benzene and petroleum products. Hence, the application of PVC is not recommended when cables are to be installed in a refinery or in the vicinity of petrochemical factories. The compound also cannot be used in places where the temperature is very low or extremely high. In cold conditions, the compound becomes hard and brittle and starts cracking. At relatively high temperatures, it starts to soften. To overcome such problems, special formulations are developed to produce a high-temperature compound withstanding PVC, which can operate at a temperature of 85°C. Some applications call for 105°C grade compounds as well.

A low-temperature PVC has also been developed, which can be used for defence cables operating at high altitudes at temperatures of –15°C.

In contrast, polythene shows better low-temperature withstanding capacity up to (–)40°C. The water absorption capacity of a PE sheath is very low. For this reason, HDPE is preferred as an outer sheath compound for defence power supply and communication cables operating at a very high altitude. HDPE is also used as an outer sheath for power cables used in nuclear power plants against radiation hazard. Pure HDPE is very hard and brittle. The polymer is also highly inflammable. The resin has to be blended with an antioxidant and ethylene vinyl acetate (EVA) to make it flexible and easy to process. The compound has to be blended with 2%–2.5% fine carbon black to protect it from sunlight and UV radiation. This black surface also protects the cables from occasional lightning discharges. Blended HDPE compound can be extruded easily. The outer protective covering of HDPE is mechanically strong and absorbs sudden impact like fallen debris.

Nowadays, cables have to be coated with materials which can withstand accidental fire. Fire may not generate from short-circuit conditions of cable, but external fire can affect the cable sheath. When a normal PVC/PE sheath catches fire, smoke and corrosive acid gas are likely to fill in the area making it difficult to evacuate the affected people and can become a cause for health hazard. Naturally, the compound should have a fire extinguishing property and should not emit corrosive acid and smoke during fire breakout.

It has been found that polyvinyl chloride is inherently flame retardant because of its chlorine content of 56%. Applications of the compound, however, require the use of plasticisers to impart the desired degree of flexibility. Plasticisers such as phthalates increase flammability by diluting the high-chlorine polymer with combustible materials. The development of phosphate plasticisers is found to replace phthalate and other plasticisers, without any loss of plasticising action, and when blended with antimony trioxide, they reduce flame propagation properties to a great extent. Further, ejection of chlorine while suppressing the fire evolves smoke and hydrochloric acid. This acid and smoke content creates problems during the evacuation of people from enclosed areas. To prepare a compound with fire-retarding low-smoke (FRLS) properties, aluminium trihydrate (ATH) and calcium carbonate are used as fillers. During an intense fire, ATH decomposes to form water in the form of steam, which in turn helps in suppressing the flame and smoke. Chlorine also acts with calcium carbonate, forming calcium chloride and water. Some carbon dioxide is also formed, which also helps in suppressing flame propagation. To make the compound heat resistant, antimony trioxide is incorporated within the compound formulation. It has to be noted that ATH and antimony trioxide must be of a very fine quality. Otherwise, the surface of the compound during extrusion will resemble that of crocodile skin. The characteristics of the material will also be impaired.

Polythene, on the other hand, is highly inflammable and easily deforms with the application of heat. To make it fire retardant, the compound is blended with an antioxidant ethylene propylene resin (EPR) to make it heat resistant and an EVA to make it fire resistant. Compounding with these ingredients is a very tedious process. A little plasticiser is added to make it flexible. Such compound, though costlier than PVC, is very effective when FRLS properties are called for.

In powerhouses and places where emergency work has to be undertaken, the cable must remain in service for a certain period of time until the fire is extinguished. In this case, the conductor must be of copper. On the conductor, one or two layers of mica tape are wrapped, on which FRLS PVC insulation is applied. The cable is wire armoured and then wrapped with a glass fibre tape. Overall, protection is provided by FRLS PVC sheaths, called fire survival cables. In case of an intense fire, these cables can remain in service for a few hours, until they are replaced. Here, the glass and mica tapes resist the fire from propagating inside and keep the insulating properties of material unimpaired for a considerable period of time.

Underground cables are subject to attack by termites, rodents, etc. The outer sheath of these cables is compounded at times with aldrin. But, in most of the cases, a copper naphthanate solution is added. But care should be taken to see that during extrusion the compound does not bloom and create a rough surface, impairing the characteristics of the sheath.

The outer sheath of HV and EHV cables is put directly on a metal sheath or a wire armour. It is necessary to keep these cables secured, even when laid in hazardous surroundings. The sheath must remain free of any pinhole or external damage. These sheaths are tested by applying a voltage between the inner metallic part and the outer surface to make sure no fault conditions exist. A coating of conductive carbon black has to be applied smoothly on the surface of the sheath, which acts as a second electrode. This layer can be applied by extruding a semi-conductive compound, or conductive black carbon can be applied as a dry powder by smearing over the sheath, or as a conductive solution prepared in a solvent. This solvent must evaporate after application and retain the conductive carbon layer on the sheath. The solvent should not react with the outer sheath compound. Carbon black also can be applied by the electrostatic precipitation method.

The outer sheath of subsea and underwater cables must not absorb water and should be tough enough to resist the attacks of fish and other aquatic life. Polythene is the preferred material for this purpose. Sometimes, a layer of bronze or stainless steel tape is applied over the sheath to keep the cables secured from the sharp teeth of crocodiles, sharks, whales and other animals. Sometimes, cables are laid on the river or seabed, when silting of mud can put pressure on the armour. Cables are also installed anchoring on floating buoys. In all such positions, the cables should be able to withstand the high current and turbulent waters during a storm and cyclone. The armour and outer coverings have to be designed to be able to withstand all such adverse conditions.

8

Electrical Parameters for Cable Design

Discussions in this book through Chapter 7 have mainly been confined to the manufacturing procedures of cables. In the course of such discussions, necessary references on electrical parameters were introduced to let one understand the basic factors related to the production system. The main focus of this chapter is on mechanical and chemical activities which are essential. However, the ultimate aim of a cable is to transmit electrical power for a regular day-to-day living.

There are two main aspects to be considered: (1) the current, that is, the amount of electricity allowed to flow through the conducting media, and (2) the pressure which develops along the route during transmission. While current flows through a conductor, resistance develops in various forms and must be overcome. In other words, by knowing the amount of various forms of resistance to be encountered, pressure is applied in the form of voltage, and the desired quantum of power is transmitted to its final destination. It is like a water pipeline having a certain bore diameter through which a definite amount of water has to be pushed through. The water current encounters resistance from the wall of the tube, the bends, constrictions, valves, etc. Naturally, the flow gets retarded, and the quantum of water coming out at the end of the tube becomes lesser than the quantum initially pumped. If the desired quantum is to be delivered at the end with a given force, then the following applies:

1. Either the size of the pipe has to be increased, that is, analogous to the increase in conductor area without altering the force of pressure (voltage).
2. Or, if the size of the delivery channel is kept constant (conductor area), then the pressure has to be increased. It means that the wall thickness and the quality of the metal have to be increased in order to accept the high pressure so that the pipe does not burst (increase in insulation thickness and quality of the material).

The previous simile is a simple form of understanding the basic transmission pattern. But in the case of the cable, many factors have to be considered, such as the condition of installation, users' demand, transmission length and environmental conditions. Since nothing can be perceived by observing a cable from its outer surface, it required to conduct various experiments and get results to understand the characteristics and behavioral pattern. Repeated cross verification of results facilitated development works and made it possible to advance realistic design parameters matching with actual operational conditions.

The conductor area depends on the amount of current which has to be pushed through. The selection in turn depends on

1. The type of metal to be used. Its resistance value, temperature coefficient, thermal resistivity and mechanical properties, such as malleability, ductility, tensile, elongation and resistance to corrosive atmosphere, all influence the transmission parameters.
2. It has to be sufficiently flexible to withstand repeated winding and unwinding. Jointing during installation should be easier, and plenty of material should also be available with a reasonable market value to become economically viable.

Considering these aspects, copper and aluminium have gained a prime position, where copper is mostly used for communication and for domestic and flexible cable construction, while aluminium is the best choice for manufacturing power cables for both overhead and underground sectors.

The resistance of metallic conductor is given by

$$R_o = \rho \frac{L}{A}$$

where

R_o is the resistance at 20°C (in ohms)
ρ is the resistivity of the metal
For copper, $\rho = 0.01724$ (Ω mm^2/m)
For aluminium, $\rho = 0.02778$ (Ω mm^2/m)
L is the length of the transmission line (m)
A is the cross-sectional area of the conductor (mm^2)

As per the International Annealed Copper Standard (IACS), the conductivity of copper is taken as 100% and that of aluminium as 63%. For a stranded conductor, an allowance of a 2% increase in resistance value is accepted for stranding. A further 2% increase is taken for laying up. This norm has been fixed by the International Electrotechnical Commission (IEC) and recognised internationally. Resistance values given in IS 8130 are based on the same parameters. The number of wire sizes for stranded conductors has not been rigidly fixed, but a guideline giving the minimum number of wires to be selected for a particular conductor area is indicated. The numbers have been fixed, taking into account the minimum amount of flexing strength which must be achieved to withstand stress and strain during winding and unwinding and bending during installation. This has given an advantage to the manufacturer, who can now select and fabricate a machine which will be cost-effective. The production process can also be rationalised, utilising the optimum capacity of the wire drawing and stranding machine. All sector-shaped and round conductors can be compacted to minimise the diameter of the cables. This will also help in releasing any internal stress developing within the conductor during stranding, facilitating the use of the rigid bobbin-stranding machine over the old ones with a back twist having a comparatively lower production rate. Naturally, while designing a conductor, all these factors have to be taken into account.

The selection of a conductor is done on the basis of its current-carrying capacity. If the cable is of longer length, the choice will depend on the voltage drop, which is calculated on the basis of its impedance value. As per IEEE regulations, the voltage drop should not be more than 2.5% of the rated voltage, irrespective of the starting condition of the load. The voltage drop is measured in millivolts and is given by mV = 2Z for single-phase conductors and mV = $\sqrt{3}Z$ for three-phase conductors, with mV = millivolt drop for 1 A flowing through 1 m length of the cable.

Z is the impedance value per kilometre length of the cable at a maximum normal operating temperature in ohms.

For a 220 V system, the single-phase drop should not exceed 5.5 V (5500 mV), and for a three-phase 440 V system, the drop should be within 9.5 V (9500 mV). Therefore, in a line of cable length, the voltage drop should be restricted as follows:

For a single-phase circuit, mV × l ≤ 5500, and for a three-phase circuit, it should be ≤9500. Selection should, however, depend on the critical value at the full load current.

Impedance is given as

$$Z^2 = R^2 + X^2 \tag{8.1}$$

where

R is the resistance
X is the reactance of the cable

$$X = 2\pi f L \times 10^{-3} \tag{8.2}$$

for three-phase or single-phase conductors, where f is the frequency (Hz) and L is the inductance of the cable (mH).

$$\text{Inductance } L = K + 0.2\ln\left(\frac{2a}{d}\right) \text{mH/km} \tag{8.3}$$

where

'a' is the axial spacing between the conductors (in trefoil formation); for flat formation, a = 1.26 times the spacing of the diameter of a phase cable
d is the diameter of the conductor

TABLE 8.1

Factors to be Added While Computing Stranded Conductor Inductance

No. Wires in a Conductor	K
3	0.0778
7	0.0642
19	0.0554
37	0.0528
≥61	0.0514

For sector conductors, corresponding round conductor diameter is to be taken.

For 'R' resistance of the conductor, AC resistance is to be taken at the operating temperature.

The value of 'K' to be same as that of the corresponding stranded conductors operating at 50 Hz (Table 8.1).

While a temperature rise, during operation, is considered, the coefficient of temperature 'α' per degree centigrade for copper and aluminium is 0.0039 and 0.0040, respectively.

Conductors carrying an AC current are influenced by an alternate magnetic field producing self-inductance. The flow of current within the conductor is influenced by this inductive field. It was found that in an AC system, the density of the current increases towards the periphery of the conductor. For a small cross-sectional conductor below 150 mm^2, this phenomenon is not very appreciable whereas for larger conductors, it is appreciable. Further, for conductors of much larger size, for HV and EHV cables, it is considerable. This effect virtually increases the resistance value. This effect is termed as the skin effect. Further, when cores and cables are laid or assembled parallelly, the adjacent core or cable-carrying current emanates a magnetic field which produces a mutual inductive current opposing the current flowing on the side and touching the adjacent cable or core. This effect also somewhat retards the flow of current and is called a proximity effect. The larger the spacing, the lesser will be the proximity effect. However, both the skin and proximity effects must be taken into account while calculating the resistance value of the conductor. Up to 120 mm^2, these values are not appreciable and are generally ignored.

The skin and proximity effects are expressed as an increase in DC resistance by an amount of ΔR, where

$$\Delta R = R_{DC}(y_s + Y_p)$$

where the DC resistance of a stranded conductor can be written as

$$R_{DC} = \frac{\rho_o}{A}\{1+\alpha_{20}(\theta_{\max} - 20°C)\}(1+s)\ \Omega/\text{m} \tag{8.4}$$

where

ρ_0 is the resistivity of the conducting metal
A is the area of the conductor
α_{20} is the coefficient of temperature at 20°C
θ_{max} is the maximum operating temperature of the conductor
s is the correction factor for stranding of the conductor, which is 0.03 for conductors below 500 mm^2 and 0.04 for those 500 mm^2 and above

Now,

$$R_{AC} = R_{DC} + \Delta R = R_{DC}(1 + Y_s + Y_p) \tag{8.5}$$

Apart from these values, conductor resistance is increased due to sheath and armour losses. Sheath loss is due to the metal sheath applied on single or assembled cable core. Armour loss is due to steel wire or strip armouring. The magnetic field emanating from the conductor induces a voltage and current within the sheath and armour. This induced field draws power from the conductor, reducing the value of current further within the conductor and virtually increasing its effective resistance. Thus, the effective resistance of the conductor is given by

$$R_{eff} = R_{AC}(1 + \lambda_{sh} + \lambda_{ar}) \tag{8.6}$$

where λ_{sh} and λ_{ar} are termed as the sheath loss factor and armour loss factor, respectively. These values are actually required to calculate the current-carrying capacity of the cable, apart from other loss factors which are to be defined hereafter. All these losses transform into heat which is given out into the surroundings. These losses in the form of heat are to be expressed in the empirical form as $I^2 R$.

The correction factor for the skin and proximity effect, Y_s and Y_p, can be determined by means of a parameter x, which in turn is based on parameter K_1:

$$K_1 = \sqrt{\omega\mu_0\mu_r\gamma}\ \text{m}^{-1} \quad \text{and} \quad x = \frac{K_1 d_c}{2} = \frac{d_c\sqrt{\omega\mu_0\mu_r\gamma}}{2} \tag{8.7}$$

When

$$\omega = 2^{\omega} = 2\pi f$$

where
f is the frequency (Hz)
$\mu_r = 1$
$\mu_0 = 4\pi \times 10^{-7}$ H/m
γ is the conductivity of metal
γ_{al} conductivity of aluminium = 38 × 10^6 S/m and for copper is γ_{cu} = 58 × 10^6 S/m
d_c is the diameter of the conductor (m)

Now, the DC resistance of the conductor for 1 m length is given by

$$R_{DC} = \frac{1}{\gamma \frac{\pi d_c^2}{4}} = \frac{4}{\gamma \pi d_c^2} \tag{8.8}$$

Taking Equations 8.7 and 8.8, the following equation can be derived:

$$X = \sqrt{\frac{2\mu_0 \mu_r f}{R_{DC}}} \tag{8.9}$$

This value is accepted for solid-type conductors. For stranded conductors, there would be a number of wires involved. With the actual lengths of each wire being larger than the actual length of the conductor, the resistance value of the conductor should naturally be more than as specified. On the other hand, the current should divide itself through parallel paths of wires and remain electrically separated from each other, somewhat reducing the summed up value of the resistance of all wires. Naturally, as anticipated, the increased value will be decreased by the reduced value of the sum of resistances of parallel wires. This means that the value of X will be somewhat lesser, for which a correction factor 'C' is to be introduced.

Taking the values of μ_r = 1 and μ_0 = 4π × 10^{-7} H/m and introducing the correction factor 'C', the value of X takes the following form:

$$X = 15.9 \times 10^{-4} \sqrt{\left(\frac{f \cdot C}{R_{DC}}\right)} \tag{8.10}$$

where
f = 50 Hz
R_{DC} is the DC resistance of the conductor (Ω/m)

TABLE 8.2

Correction Factors for Skin and Proximity Effect

Type of Conductor	Values of C	
	Skin Effect	Proximity Effect
Round compacted or stranded	1	0.8
Round or sector shaped	1	0.8
Round or segmental		
No. segments	0.435	0.370
6 segments	0.390	–

The values of C are shown in Table 8.2.

The following formulae are thus applied to calculate the increase in AC resistance of the conductor due to the skin and proximity effect and can be accepted for all practical purposes when the value of X remains within $X \leq 2.8$.

The skin effect, Y_s, is given by

$$Y_s = \frac{X_s^4}{192 + 0.8X_s^4} \tag{8.11}$$

when

$$X_s = 15.9 \times 10^{-4} \sqrt{\left(\frac{fC_s}{R_{DC}}\right)} \tag{8.12}$$

For the proximity effect, the factor Y_p is given by

$$Y_p = \frac{X_p^4}{192 + 0.8X_p^4}\left(\frac{d_c}{S}\right)^2\left[0.312\left(\frac{d_c}{S}\right)^2 + \frac{1.18}{\left(\frac{X_p^4}{192 + 0.8X_p^4}\right) + 0.27}\right] \tag{8.13}$$

and

$$X_p = 15.9 \times 10^{-4} \sqrt{\left(\frac{fC_p}{R_{DC}}\right)} \tag{8.14}$$

where

d_c is the diameter of the conductor

S is the spacing between the conductor axis (m)

TABLE 8.3

Resistivity and Temperature Coefficient of Resistivity of Metals

Metal	Resistivity, ρ (Ω m) at 20°C	Temperature Coefficient, α(°C) at 20°C
Conductors		
Copper	1.7241×10^{-8}	3.93×10^{-3}
Aluminium	2.8264×10^{-8}	4.03×10^{-3}
Sheath and armour		
Lead and its alloy	21.4×10^{-8}	4.00×10^{-3}
Aluminium	2.84×10^{-8}	4.03×10^{-3}
Steel	13.8×10^{-8}	4.50×10^{-3}
Bronze	3.5×10^{-8}	3.00×10^{-3}
Stainless steel	70.0×10^{-8}	Negligible

Values of C_p and C_s are given in Table 8.2. In case of a multicore sector-shaped conductor, the value of Y_p should be multiplied by 2/3 of the value obtained after calculating the proximity effect. These values can also be calculated by referring to IEC specification 60287.

The values of the resistivity and temperature coefficient of metals are given in Table 8.3.

By compacting the conductor, a slight change in the proximity effect can be achieved. It is a common practice to select or design a conductor by accepting the conductivity of the metal. This rationalises the consumption of the metal quantum on weight basis. But selecting conductors always on the basis of conductivity to reduce weight may lead to a bad effect. Too much restriction or close tolerance on weight will reduce the volume of the metal. This will raise the temperature by hindering the free movement of electrons in a dense, clustered form. Further, due to the reduction in the area of circumference, dissipation of heat around the conductor will be slower. On the other hand, a lower diameter of the conductor shows a lower capacitance value (when the ratio of *D*/*d* is greater), resulting in a lower dielectric loss. Hence, designing of conductors for processing must be done considering all the effects in order to obtain the best results, which is the most important but difficult task while designing a power cable (Figure 8.1).

Basically, conductors are classified into two types:

1. *For fixed installation*: In this case, the conductor is made by stranding to make it sufficiently flexible to be wound and rewound during processing, and bending, as required, during installation and fixing. These cables are laid underground, in air on supporting walls or structures, or in ducts. The nature of insulation is also selected as per installation conditions. The conductor can be made of either copper or aluminium.

FIGURE 8.1
A low-voltage cable. Stranded and compacted copper conductor, polyvinyl chloride (PVC) or cross-linked polythene (XLPE) insulation, 4-core laid-up, non-hygroscopic polypropylene fibre filler, binder polyester tape, PVC inner sheath, galvanised steel wire armour and black PVC or PE outer sheath.

2. *Flexible cables*: They are made to allow free movement of cables with repeated turns and twists and to accept strenuous working conditions. In this case, the conductor is made of fine copper wires bunched and stranded together to impart high flexibility. These cables are used as miner's working equipment, such as coal face cutting draglines, heavy-duty cranes, EOT cranes, earthmoving equipment, ship building and wiring, welding and machine tools and various other fields where flexibility is a must. Flexible cables are of numerous types. In these cases, the characteristics of insulation should also be such that the material(s) can withstand severe stress and strain during operation.

8.1 Insulation and Insulating Materials

The flow of current along the axis of the conductor produces a magnetic field around it, exerting pressure at a right angle to the axis of the conductor. This tube of force is known as voltage. The material has to withstand this magnetic pressure, that is, voltage, by protecting the conductor from coming in contact with the surrounding media. This is termed as insulation. The current, as it flows through the conductor, overcomes various components of resistance and gets weakened. The lost component transforms into heat (law of conservation of energy). As parts of the current are lost progressively on the way, components of the magnetic field also become lesser, weakening the force of pressure along the known path. This is known as the voltage drop. The quantum of loss may not be appreciable in the case of a low-voltage system, but it becomes quite high as the transmission force is raised to medium voltage (MV) high voltage (HV) and extra high voltage (EHV) systems progressively. In these cases, though the conductor remains the same, the voltage, that is, pressure, is increased (magnetic field). This is done by raising the density of the electron at a high proportion per unit area. The more is the voltage, the higher is the density level of the electron and the more is the increase of force. This can be perceived by the corona discharge around a conductor when exposed to the atmosphere, as the electrons try to get free and the current density increases substantially.

The initial loss is the heat loss known as I^2R, where I is the current (in Amperes) and R is the resistance, as defined earlier. Some amount of the generated heat is absorbed by the insulating media, and a substantial amount is set free to the surrounding media. Naturally, conductivity and the specific heat of the insulation play an important role. The conductivity of the insulating material should ideally be zero so that no current is allowed to pass through the insulation. But, in fact, none of the material on earth can be an ideal insulating substance, the reason being that every material is composed of a different conglomeration of atoms and molecules. They are bonded together to form compounds of various forms. Some of them have a loose bonding, while some have a very close composition. But in every case the electrons vibrate within the lattice space. Naturally, some amount of the remaining electrical charges cannot be detected. These electrons do not come out of their position when the pressure of the voltage is applied at certain thicknesses but absorb a part of the energy from the injected power, which increases the amplitude of the oscillating electrons within the molecular space. This absorbed power is a loss to the transmission system. Further, heat generated within the conductor also accelerates the vibration of electrons by softening the material and increasing power loss. That is the reason why measurement of losses are made at different temperatures to ascertain the stability of the insulating material.

The increased vibration of electrons is affected by the absorption of a certain amount of power (charge) which it retains in a unit area at a given temperature and is the measure of the dielectric constant or relative permittivity. This is done in comparison with the permittivity of free space (charge retained in a unit area of free space, $\varepsilon_o = 8.85 \times 10^{-12}$). The capacity of retaining the electrical charge in the insulation depends on the thickness of the insulation and the diameter of the conductor and its length. The capacitance of a single-core cable, thus, depends on

1. The dimensions of the cable, such as its length, radius of the conductor r_o and radius over the insulation R
2. Relative permittivity (dielectric constant), ε_r, of the homogenous insulation material

Thus, the measure of capacitance is given by

$$C = \frac{2\pi\epsilon_0\epsilon_r}{\ln(R/r_0)}\ \mu\text{F/km} \tag{8.15}$$

On simplification, it gives

$$C = \frac{\epsilon_r}{18\ln(R/r_0)}\ \mu\text{F/km} \tag{8.16}$$

For round three-core belted cables,

Star capacitance can be taken as $C_1 = 1.2C$ (conductor to neutral)
Conductor to conductor $C_2 = 0.6C$ (other conductors are free floating)
All conductors bunched to sheath $C_3 = 1.8C$

The relative permittivity of materials is shown in Table 8.4.

This capacity of retaining the charge per unit volume increases as the length of the cable increases, that is, the volume of insulation increases proportionately. This charge is absorbed from the flowing current, and this absorbed current is called the charging current (current absorbed by the insulation, when the cable is charged with power, i.e. current and voltage) and is given by

$$I_c = V_o\ \omega C\ \text{A/m}$$
$$\text{or}\quad I_c = V_o\ \omega C \times 10^3\ \text{A/km} \tag{8.17}$$

where
$\omega = 2\pi f$ rad/s
f is the frequency
V_o is the operating voltage

It can be seen that as the length of the cable increases, the absorption of power also increases within the insulation, which means that the value of C increases as does the charging current. As the length goes on increasing, a situation may arise when the total input current will be utilised

TABLE 8.4

Permittivity of Insulating Material

Material	Permittivity ε_r
Impregnated paper	3.4–4.3
Polythene (PE)	2.4–3.0
Cross-linked polythene (XLPE)	2.4–3.0
PVC compounds	4.0–8.0

as a charging current. This length will gradually become shorter as the voltage is increased.

For single-core HV and EHV cables, the value of C can be written as

$$C = \left\{ \frac{\epsilon_r}{2\ln(R/r_0)} \right\} \times \frac{1}{9} \times 1.1\ \mu\text{F/km} \tag{8.18}$$

Here, 1.1 is taken as an unknown factor.

It is to be noted that the value of 'ε_r' changes with the change in temperature. Hence, the value of the capacitance changes with temperature as well as with the related parameters.

Example 8.1

Let us consider a conductor area to be 400 mm² of stranded aluminium.

Voltage = 11,000 V
Diameter of conductor compacted = 24.80 mm
Thickness of the semiconducting layer = 0.60 mm
Thickness of XLPE insulation = 3.60 mm
Thickness of outer semicon = 0.60 mm
A metallic screen applied on the outer semicon
The actual dielectric constant taken to be 2.60
Capacitance as calculated 0.5856 μF/km
Charging current as calculated 2.02 A/km

Current rating in the ground at 15°C (BSS) as 535 A (trefoil formation)

The length at which the charging current will be equal to the current rating is given by 535/2.02 = 264.85 km. After this length, no current will flow through the cable. If voltage is increased further, the corresponding charging current will be higher.

At 275 kV, the charging current becomes 9.3 A/km and increases to 16.9 A for approx. 500 kV. Thus, it shows that above a 55 km length, no current will flow within a 275 kV cable. In order to keep the power line working under such conditions, shunt reactors are installed at given intervals, though it becomes a costly proposition. In case of a DC system, charging is a one-time affair during switching on or off. In this case, there is no length limitation. Hence, for long-distance EHV submarine cables and overhead lines, the HVDC system is being increasingly considered.

In case of a three-core belted cable, it is difficult to always calculate the capacitance, but the calculation shown earlier can be accepted on an approximation basis. However, for practical purposes, they are determined by actual test results, since the thickness of insulation is not uniform throughout the length of all conductors. The diameter of conductors may also vary somewhat. Laying up variations also affect the result. As per Simons' formula, a calculation can be made with fair accuracy, provided the dielectric constant of the material is found to be correct. This is very difficult in the case of polymeric insulating materials, where the dielectric constant varies considerably and with temperature as well. Simons' equation is shown in the following:

$$C_0 = \frac{0.02983 \in_r}{\log_{10}\left[1+\frac{t+t_1}{2r_0}\left(3.84-1.70\frac{t_1}{t}+0.52\frac{t_1^2}{t^2}\right)\right]} \ \mu\text{F/km} \tag{8.19}$$

where

t is the insulation thickness
t_1 is the thickness of the belt or the inner sheath

Insulation, though it helps retain charge, does not allow a free flow of current, exerting a resistance. This is known as insulation resistance and is given by

$$R_{ins} = \frac{\rho}{2\pi l}\ln\frac{R}{r_0} \tag{8.20}$$

where

ρ is the electrical resistivity of insulation
l is the length of the cable in metres
R is the outer radius of the conductor on insulation
r_o is the radius of the conductor

It may be considered that there exists a relation between relative permittivity ε_r and volume resistivity $\rho_v = 1/\gamma$, which actually determines the characteristics of the insulating media, where 'γ' is the electrical conductivity of insulation. This can be compared with the capacitor of any shape and in SI unit gives $C = \varepsilon_o\, \varepsilon_{r\lambda} \ldots F$.

Thus, an equation can be written for any shape λ of an electrode and dielectric for a capacitor

$$R = \frac{\rho_v}{\lambda} = \frac{1}{\gamma\lambda}$$

Combining the equations, $CR = \varepsilon_o\varepsilon_r\rho_v$.

Thus, the product of capacitance and volume resistivity does not depend on the geometrical form when insulation is homogeneous but depends exclusively on the quality of insulation.

The product of CR is used to determine the time constant of the dissipation of charge in the insulation, considering that the cable, as an elongated capacitor, is tested or charged with voltage and then switched off

$$V = V_o \exp\left(\frac{-t}{CR}\right) \tag{8.21}$$

TABLE 8.5
Values for Impulse Test Voltage as per IEC Recommendation

Rated System Voltage, rms Values (kV)	Max. Service Line Voltage rms Values (kV)	V Test (kV_p)		
		Minimum[a]	Normal	Reduced
33	36	154	170	–
110	123	420	550	450
132	145	500	630	550
150	170	560	750	650
220	245	820	1050	900
380	420	1360	1550	1425

[a] Insulation thickness is determined by mechanical consideration.

By drawing a curve of *CR* against V_o, the self-discharge time can be found. The longer the time, the better is the insulation.

When computing the insulation level for power cables, the following two distinct features are to be considered:

1. AC voltage
2. Impulse voltage

The cable is subjected to a long-time working voltage. Environmental stress and stress related to connected equipments, and at times voltage, are to be considered. These are the characteristics of AC voltage. Apart from these, surge voltages are also experienced due to a lightning effect. As per the IEC norm, the impulse strength of insulation must not be lower than

$$V_i = (6V_o + 40)$$

where V_o is the operating phase voltage (between the conductor and earth).

This voltage is considered after taking into account various other factors, particularly temperature conditions. Table 8.5 gives the level of the impulse voltage as given in the IEC recommendation.

8.1.1 Thickness of Insulation: Consideration

It is seen that

$$E_{max} = \frac{V_0}{r_0 \ln(R/r_0)} \tag{8.22}$$

When stress is minimum, then

$$r_0 \ln\left(\frac{R}{r_0}\right) \text{is maximum and} \frac{d}{dr_0} r_0 \ln\left(\frac{R}{r_0}\right) = 0; \text{it means} \ln\left(\frac{R}{r_0}\right) = 1$$

Or $R/r_0 = e = 2.718$ or $R - r_0 = 2.718$; now, $R - r_0 =$ is the thickness of insulation. And the stress on the conductor surface is

$$E_o = \frac{V_0}{r_0} \tag{8.23}$$

This relation is called '*e*' ratio for insulation thickness. This calculation is greatly adopted for paper-insulated cables. For high-voltage cables, this ratio does not bring insulation thickness to a practical dimension as it is too large.

For thermoplastic- and thermosetting-insulated cables, particularly for XLPE insulation, the formula adopted on the basis of a basic impulse level to decide the thickness of insulation is given by

$$t = \frac{BIL \times K_1 \times K_2 \times K_3}{E_{l(imp)}} \tag{8.24}$$

where

t is the thickness of insulation (mm)
BIL is the basic impulse level (as shown in Table 8.5)
$K_1 = 1.25$ is the thermal coefficient
$K_2 = 1.1$ is the ageing coefficient
$K_3 = 1.1$ is the coefficient of the unknown factor
$E_{l(imp)} = 50$ kV/mm

The value obtained from the Weibull plot distribution of impulse breakdown voltage is considered up to the 66–220 kV level. For 275 kV cables and above, the impulse level is taken to be 65 kV/mm.

These voltages vary from compound to compound, though for the sake of brevity, one particular value has to be fixed taking into consideration all factors related to the material and processing technique.

The second consideration to calculate the thickness is by accepting the AC breakdown voltage E_{AC} (also as per the Weibull distribution system on AC breakdown level)

$$t = \frac{\frac{E_0}{\sqrt{3}} \times K_1 \times K_2 \times K_3}{E_{AC}} \tag{8.25}$$

where

t is the insulation thickness
K_1 is the thermal coefficient = 1.1
K_2 is the ageing coefficient = 4.0
K_3 is the unknown factor coefficient = 1.1
E_o is the maximum circuit voltage (kV/mm)
E is the operating voltage

Then,

$$E_0 = \frac{E \times 1.15}{1.1}$$

Say for 33 kV

$$E_0 = \frac{33 \times 1.15}{1.1} = 34.5 \text{ kV}$$

The value of E_{AC} is taken as 20 kV/mm for rated voltage of up to 220 kV.
Above 275 kV, as per Weibull plot of minimum stress distribution, E_{AC} is taken as 30 kV/mm.

Example 8.2

When the thickness of insulation is calculated for 33, 66 and 110 kV, we get the insulation thickness as (taking the E_{AC} value to be 20 kV/mm)
For 33 kV as 4.83 mm, for 66 kV as 9.7 mm and for 110 kV as 16.1 mm
When calculated as per BIL parameters, we get the insulation level as (taking BIL as 50 kV/mm)
For 33 kV as 8.32 mm, for 66 kV as 10.60 mm and for 110 kV as 16.64 mm
In this case, whichever thickness is higher has to be taken and rounded off

For	33 kV	as	8.5 mm (IS as 8.80 mm)
For	66 kV	as	11.0 mm
For	110 kV	as	17.0 mm

Improving upon quality of the raw material, processing technique and environmental conditions, the E_L value can be increased. To do this, much needs to be done on the working front.

A transmission line or cable, as designed for a particular application, has to carry a given amount of power – combination of current and voltage. The voltage is the pressure raised to push a given amount of current through a metal conductor. To contain the pressure, a precalculated insulation thickness is applied to accept radial and longitudinal stresses. To allow a specific amount of current to flow, a metal conductor having a definite cross-sectional area is chosen. In fact, the cable is constructed with different layers of material applied in the form of concentric cylindrical shells. Each layer of material has its characteristic property and absorbs a certain amount of power, which appears in the form of heat. After a time, the absorbed heat reaches its saturation point. A temperature equilibrium is established within the cable by dissipating the excess amount of heat into the surrounding areas which may accumulate in the cable during uninterrupted operation. This means a permissible rise in temperature within the conductor is allowed for long-term performance. To achieve this, it is essential to know the characteristics of all the materials used when constructing a cable. Simultaneously, the nature of the surrounding areas has to be taken into consideration. All these factors would ultimately determine the amount of current which can be transmitted through the conductor of a cable at a given voltage:

1. The conductor in the form of a metal exerts a resistance (R_{20}) against the flow of current. With the longitudinal flow, it exerts an increasing radial pressure along with the increase in length. The rise in temperature within and in the surrounding area increases the resistance value further and needs to be considered (R_t) – the initial value has been fixed at 20°C. As the area of the conductor increases, the frequency of the transmission line starts affecting the resistance values and appears as a skin effect because of self-inductance – showing a virtual rise in the resistance value (R_{AC}). If two or more conductors are placed nearby, further rise in resistance is expected due to the magnetic field influencing each other – the effect of mutual inductance is known as the proximity effect. But to get to the initial resistance values, the conductivity of the metal must be known along with its temperature coefficient.

   Even for a small amount of increase in resistance value, applied voltage shall increase though in small (I^2R) quantum and exert a pressure in radial as well as longitudinal direction along the length of the conductor, leading to some amount of loss of transmitted power in the form of heat. Here, the thermal conductivity of metal plays a significant part to ascertain the temperature rise within.

   It is discussed that an increase in resistance of the conducting metal in various forms absorbs a certain amount of power evolving as heat, as seen with the rise in temperature, so other materials used as building blocks for a completed cable bring losses within different layers, either by obstructing dissipation of heat (thermal resistivity) or by absorbing a small amount of power through magnetic coupling (inductive influence).

2. Conductors transmitting power under a given voltage need to be kept separated from each other by an isolating media called insulation. The material for insulation as described earlier should ideally not absorb any power from the transmission line and should withstand operating voltage for a longer period without any deterioration. This being an ideal condition, selection is made considering the lowest power absorption quality, that is, the capacitance value which depends on the dielectric constant as enumerated earlier and also on the thickness of insulation.

3. The electrical strength of the insulating material is measured by considering the breakdown voltage withstanding capacity per mm basis. This is to be 15 kV/mm for 1100 V cable; for higher voltages

such as 11 and 33 kV, it should be 20 kV/mm, and for cables up to 220 kV, it should be 55 kV/mm. In case of medium- and high-voltage systems, the impulse breakdown voltage is also taken into consideration as 60 and 65 kV/mm, respectively, depending on the quality of the material. Accordingly, thicknesses are being computed. It is interesting to note that the better the insulating property, the higher the thermal resistivity. This means that heat generated within the system is retained for a longer period and dissipation is slow. Naturally, for insulation, apart from a better electrical property, lower thermal resistivity becomes a criterion for the selection of insulating media. Thermal resistivity as such changes with the rise in temperature, as well as the dielectric constant vis-à-vis capacitance value. With time, sustained temperature may affect the quality of insulation, that is, ageing property. All these factors play an important part when qualifying an insulating material.

4. For medium- and high-voltage cables, a layer of semiconducting material is used over the stranded conductor and on the insulation surface to distribute the electrical field radially because distortion of lines of force leads to uneven rise in voltage on the spiral ridges of the strand. This may create fault conditions. The thermal resistivity of these layers, though, is not of much significance but cannot be ignored totally.
5. The insulated conductor is screened by applying a closely bonded non-metallic semiconductive layer on which a copper tape wrapping or a copper/aluminium wire screen is applied to minimise stress and to accept a certain amount of fault current. An amount of voltage and current is induced within this metallic screen due to the magnetic flux generated by the flowing current within the conductor. Some amount of power is lost here. This loss is considered as the screen effect. The screen is also required to divert short-circuit current for safe performance of the cable.

   Short-circuit current for the conductor/screen can be calculated as follows (for HV and EHV cables):

$$I_{sc} = \sqrt{\left[\left\{\frac{(Q \cdot S)}{(KRt)}\right\} \ln\left\{\frac{\left(\frac{1}{K} - 20 + T_1\right)}{\left(\frac{1}{K} - 20 + T_2\right)}\right\}\right]} \text{ A} \tag{8.26}$$

   where

   I_{sc} is the maximum allowable short-circuit current (Amperes)
   Q is the specific thermal capacity of copper/aluminium (J/°C cm^3)
   S is the conductor/screen cross-sectional area (cm^2)
   K is the temperature coefficient of copper resistance (0.00393) for Al (0.004)

R is the conductor or screen resistance (copper/aluminium) at 20°C (Ω/cm)
T_1 is the maximum allowable conductor/screen temperature (250°C)
T_2 is the conductor/screen temperature before a short circuit (conductor 90°C; screen 75°C)
t is the short-circuit duration (s)

For copper screen,

$$I_{sc} = \frac{150 \times S_s}{\sqrt{t}}$$

where S_s is the screen area (mm²).

The short-circuit capacity of the metallic screen can be increased or decreased by changing the construction of the screen.

Heat develops during a short-circuit condition and is computed as follows for per metre length of the conductor:

$$\text{Heat} = \frac{\left(I_{sc}^2 t \rho_0\right)}{a}\left[1+\alpha\left\{\left(\theta_{\max}-\theta_a\right)+\frac{\theta_{sc}}{2}\right\}\right] \text{J/m} \tag{8.27}$$

where
I_{sc} is the short-circuit current (Amperes) (rms)
t is the duration of the short-circuit current (s)
a is the cross-sectional area of the conductor (mm²)
ρ_0 is the resistivity of copper/aluminium conductor at 20°C, say for copper 0.017241 and 0.028264 Ω mm²/m
α is the temperature coefficient of copper/aluminium at 20°C, for copper it is 0.00393/°C and for aluminium, it is 0.00403/°C
θ_a is the ambient temperature, say 20°C
$\theta_{\max}$ is the maximum allowable conductor temperature for XLPE insulator, 90°C
θ_{sc} is the rise in temperature due to the short-circuit condition from the initial allowable conductor temperature during operation, 250°C − 90°C = 160°C

Here, in the expression of heat, temperature $\theta_{sc}/2$ rise is linear; hence, a mean value has to be accepted. The same amount of heat is absorbed by the conductor of 1 m length and is given by

$$H' = aS_h D\theta_{sc} \times 10^{-6} \text{ J/m} \tag{8.28}$$

where
S_h is specific heat of copper, 385.2 J/kg °C
D is the density of copper 9800 kg/m³, whereas θ_{sc} = 250° − 90° = 160°C

For XLPE short circuit, the allowable temperature rise is 250°C and the conductor's allowable temperature rise during operational condition is 90°C.

On equating Equations 8.27 and 8.28, the following relationship can be obtained for a copper conductor:

$$I_{sccu} = \frac{148.5a}{\sqrt{t}} \tag{8.29}$$

Similarly, for an aluminium conductor, a short-circuit relationship can be obtained considering the following values:

ρ_0 is the resistivity of the aluminium conductor at 20°C = 0.028264 Ω mm²/m

α is the temperature coefficient at 20°C, which is 0.00403°C

S_h is the specific heat of aluminium = 920 J/kg °C

D is the density of aluminium = 2700 kg/m³

$$\theta_{max} = 90°C; \quad \theta_a = 20°C;$$

Short-circuit temperature allowable for XLPE = 250°C

$$I_{scAl} = \frac{93.6a}{\sqrt{t}} \tag{8.30}$$

The short-circuit rating for the same conductor area under the same conditions is approximately 63% lesser than a copper conductor.

To calculate the short-circuit rating at different parameters and also for different metals, it would be necessary to find out first the numerical constant of the equation as

$$I_{sc} = \frac{Ka}{\sqrt{t}}$$

The factor 'K' needs to be determined after equating Equations 8.27 and 8.28:

$$K = \sqrt{\left[\frac{S_h D \theta_{sc} \times 10^{-6}}{\rho_0 \left\{1 + \alpha\left(\theta_{max} - \theta_a + \frac{\theta_{sc}}{2}\right)\right\}}\right]} \tag{8.31}$$

6. The *inner sheath* and *outer sheath* of a synthetic polymer resist the flow of heat to a certain extent depending on its characteristics called thermal resistivity. This actually affects the rise in

conductor temperature to a certain extent vis-à-vis the affected current-carrying capacity.

7. Mechanical protection is given by steel wires or strips for a magnetic material. For a balanced three-phase system, the current flowing through the conductors rotates at 120° apart, and most of the magnetic flux thus get cancelled. But a certain amount of residual flux still acts upon the armour material. This effect is called armour loss and needs to be considered. Single-core cables are acted upon by 100% magnetic flux density causing a considerable loss. These cables are armoured with nonmagnetic material like aluminium or stainless steel wires or strips.
8. The surrounding media also play an important role. When the cable is buried underground, soil, as a thermal barrier, is considered exerting resistance to the flow of heat, the same with air when the cable is laid overhead. Thus, the thermal resistivity of soil and air, under various conditions, needs to be taken into account.

Accepting all these variations, the calculation of the current rating can be brought into a mathematical form, converting them to equivalent resistance circuits in parallel and in series.

A conductor has to carry a certain amount of current at a given voltage. The amount of current carried by the conductor is determined after taking into account the total amount of power losses within the cable system and its surroundings. When a current passes through a conductor, it generates heat due to the resistance offered by the metal. This heat has to be dissipated into the surrounding areas through different layers of construction materials, such as insulation, bedding, metal sheath, armour and an outer serving. At each stage, a certain amount of heat is absorbed untill it reaches the surrounding areas. A certain amount of heat is also given out into the surroundings untill equilibrium is reached. This can be represented by a series of resistance analogous to Ohm's law (Figure 8.2). The permissible rise in temperature within the cable is determined considering the long-term heat-enduring capacity without deformation due to an ageing process.

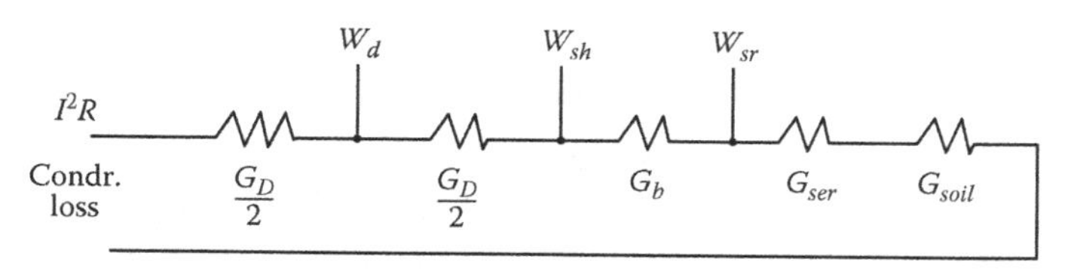

FIGURE 8.2
Equivalent thermal circuit of a single-core cable buried underground.

Accordingly, temperature rise has been fixed for a particular dielectric as indicated in the following:

Nature of Insulation	Permissible Rise in Operating Temp. °C
Poly vinyl chloride (PVC)	70
Polyethylene	70
Butyl rubber	80
Ethylene propylene rubber	85
Cross-linked polythene	90

In calculating the continuous current rating of a cable installed underground, the temperature rise of the cable conductor has to be taken into account at the rated current. The cable is constructed with different layers of materials in concentric form. Heat flowing through non-metallic protective layers experiences a certain amount of thermal resistance. Further, a metallic sheath and armour absorb a certain amount of power in the form of induced voltage under the influence of a magnetic field emanating from the conductor current. When the sheath is earthed at one or both ends, the current flows through generating a thermal condition proportional to the resistance and length of the sheath and armour. Temperature rise in all these elements is to be deducted from the conductor temperature while computing the continuous current rating of the conductor.

The losses occur through

1. Conductor losses certain amount of energy as heat due to DC resistance—an inherent property of the conductor material
2. With the rise in ambient temperature conductor resistance increases as per thermal coefficient of material. Due to this increase in resistance, some amount of energy is lost further in the form heat
3. Skin effect and proximity effect
4. Dielectric losses within insulation
5. Screening losses for MV and HV cables
6. Losses through bedding
7. Metallic sheath losses (due to circulating current and eddy current)
8. Armour losses
9. Outer sheath losses
10. Heating of soil due to heat dissipation from the cable
11. Dissipation of heat in air
12. Other special installation conditions

Heat generated within the conductor is given by I^2R. This amount of heat raises the temperature of conductor to θ_c degree. This temperature has to be reduced by the amount of rise in ambient temperatures θ_a. Further, there is a rise in temperature due to the dielectric loss factor within the insulation, which is further compounded by the rise in temperature within surrounding materials.

This total rise in temperature θ_d is also taken away from the rise in conductor temperature. These temperatures indicate that at every stage, heat is generated because of thermal resistances offered by bedding and outer sheath materials. Also, the circulating current induced within the metallic screen, sheath and armour contributed to the loss of power within the conductors. The final rise in temperature, thus, is given by the following formula:

$$\Delta\theta = \left(I^2R + ½W_d\right)T_1 + \left\{I^2R(1+\lambda_1) + W_d\right\}nT_2 + \{I^2(1+\lambda_1+\lambda_2) + W_d\} \times n(T_3 + T_4) \tag{8.32}$$

Taking this as a quadratic equation and solving for a positive route of current '*I*',

$$I = \sqrt{\frac{\theta_c - \theta_a - \theta_d}{R_{DC}(1+Y_s+Y_p)\{G_d + nG_b(1+\lambda_{sc}) + n(G_{ser} + G_{soil})(\lambda_{sc} + \lambda_{ar})\}}} \tag{8.33}$$

where

θ_c is the conductor's operating temperature
θ_a is the ambient temperature (20°C)
θ_d is the temperature in the dielectric (insulation)
$\Delta\theta$ is the temperature rise in the conductor (K) ($\theta_c - \theta_a$)
I is the current flowing through the conductor (Amperes)
R_{DC} is the DC resistance of the conductor at a maximum operating temperature (θ_{max}) per unit length (Ω/m):

$$R_{DC} = \frac{\rho_0}{A}\left\{1 + \alpha_{20}\left(\theta_{max} - 20°C\right)\right\}(1+s)\ \Omega/\text{m}$$

ρ_0 is the resistivity of the conductor material
A is the conductor area
α_{20} is the coefficient of temperature at 20°C
s is the stranding factor
W_d is the dielectric loss in insulation surrounding the conductor per unit length (W/m)
T_1 is the (G_d) thermal resistivity per unit length of insulation (km/W)
T_2 is the (G_b) thermal resistivity of bedding between the sheath and armour (km/W)
T_3 is the (G_{ser}) thermal resistivity of the outer sheath (serving) (km/W)
T_4 is the (G_{soil}) thermal resistivity of soil of the surrounding medium (km/W)
n is the number of conductors having the same cross-sectional area, each carrying an equal amount of current
λ_1 is the ratio of losses in the form of heat within the screen or metal sheath to that of all conductors in the cable (λ_{sc})
λ_2 is the ratio of losses in the form of heat within the armour to that of all conductors in the cable (λ_{ar})
Y_s is the skin effect (Equation 8.11)
Y_p is the proximity effect (Equation 8.13)

This can be rewritten as

$$I = \sqrt{\left[\frac{\theta_c - \theta_a - W_d\{\tfrac{1}{2}T_1 + n(T_2 + T_3 + T_4)\}}{R_{AC}T_1 + nR_{AC}(1+\lambda_1)T_2 + nR_{AC}(1+\lambda_1+\lambda_2)(T_3+T_4)}\right]} \text{ amp} \quad (8.34)$$

Here R_{AC}, being the AC resistance, is given as

$$R_{AC} = R_{DC}(1 + Y_s + Y_p)$$

where

R_{DC} is the direct current resistance of the conductor = $R_{20}\{1 + \alpha(\theta_{max} - 20)\}$
Y_s is the factor to be applied due to the skin effect (see Equation 8.11)
Y_p is the factor to be applied for the proximity effect (see Equation 8.13)

In the case of HV and EHV cables, the thickness of insulation being large and the depth of laying being deeper, where the number of cables in the vicinity also affects the temperature condition, the dielectric loss would be appreciable. Calculation of the current rating, thus, should be computed accordingly. This is considered also to find out the sheath and armour losses which will affect the final current-carrying capacity of the conductor.

While calculating current ratings, various other factors may also have to be taken into consideration as per surrounding and installation conditions, such as grouping factors and solar radiation, to be taken from IEC 60287.

Different loss factors are to be accounted for in the previous calculation: W_d is the dielectric loss in each phase given by

1. *Dielectric loss factor* (also refer to Equation 5.28)

$$W_d = 2\pi f C V_o^2 \tan\delta$$

and

$$C = \left\{\frac{\varepsilon}{18\ln(d_i/d_c)}\right\} \times 10^{-9} \text{ F/m} \quad (8.35)$$

where

f is the frequency of supply
V_o is the voltage between the phase and earth
$\tan\delta$ is the dielectric loss angle
C is the capacitance (F/m)
d_i is the diameter over the insulated core excluding the screen
d_c is the diameter over the screened conductor
ε is the permittivity of the insulating material (shown in Table 8.6)

TABLE 8.6

Permittivity and Loss Angle of Insulating Materials

Material	Permittivity	tan δ
PVC	4–8	0.08
Polythene	2.3	0.001
XLPE	2.5	<0.0003

The values of thermal resistivity of materials at different stages are calculated as given but have to collaborate with the IEC 60287 specification. Here, the parameters of a single-core cable have been considered to illustrate how the calculation for the current rating is evaluated in its basic form:

2. *T_1 for*
   a. *A single-core cable*

$$T_1 = \frac{\rho_t}{2\pi}\ln\left(1+\frac{2t_1}{d_c}\right) \tag{8.36}$$

ρ_t is the thermal resistivity of the insulating material (or material as considered) (km/W)

t_1 is the thickness of insulation (mm)

d_c is the diameter of the conductor (including the screen where applicable) (mm)

   b. *Multicore screened belted cable*

$$T_1 = \frac{\rho_t}{2\pi} G \times \text{screening factor} \tag{8.37}$$

G is the geometrical factor

The screening factor is calculated as follows:

$$\lambda_{sc} = \frac{\text{Power loss due to screen}}{\text{Power loss in the conductor}} = \frac{W_{sc}}{W_C} = \frac{R_{sh} \times I_{sc}^2}{R_c \times I_c^2}$$

where

R_{sh} is the resistance of the screen

I_{sc}^2 is the current induced in the screen

R_c is the resistance of the conductor

I_c^2 is the conductor current

Voltage induced within the screen

$$E_{sh} = I_c \omega M = I_c \omega 2\times10^{-7} \times \ln\left(\frac{S}{r_{sc}}\right) \text{V/m} \tag{8.38}$$

Here, $\omega = 2\pi f$ where f is the frequency

S is the distance between conductor axes, or the diameter of the conductor

r_{sh} is the mean radius of the screened conductor

The screen current is calculated as follows:

$$I_{sc} = \frac{I\omega M}{\sqrt{\left(R_{sh}^2 + \omega^2 M^2\right)}} \quad (8.39)$$

here, $\sqrt{\left(R_{sh}^2 + \omega^2 M^2\right)}$ is the impedance of the screen circuit and $M = 2 \times 10^{-7}\ln(S/r_{sh})$ H/m for a single-core cable, and between two conductors, it is approx. $M = 4 \times 10^{-7}\ln(S/r_{sh})$ H/m

3. *Thermal resistance between sheath and armour: single- and multicore cable of bedding*

$$T_2 = \frac{\rho_t}{2\pi}\ln\left(1 + \frac{2t_2}{d_s}\right) \quad (8.40)$$

ρ_t is the thermal resistance of bedding
t_2 is the thickness of bedding (mm)
d_s is the diameter over bedding (mm)

4. *Power loss due to metallic sheath.* The AC current 'I' flowing through a circuit induces a voltage in the cable sheath. This voltage depends on the amount of magnetic flux interlinking the sheath and in turn depends on the inter-axial spacing of the sheath 'S'.

   Here, the induced sheath voltage is given by $V_{sh} = IX$, where X is the mutual reactance between the conductor and sheath per metre and is given by $X = \omega M$ Ω/m. Here, 'M' is the mutual inductance per metre between the conductor and sheath (Figure 8.3). This can be shown by evaluating the following calculations:

   The magnetic field at a distance 'r' from the centre of the cable (1) in space is $H = I/2\pi r$ and the flux density at any point will be

$$B = \mu_0 \mu_r H = \frac{\mu_0 \mu_r I}{2\pi r}$$

Hence, at the distance r, the density will be $d_\phi = Bd_r = \mu_0\mu_r I d_r / 2\pi r$

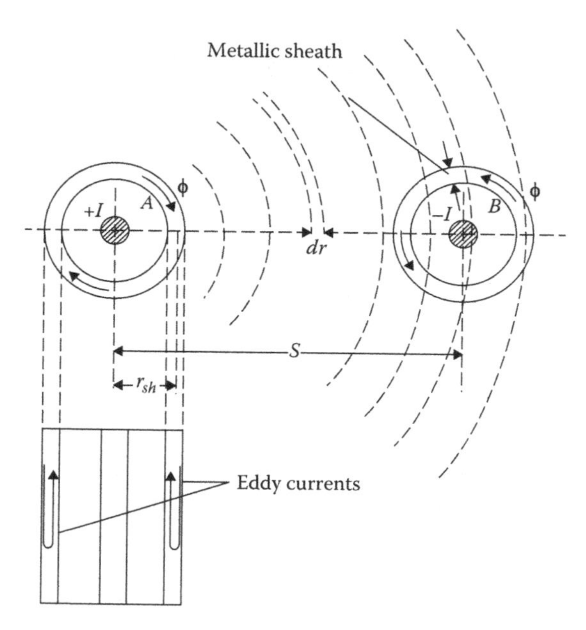

FIGURE 8.3
Loss in the cables due to metallic sheath.

Considering the direction of current in the positive direction +*I*, the flux density enclosed between two cable sheaths should be

$$\varnothing_1 = \int_{r_{sh}}^{S-r_{sh}} \frac{\mu_r \mu_0 I d_r}{2\pi r} = \frac{\mu_r \mu_0 I}{2\pi} \ln\left(\frac{S - r_{sh}}{r_{sh}}\right) \tag{8.41}$$

Here, as given earlier, '*S*' is the distance between the axes of the centre of conductors and r_{sh} is the mean diameter on the cable sheath. Now, considering the negative direction of the current, the flux density is given by

$$\varnothing_2 = \frac{\mu_r \mu_0 I}{2\pi} \ln\left(\frac{S - r_{sh}}{r_{sh}}\right) \tag{8.42}$$

Now, the total flux between the cable sheaths is

$$(\varnothing_1 + \varnothing_2) = \frac{\mu_r \mu_0 I}{\pi} \ln\left(\frac{S - r_{sh}}{r_{sh}}\right) \text{ Wb/m} \tag{8.43}$$

The mutual inductance per metre is obtained by dividing the Equation 8.43 by current '*I*'

$$M = \frac{\varnothing_1 + \varnothing_2}{I} = \frac{\mu_1 \mu_0}{2\pi} \ln\left(\frac{S - r_{sh}}{r_{sh}}\right) \text{ H/m} \tag{8.44}$$

Taking the value of $\mu_0 = 4\pi \times 10^{-7}$ H/m and $\mu^r = 1$ and r_{sh} being a very small expression, the mutual inductance can be written as

$$M = 4\times10^{-7}\ln\left(\frac{S}{r_{sh}}\right)\text{H/m} \tag{8.45}$$

And for a single cable between the conductor and sheath, it should be

$$M = 2\times10^{-7}\ln\left(\frac{S}{r_{sh}}\right)\text{H/m} \tag{8.46}$$

Therefore, for a single cable, the sheath voltage should be $V_{sh} = IX = I\omega M$ or

$$V_{sh} = I\omega^2\times10^{-7}\ln\left(\frac{S}{r_{sh}}\right)\text{V/m}$$

Now, the sheath current is

$$I_{sh} = \frac{I\omega M}{\sqrt{\left(R_{sh}^2+\omega^2M^2\right)}}\text{amp} \tag{8.47}$$

where $\sqrt{\left(R_{sh}^2+\omega^2M^2\right)}$ is the sheath impedance.

From this sheath for a single cable, the loss factor is given by

$$W_{sh} = I_{sh}^2R_{sh} = R_{sh}\left(\frac{I^2\omega^2M^2}{R_{sh}^2+\omega^2M^2}\right)\text{W/m} \tag{8.48}$$

and $W_c = I^2R_c$ W/m.

Therefore, the sheath loss factor per phase is

$$\lambda_s = \frac{W_{sh}}{W_c} = \frac{R_{sh}}{R_c}\left(\frac{\omega^2M^2}{R_{sh}^2+\omega^2M^2}\right) \tag{8.49}$$

5. *Eddy current loss.* When three single-phase cables are laid in the vicinity of each other, it may happen so that one of them running very close to the other may induce a circulating current within the conductor of the cable. Most of the magnetic flux will cut the sheath at the nearest point of the cable. This will induce an electro motive force (EMF),

causing a current within the sheath and returning along the other. This is independent of the type of earth bonding. This EMF decreases with the increase in distance between the cables. This is called localised current or eddy current and is independent of the cable length.

The sheath eddy current losses are given by

$$E_c = I^2 \left\{ \frac{3\omega^2}{R_s} \left(\frac{d_m}{2S} \right)^2 10^{-11} \right\} \text{W/m/phase} \tag{8.50}$$

where

E_c is the sheath eddy current losses
I is the current (Amperes)
ω is the $2\pi f$, where f is the frequency
R_s is the sheath resistance (Ω/m)
d_m is the mean diameter of the sheath (m)
S is the distance from the cable centre

For single-core cables, these losses are very negligible, but for aluminium sheath cables, the loss may be comparable.

6. *Armour loss.* Armouring of power cables may be considered as a secondary metal sheath with corresponding losses. When the cable is armoured with magnetic materials such as steel wires, additional hysteresis occurs. In a single-core cable, such hysteresis losses are so high that it has to be replaced by nonmagnetic wires such as hard aluminium or alloy aluminium wires.

   In case of a single-core cable, armour losses consist of

   a. Loss due to current in armouring, both in the form of circulating current and eddy current
   b. Losses due to the magnetic field around the conductor and also due to magnetic fields from nearby conductors or a group of single-core conductors, which may result in a considerable hysteresis loss

   Considering two single-core cables with steel wire armouring, open circuited and analogue, it can be established, as shown in Equation 8.41, that the flux enclosed between the sheaths due to the current $+I$ is to be expressed as (Figure 8.4)

$$\varphi_1 = \int_{r_{sh}}^{S-r_{sh}} \frac{\mu_0 \mu_r I dr}{2\pi r} = \frac{\mu_0}{2\pi} I \int_{r_{sh}}^{S} \mu_r \frac{dr}{r} \text{ W/m} \tag{8.51}$$

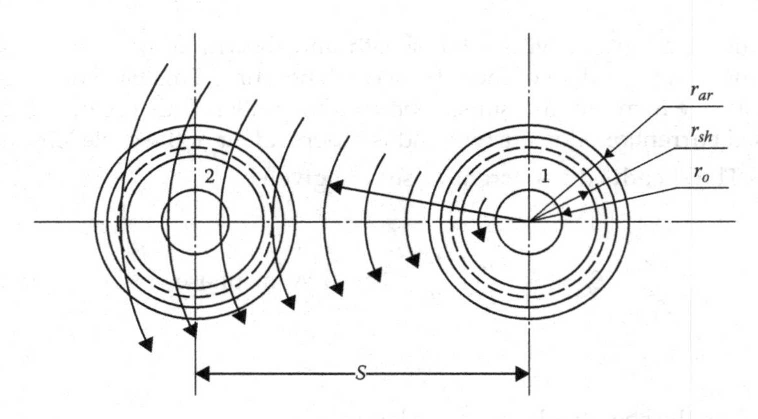

FIGURE 8.4
Armour loss in the cables.

Evaluating this integral of equations, considering the diameter over the armour as r_{ar} and diameter of the wire as p_r when 'r_{ar}' is larger than 'p_r', it is found that

$$\varphi_1 = \frac{\mu_0}{2\pi} I \left(\ln\left(\frac{S}{r_{sh}}\right) + \mu_r \frac{p_r}{r_{ar}} \right) \tag{8.52}$$

Similarly, the flux due to current $-I$ is

$$\varphi_2 = \frac{\mu_0}{2\pi} I \left(\ln\left(\frac{S}{r_{sh}}\right) + \mu_r \frac{p_r}{r_{ar}} \right) \tag{8.53}$$

Therefore, the resultant mutual inductance per metre length is

$$M = \frac{\varphi_1 + \varphi_2}{I} = \frac{\mu_0}{\pi} \left(\ln\left(\frac{S}{r_{sh}}\right) + \mu_r \frac{p_r}{r_{ar}} \right) \tag{8.54}$$

This is for a single-core cable. In the case of three-core and four-core cables, the induced magnetic field cancels within, and hence only eddy current loss and hysteresis losses have to be considered, which, being in very small quantities, can be neglected for the purpose of determining the losses.

7. *Thermal resistance of the outer sheath*

$$T_3 = \frac{\rho_t}{2\pi} \ln\left(1 + \frac{2t_3}{d_0}\right) \tag{8.55}$$

t_3 is the thickness of the outer sheath (mm)

d_0 is the diameter over the outer sheath (mm)

8. *Thermal resistance in air (not affected by solar radiation)*

$$T_4 = \frac{1}{\left[\pi d_e h(\theta_s)^{1/4}\right]} \tag{8.56}$$

d_e is the outer diameter of the cable (mm)
h is the heat dissipation factor taken from IEC 60287
$\Delta\theta_s$ is the surface temperature on the cable above ambient temperature (K)

9. *Cable laid underground at a depth of L (mm) from the ground surface (T_4).* For single and isolated cables – with no other cables nearby

$$T_4 = \frac{\rho_t}{2\pi}\ln\left\{s + \sqrt{(s^2 - 1)}\right\} \tag{8.57}$$

where
ρ_t is the thermal resistivity of oil
$s = 2L/d_e$, d_e is the outer diameter of the cable (mm)

10. *Cable laid in buried ducts (T_4).* Here, T_4 is the sum of the thermal resistance of air inside the duct, and the duct itself, plus the external thermal resistance of the material surrounding the duct.

    a. Thermal resistance of air in the duct (T_4')

    The cable diameter is restricted to 100 mm

$$T_4' = \frac{U}{1 + 0.1(V + Y\theta_m)D_e} \tag{8.58}$$

U, V and Y are constants to be taken from IEC 60287
D_e is the outside diameter of the cable (mm)
θ_m is the mean temperature of air within the duct (°C)

    b. Thermal resistance of the duct

$$T_4'' = \frac{\rho_t}{2\pi}\ln\left(\frac{D_0}{D_d}\right) \tag{8.59}$$

D_0 is the external diameter of the duct (mm)
D_d is the internal diameter of the duct (mm)
P_t is the thermal resistivity of the duct on an average of 45.2 (km/W)
Thermal resistance of the surrounding media of duct (T_4''')

This should be taken as per Equation 8.57. If the duct is surrounded by concrete and soil, then the combined resistivity of the material has to be considered.

The features described above are an example of a cable having simple constructional form. But there are various other factors to be considered, such as

- Temperature
- Cable constructional features
- Installation conditions
- Effect of the cables nearby
- Thermal resistivity of the materials

Details of various conditions and type of loss factors can be taken from IEC 60287. It is necessary to consider and study those variations properly before any calculation is made.

8.2 Thermal Resistivity

Thermal resistivity plays an important role in the construction of any type of electric cables. In the case of polymeric substances, it is found that because of different atomic structures and a combination of different blends, thermal resistivity varies from compound to compound. Everyday, a new material is introduced in the market. As per different environmental situations and requirements, materials are developed to cater to specific demands. First, the thermal resistivity of the material differs. Further, different constructions and thickness current ratings would also differ. The resistivity of soil varies from country to country, so too within the states. The thermal parameters of ducts and pipes may also differ due to the use of different construction materials. Naturally, the subject is complex. However, for guidance, a generalised approach has been established in IEC 60287. Whenever the deviations are large, the current rating should be calculated by taking into account the measured values obtained and, if possible, to get a rational figure. This would lead to gaining certain benefits of power transmission, making the system cost-effective. It would be better if the polymeric material supplier could either maintain a consistent thermal resistivity of their supplied material or provide the data of thermal resistivity of each item which they produce. It is certain that different batches of a compound with the same formulation should have different values of thermal resistivity. Compounding parameters, however small, will differ from batch to batch, hence the thermal parameters. It is not possible to calculate current ratings on this basis every time, and so an average value has to be accepted. However, in case of a major deviation, a check is essential. Therefore, it is advisable that the suppliers measure the thermal resistivity of their compound and provide the data to the user. Though, internationally, some average values have been accepted for polymeric materials, even then from time to time, a check needs to be carried out to establish its usefulness.

Similar situations can arise in the case of soil resistivity as well. Ecological parameters are changing all over the world. Deforestation, construction of dams for irrigation, hydroelectric power stations, new upcoming township, cities and roads are liable to change the thermal condition of soil from place to place. While installing cables for power supply, such changes have to be considered.

It has been shown by the author that a change in the thermal resistivity of any material or soil structure can affect the current-carrying capacity substantially.

For EHV cables, the calculation of current rating depends on the laying and installation conditions. In this case, the circulating current through the sheath and armour, as well as a rise in voltage, has to be accounted for under different earthing conditions. There are generally three types of conditions which have to be taken into account:

1. When the screen and armour are not earthed at both ends (kept floating). In this case, no current will flow through the screen. The sheath voltage will rise abnormally, damaging the cable and also becoming dangerous for working personnel. As per IEC, the sheath voltage has to be restricted to 60 V under normal operating conditions and 300 V at short-circuit conditions. Here, any loss in the operating current generates considerable heat. This system is not used in case of installation of the EHV cable system.
2. Screen and armour are earthed at one end. The other end remains floating: In this case, the circulating current starts flowing towards the end which is being earthed. Voltage rise is restricted, but if the cable length is long, then the flow of circulating current increases towards the earthing end but reduces away from the earthing point. As such, the voltage drop becomes more prominent towards the earthing end but less, away from it. This type of installation is suitable for shorter lengths.
3. Thus, for a considerable longer route length, earthing is done at both ends. But though the circulating current flows more towards both ends, at the middle portion, the voltage can rise considerably. To compensate for this increase in voltage, transposition is done after every three lengths, balancing the magnetic field. Even then, along with transposition, earthing has to be done after certain distances intermittently (if required) to reduce the intensity of residual voltage. Transposition is done through a link box, while a nominal circulating

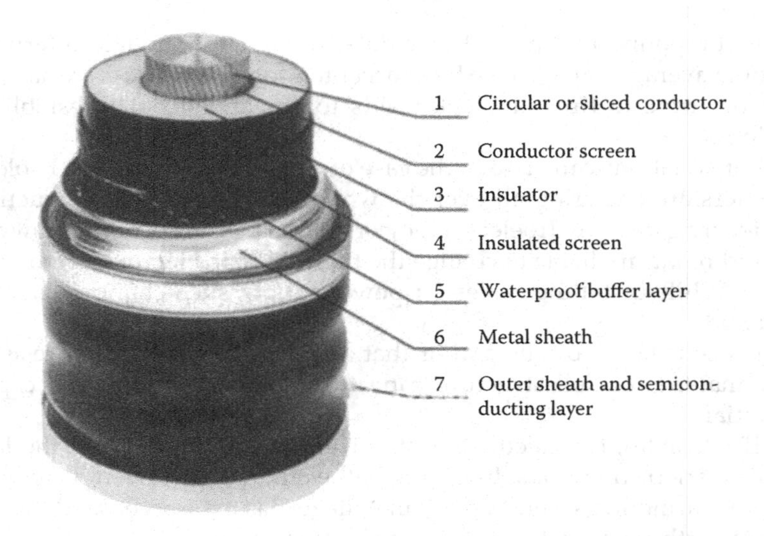

FIGURE 8.5
EHV cable with corrugated aluminium sheath.

current, as generated due to the residual magnetic field, is allowed to flow through intermediate points. Effective current rating would reduce proportionately. But this necessarily has to be accepted and accounted for.

Rigorous calculation for all the three conditions is being shown in the IEC 60287 specification (Figure 8.5).

HV and EHV cables are made as a single core. These cables are protected by a metallic sheath of either lead or corrugated aluminium sheath. Magnetic lines of force emanating from the conductor current induce a voltage within the sheath. For a longer cable length, this voltage can rise appreciably, endangering the life of working personnel and possibly damaging the monitoring of electrical equipment. As per IEC norm, this voltage must be restricted to 60 V for normal operating conditions and 300 V during short-circuit conditions. If the sheath is earthed at one end, a current starts flowing towards the earthing point reducing the intensity of the induced voltage. If the cable is short, then the rise in voltage will remain restricted. When both ends are connected to the earth, the current will flow in both directions. If the length is very long, the voltage at the middle point will rise abnormally, whereas the current flowing at the ends will help reduce the intensity of the voltage rise towards the end sections. In such longer lengths, if the cable is transposed alternately, the rise in voltage can be restricted. Since such large-diameter cables cannot be laid up as three cores, transposition is done by cross bonding of the sheath and earthing at regular intervals – say after every three lengths. This may increase a certain amount of circulating current within the transposed sections, but the voltage will be restricted to the required level. Before calculating the current-carrying capacity of an EHV cable conductor, it is necessary to determine the sheath loss factor 'λ_{sh}' at various combinations of laying and installation of cables.

Generally EHV cable is constructed as: Aluminium or copper conductor, compacted, semicon extruded, XLPE-insulated, outer semicon layer extruded, copper tape or wire screened, aluminium wire armoured or sheathed cable, PVC or HDPE outer sheathed coated with carbon black.

Hereafter, the calculations show the screen loss and AL wire armoured or metallic sheath loss factors in different laying conditions.

8.2.1 Two Single-Core Cable Sheaths Bonded at Each End

Let R_{sh} be the sheath resistance and X_c be the reactance, where $X_c = \omega M$ and $\omega = 2\pi f$.

Here, M is the mutual inductance. Self-inductance is equal to M. Now, impedance is

$$\sqrt{\left(R_{sh}^2 + X_c^2\right)}\ \Omega/\text{m} \tag{8.60}$$

from this, we calculate the sheath current as:

$$I_{sh} = \frac{I\omega M}{\sqrt{\left(R_{sh}^2 + X_c^2\right)}} \tag{8.61}$$

where

I is the current flowing through the conductor

$\omega = 2\pi f$, where f is the frequency

This value depends on the route length and spacing between conductors. Accordingly, the sheath loss for unit length is

$$W_{sh} = I_{sh}^2 R_{sh} = \frac{I^2(2\pi f)^2 M^2}{\left(R_{sh}^2 + \omega^2 M^2\right)} \tag{8.62}$$

And conductor losses are given by

$$W_c = I^2 R_c$$

where R_c is the conductor resistance.

Now, effective impedance is given by

$$Z_e = \left[R_c + R_{sh}\left\{\frac{\omega^2 M^2}{\left(R_{sh}^2 + \omega^2 M^2\right)}\right\}\right] + j\left[\omega L_c - \left\{\frac{\omega^3 M^3}{\left(R_{sh}^2 + \omega^2 M^2\right)}\right\}\right] \tag{8.63}$$

where L_c is the self-inductance.

Hence, the sheath loss factor is given by

$$\lambda_{sh} = \frac{W_{sh}}{W_c} = \frac{R_{sh}}{R_c} \times \left\{ \frac{\omega^2 M^2}{\left(R_{sh}^2 + \omega^2 M^2\right)} \right\} \tag{8.64}$$

8.2.2 HV and EHV AC Systems

Three single-core cables are used in balanced condition. In this case, cables are laid as follows:

1. In trefoil formation, forming an equilateral triangle. Cables are placed symmetrically at the corner of the triangle touching each other and clamped together.
2. In another formation in which the cables are laid equidistant parallel to each other on a horizontal plane. The sheath loss factor in this case is more for the cables on the sides than the middle one.

The following notations are accepted for three cables as I_1, I_2 and I_3 for cables 1, 2 and 3, respectively. I_{sh} is the sheath current, and r_{sh} is the radius of the cable from the centre to the mean distance of the metallic sheath, screen or armour in metres. V_{sh1}, V_{sh2} and V_{sh3} are the sheath voltages of cables 1, 2 and 3 V/m induced by the conductor current. R_c and R_{sh} are the resistance of conductor and sheath/screen/armour in Ω/m $M_{1,2}$, $M_{1,3}$ and $M_{2,3}$ are the mutual inductance in hour per metre between the conductor of cable and the sheath of the adjacent cable. 'S' is the centre distance between two adjacent cables. Every cable can be taken as the return path of the other two cables in a three-phase system. It is, therefore, shown as

$$\left\{ \begin{array}{l} (1)\ I_1 = -(I_2 + I_3) \text{ and further } I_{sh1} = -(I_{sh2} + I_{sh3}) \\ (2)\ I_2 = -(I_1 + I_3) \text{ and further } I_{sh2} = -(I_{sh1} + I_{sh3}) \\ (3)\ I_3 = -(I_1 + I_2) \text{ and further } I_{sh3} = -(I_{sh1} + I_{sh2}) \end{array} \right\} \tag{8.65}$$

and the sheath voltage can be expressed as

Assuming that eddy currents are not involved in these calculations and capacitive currents are very low and can be neglected

$$\left\{ \begin{array}{l} (1)\ V_{sh1} = I_{sh1}R_{sh} - j\omega[M_{1,2}(I_2 + I_{sh2}) - M_{1,3}(I_3 + I_{sh3})] \\ (2)\ V_{sh2} = I_{sh2}R_{sh} - j\omega[M_{1,2}(I_1 + I_{sh1}) - M_{2,3}(I_3 + I_{sh3})] \\ (3)\ V_{sh3} = I_{sh3}R_{sh} - j\omega[M_{1,3}(I_1 + I_{sh1}) - M_{2,3}(I_2 + I_{sh2})] \end{array} \right\} \tag{8.66}$$

Now, it follows:

i. *Three single-core cables in a three-phase system installed in trefoil formation at an equal axial spacing.* In this case, the mutual inductance of cables 1, 2 and 3 should be equal because of a symmetrical arrangement:

$$M_{1,2} = M_{1,3} = M_{2,3} = 2\times 10^{-7} \ln\left(\frac{S}{R_{sh}}\right) \text{H/m} \tag{8.67}$$

where

r_{sh} is the mean radial distance of nonmagnetic screen/sheath/wire armour from the screen in metre

S is the axial spacing between cables in metres

Considering the balanced load in the three cables, the induced voltage and circulating current can be computed from the two sets of Equations 8.65(1) and 8.66(1) as follows:

$$\begin{aligned} V_{sh1} &= I_{sh1}R_{sh} - j\omega M(I_2 + I_{sh2}) - j\omega M(I_3 + I_{sh3}) \\ &= I_{sh1}R_{sh} + j\omega M(I_1 + I_{sh1}) \end{aligned} \tag{8.68}$$

1. *Case 1:* Cable sheath bonded and earthed at one end
   a. In this case, the sheath current should be $I_{sh} = 0$
   b. Sheath voltage = ωIM V/m, where $\omega = 2\pi f$

      I is the rated current flowing through the conductor

   Therefore,

$$V_{sh} = 2\times 2\times \pi\times 50\times I\times 10^{-7}\times \ln\left(\frac{S}{r_{sh}}\right) \text{V/m} \tag{8.69}$$

or

$$V_{sh} = 0.62832\times 10^{-4}\times I \ln\left(\frac{S}{r_{sh}}\right) \text{V/m} \tag{8.70}$$

2. *Case 2:* Cable sheath bonded at both ends. There will be current flowing through the sheath, and the sheath voltage should be

$$V_{sh1} = V_{sh2} = V_{sh3} = 0$$

   a. Then, the sheath current should be I_{sh}

$$I_{sh} = I\frac{\omega^2 M^2}{\sqrt{R_{sh}^2 + \omega^2 M^2}} \text{A/m} \tag{8.71}$$

   R_{sh} is the screen or metallic sheath or armour resistance.

Now,

$$\omega M = 0.62832 \times 10^{-4} \ln\left(\frac{S}{R_{sh}}\right)$$

Hence,

$$I_{sh} = I\left[\frac{\left\{0.62832 \times 10^{-4} \times \ln\left(\frac{S}{r_{sh}}\right)\right\}^2}{\sqrt{R_{sh}^2 + \left\{0.62832 \times 10^{-4} \times \ln\left(\frac{S}{r_{sh}}\right)\right\}^2}}\right] \text{A} \quad (8.72)$$

b. Even though it is assumed that there will be no sheath voltage, that is, $V_{sh} = 0$, a certain amount of residual voltage will be there.
c. Sheath/screen loss $I_{sh}^2 R_{sh} = W_{sh}$ W/m

Therefore,

$$W_{sh} = I_{sh}^2 R_{sh} = I^2 R_{sh}\left[\frac{\left\{0.62832 \times 10^{-4} \times \ln\left(\frac{S}{R_{sh}}\right)\right\}^2}{R_{sh}^2 + \left\{0.62832 \times 10^{-4} \times \ln\left(\frac{S}{R_{sh}}\right)\right\}^2}\right] \text{W/m} \quad (8.73)$$

d. Sheath loss factor $\lambda_{sh} = I_{sh}^2 R_{sh} / I^2 R_{AC}$, where R_{AC} is the AC resistance of the conductor at the operating temperature (Ω/m)

Hence,

$$\lambda_{sh} = \frac{R_{sh}}{R_{AC}} \times \left[\frac{\left\{0.62832 \times 10^{-4} \times \ln\left(\frac{S}{r_{sh}}\right)\right\}^2}{R_{sh}^2 + \left\{0.62832 \times 10^{-4} \times \ln\left(\frac{S}{r_{sh}}\right)\right\}^2}\right] \quad (8.74)$$

All the earlier calculations relate to symmetrical trefoil circuit formation.

8.2.3 Three-Phase Circuit with Flat Formation

When cables are equidistant (S) and laid parallel to each other, they are identified as cables 2 and 3. The distance between cables is 'S'. Current is taken to be in a balanced condition; then,

$$M_{1,2} = M_{2,3} = M = 2\times10^{-7}\ln\left(\frac{S}{R_{sh}}\right) \text{ H/m}$$

And, it will be shown for a balanced three-phase system,

$$\left.\begin{aligned} I_1 &= I_2\left(-\frac{1}{2}+j\frac{\sqrt{3}}{2}\right) \\ I_3 &= I_2\left(-\frac{1}{2}-j\frac{\sqrt{3}}{2}\right) \end{aligned}\right\} \tag{8.75}$$

and

$$M_{1,3} = 2\times10^{-7}\ln\left(\frac{2S}{r_{sh}}\right) = 2\times10^{-7}\ln 2 + 2\times10^{-7}\ln\left(\frac{S}{r_{sh}}\right) = M_n + M$$

where

$$M_n = 2\times10^{-7}\ln 2 = 1.386\times10^{-7} \text{ H/m}$$

and

$$M = 2\times10^{-7}\ln\left(\frac{S}{R_{sh}}\right) \text{ H/m}$$

8.2.3.1 Sheath Bonded at One End

Then, the sheath current on all the three cables should be

$$Is_1 = Is_2 = Is_3 = 0$$

Taking

$$X = \omega M \quad \text{and} \quad X + X_m = \omega M + \omega M_n$$

then the sheath voltage is given by (taking the help of Equations 8.65 and 8.75)

$$\begin{bmatrix} V_{sh1} = -\frac{I_2}{2}\left\{\sqrt{3}(X+X_m)+j(X-X_m)\right\} \\ V_{sh2} = jI_2X \\ V_{sh3} = -\frac{I_2}{2}\left\{\sqrt{3}(X+X_m)+j(X-X_m)\right\} \end{bmatrix} \tag{8.76}$$

Thus, in absolute value,

$$V_{sh1} = V_{sh3} = [I]\sqrt{(X^2 + XX_m + X_m^2)} \tag{8.77}$$

and $Vs_2 = [I]X$.

Considering $I_1 = I_2 = I_3 = I$,

the sheath loss factor here is

$$\lambda_{sh} = \frac{R_{sh}I_{sh}^2}{R_{AC}I^2} = 0 \quad \text{since } I_{sh} = 0 \tag{8.78}$$

This type of arrangement is suitable for a short cable length, where the sheath voltage can be limited within a specified value. It is seen that V_{sh} increases with the increase in cable spacing.

8.2.3.2 Sheath Bonded at Both Ends Together

Sheath current will flow out through the earthing points of all cable ends; hence, no sheath voltage is expected to rise. However, a residual voltage, though small, is retained within the system. The magnitude of this voltage is to be determined as follows without going much into the details of the calculations.

Sheath current on all three cables should be flowing.

Only a residual sheath voltage as $V_{sh1} = V_{sh2} = V_{sh3} = V_o$ could be determined, where

$$I_{sh1} + I_{sh2} + I_{sh3} = 0$$

Now, taking the help of a general equation (Equation 8.66), it can be shown that

$$V_{sh1} = V_0 = I_{sh1}(R_{sh} + jX) - \frac{1}{2}jI_2(X - X_m) - \frac{\sqrt{3}}{2}I_2(X - X_m) - jI_{sh3}X_m$$

$$V_{sh2} = V_0 = I_{sh2}(R_{sh} + jX) + jI_2X \tag{8.79}$$

$$V_{sh3} = V_0 = I_{sh3}(R_{sh} + jX) - \frac{1}{2}jI_2(X - X_m) + \frac{\sqrt{3}}{2}I_2(X - X_m) - jI_{sh1}X_m$$

Now, it is deduced by referring to the previous equations:

$$V_0 = V_{sh1} + V_{sh2} + V_{sh3} = j(I_2 + I_{sh2})X_m \tag{8.80}$$

Again,

$$3V_0 = 3V_{sh2} = 3I_{sh2}(R_{sh} + jX) + 3jI_2X \tag{8.81}$$

$$I_{sh2} = -I_2 \frac{j\left(X - \frac{X_m}{3}\right)}{R_{sh} + j\left(X - \frac{X_m}{3}\right)} \tag{8.82}$$

1. Where $I_1 = I_2 = I_3 = I$, and the residual voltage is given by

$$[V]_0 = \frac{1}{3}\left\{\frac{IX_mR_{sh}}{\sqrt{R_{sh}^2 + Q^2}}\right\}$$

or

$$V_0 = \frac{1}{3}\left\{\frac{I \times 0.43542 \times 10^{-4} \times R_{sh}}{\sqrt{\left(R_{sh}^2 + 0.62832 \times 10^{-4} \ln(S/r_{sh}) - 0.14514 \times 10^{-4}\right)^2}}\right\} \text{V/m} \tag{8.83}$$

2. By transposing and transforming, the circulating current flowing through the sheaths is given by
   a. For cable 1,

$$I_{sh1} = I\sqrt{\left\{\frac{Q^2}{4\left(R_{sh}^2 + Q^2\right)} + \frac{3P^2}{4\left(R_{sh}^2 + P^2\right)} + \frac{\sqrt{3}}{2}\frac{PQR_{sh}(Q-P)}{\left(R_{sh}^2 + Q^2\right)\left(R_{sh}^2 + P^2\right)}\right\}} \text{ A} \tag{8.84}$$

   b. For cable 2,

$$I_{sh2} = I\frac{Q}{\sqrt{\left(R_{sh}^2 + Q^2\right)}} \text{ A} \tag{8.85}$$

   c. For cable 3,

$$I_{sh3} = I\sqrt{\left\{\frac{Q^2}{4\left(R_{sh}^2 + Q^2\right)} + \frac{3P^2}{4\left(R_{sh}^2 + P^2\right)} - \frac{\sqrt{3}}{2}\frac{PQR_{sh}(Q-P)}{\left(R_{sh}^2 + Q^2\right)\left(R_{sh}^2 + P^2\right)}\right\}} \text{ A} \tag{8.86}$$

Here,

$$Q = \left\{0.62832 \times \ln\left(\frac{S}{r_{sh}}\right) - 0.14514\right\} \times 10^{-4}$$

$$P = \left\{0.62832 \times \ln\left(\frac{S}{r_{sh}}\right) + 0.43542\right\} \times 10^{-4}$$

I is the rated current at the operating temperature of the conductor
R_{sh} is the metallic sheath or screen resistance (Ω/m)
S is the axial spacing between two cables (m)
r_{sh} is the mean radius of the metallic sheath or screen (m)

8.2.4 Metallic Sheath or Screen Loss Factors

For cable 1,

$$\lambda_{sh1} = \frac{R_{sh}}{R_{AC}}\left\{\frac{Q^2}{4\left(R_{sh}^2 + Q^2\right)} + \frac{3P^2}{4\left(R_{sh}^2 + P^2\right)} + \frac{\sqrt{3}}{2}\frac{PQR_{sh}(Q-P)}{\left(R_{sh}^2 + Q^2\right)\left(R_{sh}^2 + P^2\right)}\right\} \quad (8.87)$$

$$\lambda_{sh2} = \frac{R_{sh}}{R_{AC}}\left(\frac{Q^2}{R_{sh}^2 + Q^2}\right) \quad (8.88)$$

$$\lambda_{sh3} = \frac{R_{sh}}{R_{AC}}\left\{\frac{Q^2}{4\left(R_{sh}^2 + Q^2\right)} + \frac{3P^2}{4\left(R_{sh}^2 + P^2\right)} - \frac{\sqrt{3}}{2}\frac{PQR_{sh}(Q-P)}{\left(R_{sh}^2 + Q^2\right)\left(R_{sh}^2 + P^2\right)}\right\} \quad (8.89)$$

Sheath loss thus is given by

$$\left.\begin{array}{l}\text{For cable 1}: I_{sh1}^2 R_{sh} \\ \text{For cable 2}: I_{sh2}^2 R_{sh} \\ \text{For cable 3}: I_{sh3}^2 R_{sh}\end{array}\right\} \quad (8.90)$$

and is proportional to the ratio of the conductor to sheath resistances. This is dependent on the reactance values of the frequency of supplies (f).

And total sheath loss factor of three cables combined

$$\lambda_{sh} = \frac{1}{2}\frac{R_{sh}}{R_{AC}}\left(\frac{Q^2}{R_{sh}^2 + Q^2} + \frac{P^2}{R_{sh}^2 + P^2}\right) \quad (8.91)$$

To calculate I first λ_{sh} of individual cable to be determined. It is desirable to take the highest value of λ_{sh}, as calculated, out of the three cables.

8.2.5 Single-Core Cable Sheath Bonding

If the sheath is bonded at both ends, a circulating current will flow, which should actually cause a considerable power loss. If three cables are laid at a distance S from each other, then an induced voltage will act on each cable separately (Figure 8.6). The induced voltage can be calculated by knowing the reactance of each sheath as follows:

I_a, I_b and I_c are the currents flowing through each of the cables a, b and c, respectively. Then, the magnetic field at a radial distance 'r' from the conductor centre is given by $H = I_a/2\pi r$ or the flux density is $B = \mu_o H = \mu_o I_a/2\pi r$.

The flux linking the sheath of cable 'a' is given by the current of all the three cables which are $I_a = I$ and $I_b = h^2 I$, due to both cables from the sides, and $I_c = hI$ for cable 'c' due to the current of cable 'b' (as cable 'c' is far away from cable 'a' and its influence is almost negligible), where 'h' is the factor which rotates the phasor counterclockwise. Naturally, $h = e^{j2\pi/3}$ and $h^2 = e^{j4\pi/3}$.

Now solving for the equation of total flux density on cable 'a'

$$\varphi_{sha} = \left(\frac{\mu_0 I}{2\pi}\right)\ln\left(\frac{2R}{D_{sh}}\right) + \left(\frac{\mu_0 h^2 I}{2\pi}\right)\ln\left(\frac{R}{S}\right) + \left(\frac{\mu_0 h I}{2\pi}\right)\ln\left(\frac{R}{2S}\right) \tag{8.92}$$

where

R is the radius of distance from the sheath around cable 'a' on which flux is concentrated

S is the distance of middle cable 'b' from the cable 'a'

$2S$ is the distance of cable 'a' from cable 'c'

D_{sh} is the diameter of the sheath

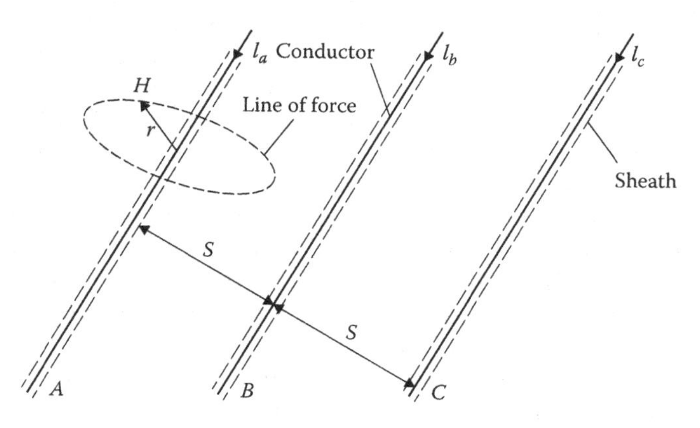

FIGURE 8.6
Three cables laid parallel.

Now, reactance is given by

For cable 'a', $X_a = \omega\varphi_{sha}/I_a$

or

$$X_a = 2\omega\times10^{-7}\left(\ln\frac{2S}{D_{sh}} + 0.5\ln 2 - j\frac{\sqrt{3}}{2}\ln 2\right)\Omega/\text{m} \tag{8.93}$$

For cable 'b',

$$X_b = 2\omega\times10^{-7}\ln\frac{2S}{D_{sh}}\,\Omega/\text{m} \tag{8.94}$$

And for cable 'c',

$$X_a = 2\omega\times10^{-7}\left(\ln\frac{2S}{D_{sh}} + 0.5\ln 2 + j\frac{\sqrt{3}}{2}\ln 2\right)\Omega/\text{m} \tag{8.95}$$

The equations can be rewritten as

$$X_a \text{ and } X_c = 0.6283\times10^{-4}\sqrt{\left\{\ln\left(\frac{2S}{D_{sh}}\right)^2 + 0.69135\ln\left(\frac{2S}{D_{sh}}\right) + 0.48054\right\}}\;\Omega/\text{m} \tag{8.96}$$

and

$$X_b = 0.6283\times10^{-4}\ln\left(\frac{2S}{D_{sh}}\right) \tag{8.97}$$

In case of longer HV and EHV transmission lines, the sheath voltage rises considerably in the middle portion of the cable, even though the cable ends may be connected to earth. To keep the voltage within the specified limit, cross bonding has to be done as shown in Figure 8.7. After every three lengths, the sheath has to be connected to earth through a link box to allow the induced current to drain out the limiting sheath voltage to approach zero value. This is similar to assembling three core cables in parallel formation. It is actually comparable to a virtual transposition (rotation of each phasor by 120° with respect to the other in a regular sequence). In this system, the magnetic flux of each cable is neutralised, except certain stray magnetic fields and eddy currents which give the effect of residual voltage because of some manufacturing imperfections, however small it may be.

In these cases, in two-thirds of cables, the joint sheaths are isolated and bonded through the link box, whereas at the beginning and at the end of each of the three parallel lengths, the sheaths are connected to the earth. This minimises the circulating current and contains the rise in sheath voltage

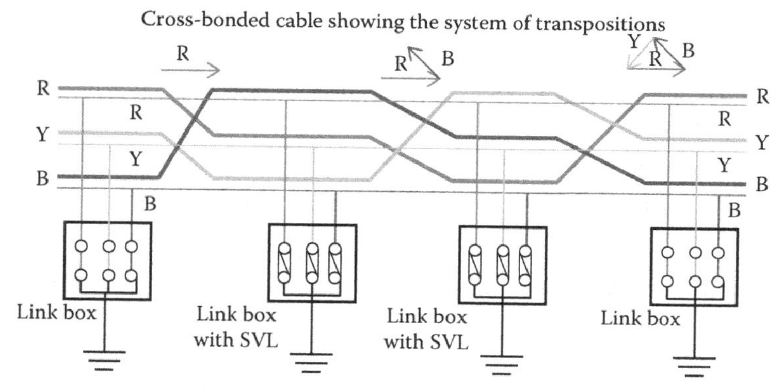

FIGURE 8.7
Cross bonding of cables.

(as per IEC 60287 norm, the sheath residual voltage should not rise more than 60 V during operating conditions, and during short-circuit conditions, it should not rise beyond a specified limit).

In case cables are installed in the trefoil formation, they are placed symmetrically at an equidistant position. In this case, the circulating current is

$$I_{sha} = I_{shb} = I_{shc} = 0$$

No sheath current flows, hence no sheath loss.

In case cables are laid in flat parallel formation, the reactance in each cable sheath is not equal and symmetrical. Even after cross bonding, a certain amount of sheath current will flow, inducing residual voltage, which can be shown as follows:

The voltages of cables *'a'*, *'b'* and *'c'* are to be given by V = reactance × current.

The length of cross bonding is 'l_1', 'l_2' and 'l_3', but in actual practice, $l_1 = l_2 = l_3 = l$.

The voltages are neglecting an imaginary part of the phasor:

a.

$$|V_{sh}| = I_c \times l \times 0.31416 \times 10^{-4}\left[\left\{\sqrt{(\ln(2S/D_{sh})+0.3466)^2 - 0.36}\right\} + \left\{\sqrt{(\ln(2S/D_{sh})+0.3466)^2 - 0.36}\right\} - 0.7071\ln(2S/D_{sh})\right]$$

volts per section of length 'l' (8.98)

where

S is the axial spacing of the cable

D_{sh} is the mean diameter over the metallic screen/sheath/armour

I_c is the rated current at operating temperature (°C)

b. Impedance of the circuit length: [Z] of a, b and c is equal to

$$l\sqrt{9R_{sh}^2+0.0987\times10^{-8}\left[\left\{\sqrt{(\ln(2S/D_{sh})+0.3466)^2-0.36}\right\}+\left\{\sqrt{(\ln(2S/D_{sh})+0.3466)^2+0.36}\right\}+\ln(2S/D_{sh})\right]^2} \quad (8.99)$$

Taking Equations 8.98 and 8.99, the sheath current can be calculated as

c.

$$I_{sh}=\frac{V_{sh}}{|Z|} \quad (8.100)$$

d. Sheath loss factor

$$\lambda_{sh}=\frac{I_{sh}^2R_{sh}}{I_c^2R_{AC}}=\frac{R_{sh}\times0.0987\times10^{-8}\left[\left\{\sqrt{(\ln(2S/D_{sh})+0.3466)^2+0.36}\right\}+\left\{\sqrt{(\ln(2S/D_{sh})+0.3466)^2-0.36}\right\}-0.7071\ln\frac{2S}{D_{sh}}\right]^2}{R_{AC}\times9R_{sh}^2+0.0987\times10^{-8}\left[\left\{\sqrt{(\ln(2S/D_{sh})+0.3466)^2+0.36}\right\}+\left\{\sqrt{(\ln(2S/D_{sh})+0.3466)^2-0.36}\right\}+\ln\frac{2S}{D_{sh}}\right]^2} \quad (8.101)$$

where

R_{sh} is the sheath resistance per metre

R_{AC} is the AC resistance of the cable conductor per metre

If the balancing and cross bonding of cables are done properly, the loss factor will be very low. Since all sections are almost equal in length, R_{sh}, R_{AC}, Z and λ_{sh} will all be almost the same in all cases.

Similarly, armour and eddy current losses can also be determined to calculate the final current-carrying capacity of a conductor in a cable. Detailed treatments are shown in IEC specification No. IEC 60287-1-1, 'Electric Cables – Calculation of the Current Rating'.

The dielectric loss in HV and EHV cables becomes a prominent feature. For a single-core cable, the total temperature rise above ambient temperature is also due to losses, corresponding to the maximum conductor temperature flow through one-half of the internal thermal resistance of the cable. The temperature rise above the ambient temperature θ_a is only due to the dielectric loss W_d. The loss is independent of the load current and causes an

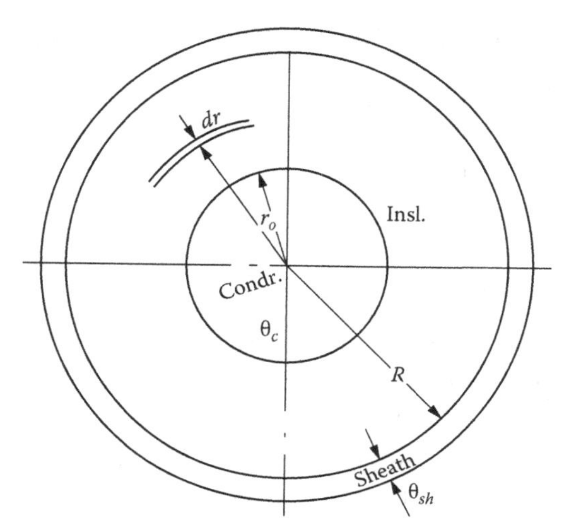

FIGURE 8.8
Dielectric loss in insulation.

increase in the temperature of the dielectric, which must be accounted for in all calculations of EHV cables' current rating.

Normally, cable dielectric loss is computed at the maximum permissible operating temperature of the conductor, considering that the loss is distributed evenly over the total thickness of insulation. The additional temperature rise has to be computed, taking into account the ground ambient level due to dielectric loss only. Naturally, the rise of temperature θ_d' is determined in the cable insulation itself.

Figure 8.8 shows a single-core cable having a dielectric permittivity ε_r.

In the figure, let dV_0 be the voltage across the annular cylinder d_r and d_C be the capacitance having a length of 1 m.

The capacitance of the annular cylinder will be

$$d_C = \frac{\varepsilon_0 \varepsilon_r 2\pi r.1}{d_r}$$

Therefore, the dielectric loss within the annular ring is

$$(dV_0)^2 \omega d_C \tan\delta = (dV_0)^2 \omega \frac{\varepsilon_0 \varepsilon_r 2\pi r}{dr} \tan\delta \text{ W/m}$$

where

$$\tan\delta = \text{power factor} = \text{dielectric loss angle}$$

The dielectric loss per volume is W_{cd}:

$$W_{cd} = (dV_0)^2 \omega \frac{\varepsilon_0 \varepsilon_r 2\pi r}{dr} \tan\delta \frac{1}{2\pi r.1.dr} = \left(\frac{dv_0}{dr}\right)^2 \omega \varepsilon_0 \varepsilon_r \tan\delta \text{ W/m} \quad (8.102)$$

Now, the potential gradient is given by $dV_0/dr = E_r$ at the distance of radius 'r'.

As such,

$$W_{cd} = E_r^2 \omega \varepsilon_0 \varepsilon_r \tan\delta \text{ W/m} \tag{8.103}$$

The electrical stress is given by

$$E_r = \frac{V_0}{r \ln(R/r_0)}$$

where

V_o is the voltage between the conductor and earth
R is the radius of the finished cable
r_0 is the radius of the conductor

Now, the dielectric loss per unit length of the annular cell in the cable insulation at a distance r having a thickness dr is given by

$$dW_{d(r)} = W_{cd} 2\pi r\, dr = \omega \varepsilon_0 \varepsilon_r \tan\delta E_r^2\, 2\pi r\, dr$$

$$= \omega \varepsilon_0 \varepsilon_r \tan\delta\, 2\pi \frac{V_0^2}{(\ln(R/r_0))^2} \frac{dr}{r} \tag{8.104}$$

Taking

$$p = \omega \varepsilon_0 \varepsilon_r \tan\delta\, 2\pi \frac{V_0^2}{(\ln(R/r_0))^2}$$

the equation becomes

$$p \frac{dr}{r} \tag{8.105}$$

Integrating

$$dW_{d(r)} = \int_{r_0}^{r} dW_{d(r)} = \int_{r_0}^{r} p \frac{dr}{r} = p \ln \frac{r}{r_0}$$

Thus, the dielectric loss of the insulated conductor over a unit length will be $W_d = p \ln(R/r_0)$ and the thermal resistance of a thin cylindrical shell at distance 'r' with a thickness of dr should thus be $dG_{d(r)} = g_d(dr/2\pi r)$; hence, the total thermal resistance of the cable on unit length is

$$G_d = \frac{g_d}{2\pi} \ln\left(\frac{R}{r_0}\right)$$

where g_d is the thermal resistivity of dielectric.

Now, the heat flow due to dielectric loss within the cylindrical shell at a distance '*r*', when crossing through the thin shell *dr*, is given by $dG_{d(r)}$ and experiences a temperature rise of

$$d\theta'_{d(r)} = W_{d(r)}G_{d(r)} = p\left(\ln\left(\frac{R}{r_0}\right)\right)\frac{g_d}{2\pi}\frac{dr}{r} \tag{8.106}$$

Therefore, the total temperature rise within the above ambient ground temperature due to dielectric loss within a unit length within insulation will be

$$\theta'_d = \int_{r_0}^{R} p\frac{g_d}{2\pi}\left(\ln\frac{R}{r_0}\right)\frac{dr}{r} = p\frac{g_d}{2\pi}\cdot\frac{1}{2}\left(\ln\frac{R}{r_0}\right)^2 = \frac{1}{2}\left(p\ln\frac{R}{r_0}\right)\left(\frac{g_d}{2\pi}\ln\frac{R}{r_0}\right) = \frac{1}{2}W_dG_d \tag{8.107}$$

Now, the temperature rise above ground temperature and within the dielectric loss alone brings an additional rise of conductor temperature equal to half of the product of the dielectric loss W_d and thermal resistance G_d of the cable insulation.

The final temperature rise of the cable conductor surface above ground temperature due to dielectric loss is

$$\theta_d = \theta'_d + W_d(G_b + G_{ser} + G_s) = \frac{1}{2}W_dG_d + W_d(G_b + G_{ser} + G_s)$$

$$= W_d\left(\frac{1}{2}G_d + G_b + G_{ser} + G_s\right) \tag{8.108}$$

This amount of rise in temperature has to be reduced from the conductor temperature rise due to load current.

Further, the values of thermal resistivity calculations are to be referred as per Equations 8.33 through 8.59 as appropriate for the calculations of current rating. It is to be noted that previous calculations are shown as a guideline when cables are laid underground. For other types of installation and different environmental conditions, refer to the different parts of IEC 60287.

The voltage and the current induced within the screen of an MV aerial bunch cable need to be considered while installing these cables:

Frequent failure of the AB cable is being reported during service conditions. It has been found that most of the failures are near the end termination, where termination of the screen is not properly done, and the earthing of the point is casually terminated. The failure is more in the case of aluminium tape screened cables.

Induced voltage within the screen is calculated with the help of following formula. (To consider that there are three phases around the messenger conductor which are running parallel in spiral formation—and are not transposed)

When the cable sheath is bonded and earthed at one end

1. The case sheath current will be $I_{sh} = 0$.
2. Sheath voltage = ωIM V/m, where $\omega = 2\pi f$ and I = rated current flowing through the conductor.

Therefore V_{sh} volts per meter, is to be calculated as per Equations 8.76 and 8.77.

$$V_{sh1} = V_{sh3} = [I]\sqrt{(X^2 + XX_m + X_m^2)} \quad \text{and} \quad V_{s2} = [I]X$$

considering $I_1 = I_2 = I_3 = I$

Example 8.3

Voltage induced within the screen of each phase
AB cable 120 mm² × three cores × 11 kV + 70 mm² messenger conductor
AB cable 120 mm² × three cores + 70 mm² Messenger.
Stranded, compacted, aluminium conductor, inner semicon, XLPE-insulated, outer semicon, copper tape screened Black polythene sheathed, three core laid around a messenger conductor.

Construction Details	Unit	Values	Unit
Area of condr.	mm²	120.00	
Diameter of condr.	mm	13.18	
Thickness of inn. semicon	mm	0.50	
Diameter over screen	mm	14.18	
Thickness of insl	mm	3.80	
Diameter over insl	mm	21.78	
Thickness of outer screen	mm	0.60	
Diameter over outer screen	mm	22.98	
Thickness of copper screen	mm	0.045 × 2 nos.	
Diameter over copper screen	mm	23.16	
Thickness of outer sheath PE	mm	2.20	
Diameter over outer sheath	mm	27.56	
Apx.	mm	28.00	
Resistance of condr	Ω/km	0.253	
Current rating	A	260.00	
Resistance of screen			
Area of screen	mm²	6.52	
Resistance of screen	Ω/km	2.64	
Short-circuit current	kA	11.32	
$X =$	ω M	0.0863	
$X_m =$		0.0436	
$V_{sh} =$	Per km	29.7656	V/km
$V_{sc} =$	Per km	1295.95	V/km

In this case, a one-point earthing gives a voltage rise in voltage as 30 V/km. If the length is longer, it will be multiplied by the quantum of length. Without proper earthing at both ends, this voltage will erode the insulation at the termination point creating fault conditions. In, the case of an aluminium tape screen, there are more problems as the earthing contact with the metal cannot be achieved properly unless a proper jointing technique is adopted, which will be discussed in Chapter 11 in more detail.

Current ratings for cables laid and installed under different conditions, and in different configurations, are to be determined by consulting the calculations given in IEC 60287. The aforementioned calculations are shown only as a guideline for beginners.

8.3 Voltage Induced in the Conductor of a Control Cable When Laid alongside an Electrified Railway Track

It is seen that a power line emitting a magnetic field influences a conductor within a cable running alongside such power lines. An induced current and voltage when generated within the conductor of a signalling or communication cable can distort transmitted signals. This ultimately can cause malfunctioning of railway communication systems and signal lamps by the sides of the railway tracks. This may lead to unwarranted accidents. To eliminate such problems, the cable has to be covered with a metallic sheath or screen. The magnetic field, as seen earlier, should induce a voltage within the metallic sheath. If the sheath is solidly earthed, the induced current will flow into the earth and the voltage should be reduced theoretically to a zero level. To create such an ideal condition, the metallic covering has to be infinitely thick. However, for practical applications, it has been specified in Verinigte Deutsche Elektrotechnik (VDE) 0227 that induced voltage within the sheath should be restricted to 60 V at 50 Hz at normal operating conditions and 300 V during short-circuit conditions. The cable has to be tested as per VDE 0472 at 50 Hz. This voltage increases as the frequency increases. The reduction in voltage has been specified by defining the reduction factor, which is given by

$$r_k = \frac{\text{Voltage induced within the conductor after a metallic sheath covers the cable}}{\text{Voltage induced within the conductor without any metallic covering}}$$

As per Indian railways, the reduction factor has been specified and should be below $r_k \leq 0.4$ at 50 Hz. In order to achieve this value, the thickness of the metallic sheath has to be determined as follows:

The following information is required to proceed with the calculation (Table 8.7): (Taken as per Indian railway standard requirements: the over head (OH) transmission line is of 25 kV AC at 50 Hz, and on two-track basis).

During operation of an electric locomotive, the change of AC power to DC power produces a series of harmonics. These harmonics interferes with the signalling frequency transmitted through the conductor due to unbalance in the capacitance and magnetic field. A little unbalance can be amplified by the induced voltage. In order to keep this influence/voltage within a permissible limit, a metallic covering is required, which gives an effective result. This metallic sheath should have low resistance. Lead, aluminium or copper is the preferred material for screening or sheathing. To determine the thickness of the sheath, the reduction factor, as stated earlier, has to be known and specified. The reduction factor can also be defined as

$$r_k = \frac{R}{R^2 + \omega L^2} = \frac{\text{DC resistance}}{\text{AC resistance}} \tag{8.109}$$

where

R is the DC resistance of the sheath
$\omega = 2\pi f$, where f is the frequency
L is the induction of the cable, normally taken as 2 mH/km

To achieve a low reduction factor, 'R' should be as low as possible or the frequency f can be increased. Or $(R^2 + \omega L^2)$ has to be increased. This can be achieved by the armour of the cable with a steel tape having high permeability.

The voltage induced in a cable is determined by

$$U_I = 2\pi f \times M_e \times I_f \times l \times r \tag{8.110}$$

$$U_K = 2\pi f \times M_e \times I_K \times l \times r \tag{8.111}$$

where

E_{IF} is the induced E within the sheath due to normal operating current in the overhead contact wire
E_{Ik} is the induced E within the sheath due to short-circuit current
F is the frequency
I_f is the operating current (A) (track being the return path)
I_k is the short-circuit current (A)
M_e is the mutual inductance between sheath and railway track
r is the reduction factor of rail (or number of rails running parallel).
l is the length of the cable

TABLE 8.7

Data Required for Calculating the Reduction Factor of Traction Cables

	Required Information	Data Obtained
[1] (a)	Normal operating current	600 A for two tracks (300 A per track).
(b)	The short-circuit current	3000 A.
(c)	How many tracks are there? How many engines run at a time?	Two tracks. Maximum current used is 600 A. For double track, i.e. 300 A per track. The number of engines at a time will consume 300 A per track.
(d)	Earth conductivity in microsiemens (µS)	The same is not available but resistivity (ρ) is = 25,000 Ω/cm³.
(e)	Permissible voltage in the cable (according to German Standard, the max. voltage permitted in the cable is 300 V without transformer and 1200 V or 60% of the testing voltage with transformer in a short-circuit condition and 65 V on normal operating condition)	Permissible cable in voltage 35 V/km. The maximum voltage under condition as is governed by the CCITT specification.
[2] (a)	The distance of the cable from the railway line.	In between the two tracks, the distance from the midpoint of a track to the cable should not be less than 1.52 m.
(b)	The depth of laying	The depth of laying of cable is normally 0.8 m, but the cable may at times run on the surface too.
(c)	If the cable is not terminated with transformers, the maximum length of the cable to be laid	The cables are sometimes terminated with transformers, in case the length of the cable exceeds 3 km, but whenever the cable is not terminated with transformers, the maximum length of the cable is restricted to 3 km.
(d)	Minimum height of the overhead traction line	The minimum height of the contact wire from the rail level is 5.5 m and the maximum sag of the catenary from the point of suspension is 0.9 m. The rail level from the datum line (ground) is 0.5 m.
(e)	Reduction factor of rail	Reduction factor of rail is = 0.4.
(f)	Supply frequency (AC)	50 Hz.

To determine the value of M_e, it is necessary to know the distance 'a' of the cable from the rail, as it is installed underground in metres. Here, the conductivity of soil σ is to be known (μS/cm) by knowing the resistivity of soil as defined by $\sigma = 1/\varrho$ and is given as

$$1\,\mu\text{S/cm} = \frac{1}{10^4\,\Omega\text{m}};\ \text{or}\ 1\ \Omega\text{m} = \frac{1}{10^4\,\mu\text{S/cm}}$$

To obtain the value of M_e, the factor 'x' is to be determined first

$$x = a/m\sqrt{\frac{\sigma f}{(1000\,\mu\text{S/cm})\ \text{at a given Hz}}} \tag{8.112}$$

Once 'x' is determined, the corresponding value of M_e in μH/km is to be taken from the graph as given in specification No. VDE 0227.

While calculating the thickness of the metal sheath, it is advisable to take the short-circuit value for safety reasons.

Example 8.4

a. The single track rail line reduction factor 0.7
b. Short-circuit current 3000 A
c. Conductivity of soil (σ) 2000 μS
d. Frequency of OH supply 50 Hz
e. Working voltage 25 kV
f. Distance from the cable track (a) 30 cm
g. Cable length (l) 1 km

Voltage has to be induced in the sheath, taking into account the short-circuit current only

$$U_K = 2\pi f \times M_e \times I_k \times l \times r$$

$$= 2 \times \pi \times 50 \times 1.300 \times 10^{-3} \times 1000 \times 0.7$$

$$= 825\ \text{V}$$

To get M_e, the value of x to be determined is

$$x = \frac{30}{100} \times \sqrt{\frac{2000 \times 50}{1000}} = 3\ \text{(Ref. 8.112)}$$

From the graph, the value of M_e is = 1300 μH/km.

The permissible voltage rise of 300 V should be allowed for short-circuit ratings.

Hence, the reduction factor obtained is to be

$$r_k = \frac{300}{825} = 0.36 \text{ apx. is lower than required } 0.4$$

Now, referring to Equation 8.109

$$R = r_k \sqrt{\frac{\omega^2 L^2}{1 - r_k^2}} = 0.36 \sqrt{\frac{0.887}{1-(0.36)^2}} = 0.363\ \Omega/\text{km (where } L = 3 \text{ mH)}$$

This is the resistance required for the metal sheath to get the desired reduction factor.

For lead, ρ = 0.208, and for aluminium, ρ = 0.0287. By knowing the diameter of the cable under sheath, the thickness can be calculated.

The diameter of the cable under sheath is 16.2 mm.

The area, 'Q', required for the lead sheath to get the desired resistance is

$$0.36 = \frac{0.208 \times 1000}{Q}\text{; hence, } Q = 578 \text{ mm}^2$$

The area of the annular ring is $Q = \pi\,(D + t)t$, where D is the diameter of cable under sheath and t is the thickness of the sheath (this is a quadratic equation); by solving, it comes to 't' = 7.7 mm apx. for lead, whereas for aluminium, 't' should be = 2.21 mm apx.

8.4 Transmission of High-Frequency Signal through HV and EHV Cables

An HV or EHV cable can be taken as a coaxial cable for the transmission of carrier frequency for sending signals from end point to end point. With the inner conductor being the carrier of power and screen or metallic sheath acting as a wave guide, the attenuation factor is higher in this case, as the dielectric is solid and the value of 'ϵ_r' is higher, that is, 2.5. Even then, attenuation and impedance values can be calculated at a given frequency.

Required parameters as mentioned above can be calculated considering the cable to be a coaxial one having frequency range exceeding 50 Hz. Due to increase in frequency, active resistance R_a consists of R_d internal resistance of wire and R_D external resistance of the concentric conductor (metal screen/sheath). Hence,

$$R = R_d + R_D = \left\{ \sqrt{\left(\frac{\mu_d f}{\sigma_d}\right)} \times \left(\frac{2}{\sqrt{10}}\right) \times \frac{1}{d} + \sqrt{\left(\frac{\mu_D f}{\sigma_D}\right)} \times \frac{2}{\sqrt{10}} \times \frac{1}{d} \right\} \Omega/\text{km} \quad (8.113)$$

where

σ_d, σ_D and μ_d, μ_D are the electrical conductivities and magnetic permeability of internal and external conductors, respectively
d is the diameter of the internal conductor (mm)
D is the diameter of the external concentric conductor (mm)

The magnitude $\sqrt{(\mu f/\sigma)}$ for various metals is shown in the following:

For copper wire coaxial cables, it can be expressed in the following form:

$$R = R_d + R_D = 0.0835\sqrt{f} \times \left(\frac{1}{d} + \frac{1}{D}\right) \quad (8.114)$$

where d and D are in millimetres.

The inductance of the circuit is composed of an internal self-inductance 'L' of both the conductors (R_d and R_D) and an external mutual inductance 'L_e' and is given by

$$L = L_d + L_D = \left[2\ln\frac{D}{d} + \left\{ \left(\frac{133.3}{\sqrt{f}}\right) \times \left(\frac{1}{D} + \frac{1}{d}\right) \right\} \right] \times 10^{-4}\ \text{H/km} \quad (8.115)$$

Since internal inductance is considerably less than the external one and with the increase in value of 'f', it reduces more and more. Therefore, the inductance of coaxial cable can be given by the following formula to a sufficient degree of accuracy:

$$L = \left[2\ln\left(\frac{D}{d}\right) \right] \times 10^{-4}\ \text{H/km} \quad (8.116)$$

Capacitance is given by the following formula: $C = \{\varepsilon/[18 \ln (D/d)]\} \times 10^{-6}$ F/km where 'ε' is the relative permittivity.

The leakance 'G' is calculated as

$$G = \omega C \tan \delta \quad (8.117)$$

where $\tan \delta$ is the tangent of the equivalent power factor.

From the previous attenuation factor, getting the primary parameters as before is given by

$$\alpha = \alpha_m + \alpha_{ins} = \frac{8.35\sqrt{f\varepsilon}\left(\dfrac{D}{d}+1\right)\times 10^{-3}}{12D\ln\dfrac{D}{d}} + 10\pi f\sqrt{\varepsilon}\tan\delta\times 10^{-6}\times\frac{1}{3}\,\text{neper/km} \tag{8.118}$$

In this case, α_m is the attenuation due to the metal conductor which is proportional to $\sqrt{f}$ and α_{ins} is attenuation due to the dielectric, where it increases rapidly with the increase in frequency 'f'.

Phase coefficient β is determined by the following equation:

$$\beta = \omega\sqrt{LC}\ \text{rad/km} \tag{8.119}$$

or it can be expressed by the following equation:

$$\beta = \omega\sqrt{\frac{\varepsilon}{c}}\ \text{rad/km} \tag{8.120}$$

where c is the velocity of light (c = 300,000 km/s).

The electromagnetic energy propagation velocity over the coaxial circuit is given by

$$V = \frac{\omega}{\beta} = \frac{c}{\sqrt{\varepsilon}}\ \text{km/s} \tag{8.121}$$

Here, the characteristic impedance is given by

$$Z_w = 60\frac{\ln(D/d)}{\sqrt{\varepsilon}}\ \Omega \tag{8.122}$$

Nowadays, signal and communication frequencies are transmitted through optic fibre tubes inserted within the cable during the manufacturing of an underneath cable sheath (Tables 8.8 and 8.9).

TABLE 8.8

Magnitude $\sqrt{(\mu f/\sigma)}$ for Various Metals

Metal	$\sqrt{(\mu f/\sigma)}$
Copper	0.132
Aluminium	0.171
Lead	0.47
Steel	3.72

TABLE 8.9

Thermal Resistivities of Materials

Material	Thermal Resistivity (km/W)
Insulations	
Polythene (PE) and XLPE	3.5
PVC up to 3 kV	5.0
PVC above 3 kV	6.0
EPR up to 3 kV	3.5
EPR above 3 kV	5.0
Butyl rubber and natural rubber	5.0
Protective coverings	
Polychloroprene (PCP)	5.5
PVC covering up to 35 kV grade cables	5.0
PVC coverings above 35 kV cables	6.0
PE	3.5
Materials for ducts	
Concrete	1.0
Fibre	4.8
Asbestos	2.0
Earthenwares	1.2
PVC	7.0
PE	3.5

9

Quality Systems, Quality Control and Testing

Every product made needs to be checked for its quality.

9.1 What Is Quality?

Quality is the characteristic of a product, attributed to it during manufacturing, to be able to achieve a given function under certain defined circumstances, economically and in a rationalized and specified manner, to give maximum benefit to the user on a long-term basis.

Its explanations are as follows:

1. A product: such as a power cable.
2. Characteristic to be attributed: Design parameters such as
   a. Conductor material: to be selected for efficient transmission of power
   b. Manner of formation: impart flexibility and mechanical stability
   c. Insulation: the type and strength to endure electrical pressure and variable thermal conditions
   d. Formation of cable core and a protective layer: protect against magnetic and thermal behaviour
   e. Armouring: for mechanical protection and safety against short-circuit conditions
   f. Final protective covering: against the environment, mechanical abrasion and corrosive actions

   All these aspects are to be considered for a long-term safe and reliable performance.
3. Compatible: Design and manufacturing should be such that under different conditions, the cable should be able to withstand all hazardous environments, both during installation and operation.

4. Specified manner: Design and manufacturing should conform to national and international standards.
5. Rationalised manner: Overall cost should be kept to a minimum while accepting given manufacturing tolerance; utilising minimum machinery, equipment and raw material; and generating a minimum amount of scrap. There should also be an efficient utilisation of power, manpower, cost of handling and transportation.

In this context, attention should be paid to the following actions:

1. The quality of available resources should be compatible to manufacturing techniques to be able to yield the desired result with minimum expenditure.
2. Input material should be as per given standards. The supplier has to ensure consistent quality throughout.
3. A statistical evaluation has to be brought into the system to ascertain a consistent quality of the finished product.
4. Long-term analysis should bring forward the level of quality trend of the institution. Accordingly, factory standards are to be established based on the acquired input. Ongoing improvement needs to be initiated in all spheres of activity.

Available resources include the following:

1. Land and building
2. Machinery
3. Manpower
4. Raw materials
5. Power
6. Water
7. Road and transportation
8. Communication

To initiate the aforementioned working policy and philosophy, stage-wise inspection and testing have to be carried out on every batch of materials, both during production and final testing of finished products. The obtained result has to be documented and studied for evaluation. The evaluated result should be judicially incorporated into the factory's manufacturing standards for implementation to improve results progressively.

9.2 Testing of Cable

Testing has to be carried out in three stages:

1. Raw material. Input material should be as per the given standard. The supplier has to ensure quality and consistency throughout. The factory has to formulate specifications as per national, international or factory requirement.
2. In process. A statistical evaluation has to be brought into the system to ascertain a consistent quality of the finished product. In-process standard has to be worked out based on machine capacity, human ability and given design parameters. These have to be documented, studied and utilised for continuous improvement.
3. Final testing of the finished product. Long-term analysis will bring forward the trend in the level of quality of the organization. Accordingly, factory standards have to be established and revised based on whatever continuous improvements are being carried out in all spheres of activities.

Three types of tests are carried out:

1. Physical tests – mechanical
2. Material composition and behavioural tests – chemical
3. Electrical tests – for electrical strength

Electric cables are normally laid and installed outdoors, underground, in the open and on rough terrains near refineries and factories, where chemical affluents remain in the surrounding area. Cables have to be bent near corners and installed vertically. Cables have to be laid along the streets, on bridges and near railway tracks where they are subjected to constant vibration. Sometimes, cables may have to be laid across water drains and at times may even remain submerged in water in lowland areas during the rainy season. During laying, cables are dragged over abrasive grounds. Naturally, the conductor along with protective materials like outer sheaths, and insulation enclosing conductors, must withstand all types of abuses – stress and strain. Fluctuations in humidity and temperatures during day and night constantly affect the materials. Cold climates and hot and humid atmospheres also affect the conductor, insulation and cable sheath.

Furthermore, during operation, a temperature rise within the conductor will affect all construction materials and most likely affect their physical characteristics in the long run. A sudden surge in current and short-circuit condition will also affect the parameters of a cable.

It is important to ascertain and ensure that all the materials used, or to be used, have the capacity to withstand such stress and strain. Since it is not possible to test all raw materials and finished products, tests have to be conducted on prototype from each batch. The mechanical characteristics, thus, have to be examined by testing tensile and elongation on samples of raw materials and finished cables.

In today's production system, though automation has been introduced to eliminate human error in most areas, some amount of variation, however small, may still come into play. Naturally, a batch-wise quality check has to be conducted compulsorily. The results obtained should lie within the given limit of tolerance. A variation in results may become more pronounced in case of polymeric materials, where the basic polymer resin quality may vary batch to batch, however small it may be. Therefore, the characteristics of compounded materials are restricted to vary within a specified limited range.

9.2.1 Tests on Raw Materials

As far as possible, input materials have to be identified as follows:

1. 'A' grade materials. They are procured in bulk and are considered essential items for manufacturing:
    a. Conductor material: Electrochemical (EC) grade copper wire rods
    b. Conductor material: EC grade aluminium wire rods
    c. Insulating materials: polyvinyl chloride (PVC), polythene, cross linked polythene (XLPE), etc.
    d. Armouring materials: Galvanised steel wires, steel tapes, galvanised steel strips
    e. Outer sheath material: PVC, high density polythene (HDPE), low density polythene (LDPE), etc.
    f. Semiconducting materials
2. 'B' grade materials: Like copper tape for screening, filler tapes and materials, binder threads, polyester, PVC and other types of tapes, water blocking tapes, and so on.
3. 'C' grade materials: Consumables, lubricants, etc.

Of these three categories, the materials specified under 'A' have to be tested frequently. In these cases, though the supplier's certificate is considered to be in order, to ensure quality, occasional testing has to be carried out to establish the proper credentials of the supplier. Furthermore, the test can also direct design and manufacturing engineers to regularly monitor raw material consumption.

When testing copper wire rods

1. Check the average diameter
2. Check the density of the rod
3. Check the conductivity of the material

Tests (2) and (3) can influence the production speed and consumption of raw materials.

Similar tests on aluminium rods will determine the production speed and raw material consumption parameters. Whereas copper consumption can be controlled by an annealing process, aluminium consumption will mostly depend on the quality of the wire rod received.

9.3 Tensile Strength

Tensile strength ascertains the capacity of the material to withstand mechanical abuses during winding and unwinding and to accept a pulling force while laying on rough terrain.

9.4 Elongation

Elongation is the capacity to withstand strain during bending, unbending and twisting.

The following are tests conducted on copper and aluminium wire rods and wires:

1. *Wire rod stage*: A wire rod of either copper or aluminium as received from the supplier must be tested for tensile strength and elongation capacity to ensure that the wire rod breakdown machine is capable of drawing the required sizes of wires without getting overloaded. The breaking load and elongation determine the degree of softness of wire rods. Accordingly, the machine load, wear and tear and lubricant temperature and viscosity have to be adjusted to get the best possible speed of the wire drawing machine. If the wire rod is too hard, it has to be annealed before drawing. Nowadays, wire rods received from the Properzi mill or from the up-cast machine give consistent results. Even then a check on the quality should be ensured and a record kept to shortlist the best supplier. During drawing, due to working hardness, the value of resistance changes. Hence, annealing has to be done to regain the properties. Hence, an annealing test is done by ascertaining the elongation of the drawn wires.

FIGURE 9.1
Wire rod, wire tensile and elongation testing unit.

2. Tensile and elongation tests are carried out on the same tensile testing machine (Figure 9.1).
3. After drawing wires as per the required sizes, they have to be checked for tensile and elongation. To ensure that the correct compacted stranded conductors have the specified electrical resistance values, it is necessary to measure the resistance of individual wires. Mechanical stress and strain values are obtained by testing for breaking load and elongation. This is to make sure that after stranding and compacting, the final result satisfies the designed values. At the starting stage, 1 m of the stranded conductor has to be taken for measurement. The value of resistance of 1 m being very low, a standard double Kelvin bridge has to be used to get the accurate result (see electrical tests). Before testing, a dimensional check has to be carried out on the samples such as diameter (sector height and width in case of a sector conductor) and weight. At least five to six samples have to be checked from every lot to ascertain the consistency (Figure 9.2).
4. The quality of PVC and polythene (poly ethylene) (PE) can be checked in the following manner:

   Raw material polymeric compounds are delivered by the suppliers in the form of granules in watertight packing. Certain quantities of the granule have to be taken from a batch and masticated in a lightly heated laboratory rolling mill. After masticating, the material has to be taken out in a sheet form. This sheet is then placed

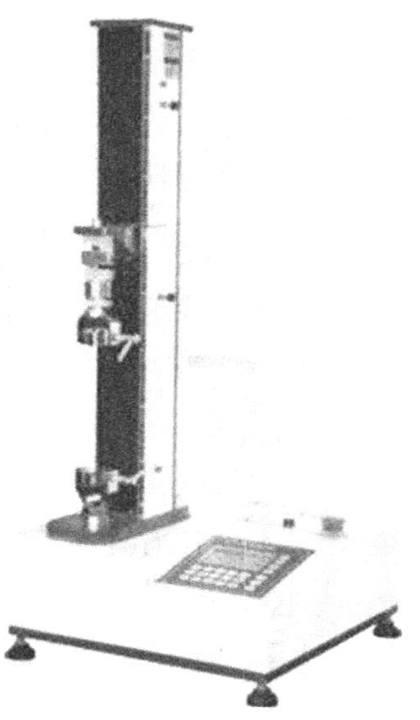

FIGURE 9.2
Electronically controlled computerised tensile and elongation testing machine for polymeric compounds.

in a grooved dice with a fixed thickness, length and width. This dice hot-pressed into a sheet with the given dimensions. The sample strips are then cut into a dumb-bell (Figure 9.3), lengthwise and widthwise. These strips are measured carefully after cooling down to room temperature and then tested for tensile and elongation. As for grafted XLPE polymers, normally the supplier's test certificate is considered authentic. Even then the polymer can be tested in a mooney viscometer to ascertain the curing characteristics in relation to time and temperature.

The main criterion is the cleanliness of the XLPE compound. The higher the voltage, the greater is the demand for a contamination-free compound. The contamination level is specified by the supplier of the materials.

5. Tensile and elongation tests on polymers such as PVC, polythene, XLPE and other polymeric compounds.

   The present polymeric tensile and elongation testing machine, as designed, is operated by an electronically computerised system. The results and variations are automatically recorded and statistically evaluated to indicate the suitability of materials.

When all the testing is finished, samples are drawn from every lot. Strips are taken from the insulated conductor and also from the outer sheath material.

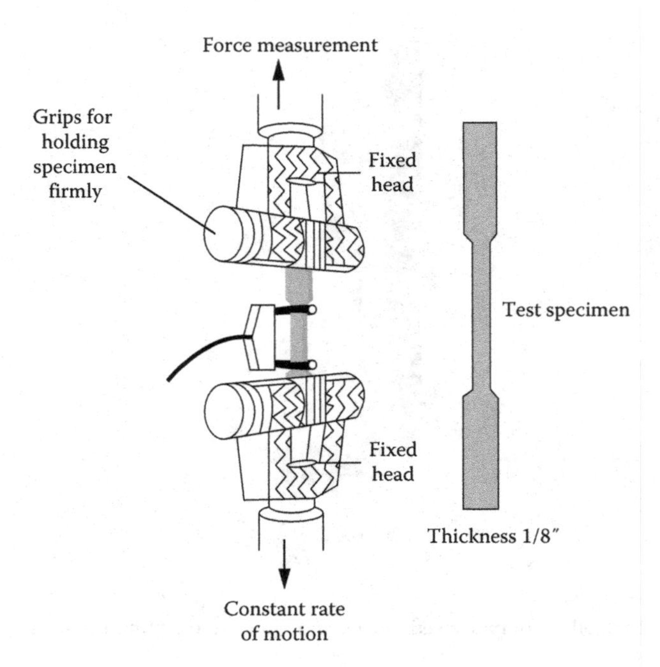

FIGURE 9.3
Test specimen and grips.

These samples are tested for tensile and elongation. The results are to be corroborated with the results obtained at the raw material stage. Normally, suppliers have to supply the relevant test results for initial approval.

9.5 Ageing Test

When current flows through the conductor, heat is developed as per the equation $W = I^2R$, where I is the current (A), R the resistance of the conductor (Ω) and the heat developed 'W' is in watts. The temperature rise within the conductor is given by

$$T = \frac{0.24VIt}{Mc}, \quad \text{taking} \quad t = 1\text{s}\ T = \frac{0.24VI}{Mc}\ {}^{\circ}\text{C}$$

T is the temperature rise over ambient (°C)
V is the voltage (V)
I is the current (A)
M is the mass of the conductor metal (g/cm³)
c is the conductivity of metal
t is the time (s)

This amount of rise in temperature during operation would act constantly on the polymeric insulation and also on the bedding and outer sheath materials. At times, a short-term overload current and short-circuit current develop transient heat, which takes time to dissipate. This transient heat acts upon the polymeric materials used for the construction of electric cables.

Cables are normally installed underground or in an open atmosphere. These cables may also pass through hot zones of factory premises. Naturally, outside environmental factors will react on the polymer and degrade its life, in a slow process. In such conditions, molecular structures start reorienting and dissociation comes into play within the structure of the materials.

In order to ascertain the durability of the materials for long-term performance, ageing tests are simulated artificially, by raising the temperature at a predetermined level and for a certain time. Thereafter, tensile and elongation tests are done on the aged samples. The percentage of retention of tensile and elongation is noted and compared with specified values.

The ageing process is carried out by keeping the conditioned samples in an enclosed chamber (Figure 9.4), with electrical heating and hot air circulating within to keep the temperature constant throughout.

FIGURE 9.4
Ageing oven.

9.6 Hot Set Test (A Measure to Determine the Amount of Cross-Linking Achieved) for Cured Materials such as XLPE and Elastomers

The test comprises of two parts: elongation test and set test.

1. *Elongation test*: A specimen is subjected to a constant load stress while suspended in an air oven, at a specified temperature and for a specified time period. At the end of the time period, the increase in elongation of the specimen is determined.
2. *Set test*: Immediately after the elongation test has been completed on a specimen, the same specimen with the load stress removed is subjected to an additional time period in the oven at the same elevated temperature. The specimen is then removed and allowed to cool. The set of the specimen, based on the original length, is then determined (Figure 9.5).

FIGURE 9.5
Hot set test apparatus.

The test consists of preparing a tubular specimen for insulations of 60 mils, or less, or a die-cut specimen for larger thickness insulations. For larger thicknesses, the specimen shall be taken from the inner 25% of the insulation. The cross-sectional area is calculated to determine the weight that has to be suspended from the specimen. The specimen is placed in an oven at 150°C. After 15 min exposure, without removing the specimen from the oven, the elongation is measured and a % change is recorded from the original 25 mm gauge marks. The set test is performed on the same specimen while in the oven and left for 5 min. It is then removed and allowed to cool to room temperature. The distance between the original 25 mm gauge marks is again measured, now that the specimen is cooled, % change to the original is calculated as the 'set' value. This is to check the amount of cross bonding being accomplished within the material.

9.7 Solvent Extraction

At the initial state, the solvent extraction test was developed to determine the amount of cross-linking within the cured XLPE material so as to ensure quality of insulation.

This test determines the precise amount of cross-linking that has taken place in a polymer by measuring the gel content (insoluble fraction). The procedure involves grinding a very fine sample from the insulation and boiling it in a solvent that dissolves the un-cross-linked portion. The sample is then placed in a vacuum chamber and heated to drive off any residue of the solvent. The sample is then weighed. By measuring the before and after weights, the percentage of the extractable can be determined (soluble portion). This test method serves as a referee test to the hot creep test with a maximum of 30% extractable (weight loss), or a minimum 70% gel content, and after a 20 h drying time in the vacuum chamber.

9.7.1 Test Evolution

It was the solvent extraction test that was used by the early developers of thermoset compounds. With the introduction of cross-linked extruded semi-conducting shields, the specification was then expanded to include these materials in 1974. It wasn't until 1987 when the solvent extraction test for extruded semiconducting shields was renamed the wafer boil test with the same requirements. Also, the hot set test replaced the solvent extraction test.

Solvent extraction testing took many hours to perform, which could result in the scrapping of possibly thousands of kilometres of cable being produced if the test failed. It was determined in the early 1980s that a faster test method was needed which could determine if the cable was sufficiently cured. Thus, the hot set test was developed. With a good correlation between the two, hot set replaced solvent extraction as the main test. The hot set test was used for a short time in 1994 to determine the cure state for semiconducting insulation shields. However, it was discovered that insulation shields could be formulated to meet hot creep requirements but had not been sufficiently cured when tested by the solvent extraction test. The use of the hot creep test was discontinued for insulation shields by 2000. This left the wafer boil test as the main test for this material.

The previous tests are routinely carried out during production for in-process control and also on finished cables.

The following routine and acceptance tests were carried out to confirm the integrity of the produced cables before being supplied to the customers (samples are drawn as specified in the relevant standards):

1. Tests on finished products: Initially, the external characteristics are examined.
   a. Cables in every container drums are examined visually for any external fault or damage. This is to be done during winding on the final shipping drum. The drums are checked for quality before winding.
   b. Constructional tests on specimen: A sample length is taken from a lot and checked as follows:
      i. External diameter of the cable. Measurement is taken over the cross section of the cable. Three readings are taken of the same circle rotating 120° on the circumference. Readings are taken on five different points and the average calculated.
      ii. Thickness of sheath, insulation and semiconducting layer. Measurement is taken with a vernier caliper or micrometre. To get an accurate measurement, a travelling microscope can also be used.
      iii. Measure the diameter over the armour, under the armour and insulated conductor dimensions.
      iv. Armour wire diameter and torsion tests have to be conducted along with tensile and elongation. This is required to understand that the armour wires and strips are flexible enough to be applied on the cable by an armouring machine without any twist or breakage.

It has to be understood that apart from controlling the weight and electrical stability, all these measurements facilitate

a. Selection of packing length and drum dimension
b. Selection of jointing kits and accessories
c. Selection of equipments, manpower and implements for laying and installation of cables

Electrical tests:

Test conditions

2. Electrical source: Power frequency voltage test. The frequency of the voltage should be either 50 or 60 Hz. The waveform should be in the form of a sine wave, if not specified as to be a pure sine wave.
3. Application of voltage: When the voltage is applied at the start, it has to be 20% of the maximum value. The voltage is then raised slowly within 1 min at a uniform rate to the specified value.
4. Ambient temperature is to be set to 20°C ± 15°C.

   The number of lengths to be tested may be reduced, or an alternative test method adopted, according to the agreed quality control procedures. The routine tests required by this standard are as follows:

   a. Measurement of the electrical resistance of conductors.
   b. Partial discharge test on cables having cores with conductor screens and insulation screen.
   c. Voltage test.

      The sample length must be conditioned at room temperature before measurement is being carried out. A temperature correction factor has to be applied to maintain it at 20°C.

9.7.1.1 Insulation Resistance Test

There is no material which does not absorb heat from the available field of energy surrounding it. When the amount of energy absorption is very limited, then the material falls under the category of insulation. As explained earlier, the vibration of atoms within the bonded molecular structure does not allow electrons to get dissociated, but the intensity increases with the increasing thermal stress and electrical field surrounding it. The quantum of resistance exerted to resist the electrical stress against the conduction of electron flow is the measure of insulation resistance. The higher the value, the better is the material. But this value deteriorates with the rise in temperature and moisture content. The temperature allows the electrons to absorb more energy, increasing the frequency of vibration. Water being a conductive element allows free passage of electrical energy.

While processing, polymers undergo many changes. Internal molecular structures reorient as the flowing material from the extruder is drawn forwards longitudinally. Further, the evolution of internal gas generates microporosity within the insulation. Some variation in contour and thickness may occur. Even after cooling, the ageing process starts within the insulation but very slowly. Fluctuation of environmental conditions and water absorption influences the quality of insulation. It has been discussed earlier that during operation, a small amount of power is absorbed by the structural molecules of insulating compounds. Measurement is done with a Million Megohm Meter. In the case of armoured cables, measurement is done by connecting the armour to the earth and the conductor to the charging terminal. In the case of single-core cables, the insulated core is immersed in water at 20°C, keeping both ends above water. In this case, the water is connected to the earth. The conductor is charged with 500 V DC for 1 min, until the indicator dial reaches a steady state. This gives the measure of the insulation resistance value. It is then calculated to convert on a per km basis using the following formula:

$$R_{ins} = \frac{\rho}{2\pi l} \ln \frac{d_2}{d_1} \; \mathrm{M\Omega / m}$$

where

ρ is the resistivity of insulation
l is the length of the cable or conductor
d_2 is the diameter of insulation
d_1 is the diameter of the conductor

Generally, the calculation is converted to volume resistivity for better comparison and understanding.

The volume resistivity of the outer sheath is also determined in a similar manner. While testing is carried out, a certain length is immersed in water. Water is connected to the earth. The armour or screen is connected to a DC terminal. In extra-high-voltage cables, a conductive layer is applied over the outer sheath. The inner screen or metal sheath is connected to the DC terminal and the outer sheath is connected to the earth:

$$R_{ins} = \frac{\rho}{2\pi l} \ln \frac{d_4}{d_3} \; \mathrm{M\Omega / m}$$

where

ρ is the resistivity of the sheathing material
l is the length of the cable
d_4 is the diameter over the sheath
d_3 is the diameter under the sheath

The integrity of insulation is determined by this test.

Insulation resistance is measured at a high temperature, that is, at the operating temperature of the conductor.

1. A water absorption test is performed as per NEMA WC-5 specification.

   In this case, the insulated conductor is immersed in a water bath keeping the ends above the water surface. The temperature of the bath is maintained at 60°C. The initial capacitance is then measured and noted. The conductor is kept in this bath at this constant temperature for 14 days. The capacitance reading is noted every day and the changes are studied. After 14 days, if the capacitance does not deteriorate beyond a specified limit, it is understood that the insulation did not absorb any water beyond the specified limit.
2. It is also necessary to protect the cable from ingress of moisture or water at all times.
3. During operation, the insulation remains constantly under a hot condition under the pressure of voltage and current. With time, the molecular structure may become weak, particularly for synthetic polymers. In such cases, the insulation resistance (IR) value gradually drops down to a lower level. This needs to be monitored periodically to ascertain the health of the cable.
4. Repeated short circuits and overloading of the cables also affect the characteristics of the insulation.
5. Any faulty constriction, pinholes, foreign particle and burnt materials affect the performance of the dielectric, which may initiate failure of the cable.

This makes it necessary to look for the best quality materials and to follow a strict working principle during the application of insulation on the conductor by the extrusion process or otherwise.

9.7.1.2 Capacitance and Tan δ Test

It is seen that a certain amount of energy (electric charge) is absorbed by the dielectric of the cable during operation. The capacity of retaining this charge within certain lengths of the dielectric depends on the measure of retaining the quantum of charge within a unit volume at a given temperature, called dielectric constant 'ε' in relation to the relative permittivity of free space 'ε_0' = 8.85×10^{-12}. The electrical properties of insulating systems change due to ageing and a continuous electrical stress. By measuring electrical properties such as capacitance and tan δ regularly, it is possible to ensure the operational reliability of high voltage (HV) insulating systems and to avoid costly breakdowns.

This is particularly important for high-voltage and extra-high-voltage cables, cable bushings, power transformers, generators, power capacitors, etc.

A capacitance and tan δ test set comprises a C and tan δ bridge, high-voltage power supply with built-in standard capacitor and a set of cables. The compact design of the bridge uses the principle of three winding differential transformers on a high-permeability μ metal core. The bridge is contained in a sturdy metallic housing with μ metal lining, which shields it from external electromagnetic and electrostatic influences. A built-in battery-powered null indicator makes the bridge suitable for operation in workshops, factories, high-voltage substations, switch yards, etc. The high-voltage power supply is provided with a built-in SF6 gas-filled standard capacitor. It is suitable for both grounded and ungrounded objects. Capacitance and tan δ are measured directly, and no further calculations are required.

9.8 Null Detector

A built-in battery-operated electronic null detector is most suitable for balancing the bridge. It gives high sensitivity and accuracy for most of the applications and is very user-friendly. A phase-sensitive null detector can be used in place of the electronic null detector. For higher sensitivity and accuracy in critical operations, an oscillograph null detector can also be used.

9.9 Interface Suppression

The C and tan δ bridge has been specially shielded with μ metal sheets to avoid the effect of external interferences, making the measurement accurate. The phase reversal switch provided in the HV power supply effectively cancels interference/pick up by the object under test in an energised environment. For a situation where the induction is excessive and cannot be cancelled by the phase reversal switch, a separate three-level interference suppression unit can also be incorporated with separate C and tan δ adjustments.

9.10 Protection

The bridge is provided with built-in, high-voltage protection devices which protect it and operate against the failure of the test object or standard capacitor.

The amount of power lost due to the absorption of energy in a dielectric is determined by measuring the value of tan δ. For lower-voltage cables and below 66 kV, the value is not of much significance, but above 66 kV, the values of tan δ cannot be ignored. Calculation in determining tan δ is shown in Equation 5.27.

Tan δ is measured as the voltage U_o on the full drum at ambient temperature. An increment is noted after applying voltages of 0.5 and 2 U_o.

The measurement has to be done as a function of temperature, by raising the temperature on a 6 or 8 m cable sample and bending it around a mandrel 10 times its diameter, over the metallic shield with a half turn, and then unwinding it, repeating the same treatment and rotating it to 180° about its axis. After this, the bending test voltage should be applied to the cable between the conductor and the metallic sheath at a specified temperature.

After finishing this test, the specified voltage should again be applied at 105°C. In both cases the result should not vary by more than 0.1%.

9.11 Resistance of the Conductor

The electrical resistance of the conductors should be tested for all cable lengths that are submitted to routine tests, including the concentric conductor, if present. The complete cable length, or a sample from it, should be placed in the test room with a reasonably constant temperature, for at least 12 h before the test. In case of doubt as to whether the conductor temperature is the same as the room temperature, the resistance can be measured after the cable has been in the test room for 24 h. Alternatively, the resistance can be measured on a sample of conductor conditioned for at least 1 h in a temperature-controlled liquid bath. The measured value of the resistance should be corrected to a temperature of 20°C and 1 km length in accordance with the formulae and factors given in IEC 60228. The DC resistance of each conductor at 20°C should not exceed the appropriate maximum value specified in IEC 60228.

For concentric conductors, the resistance should comply with national regulations and/or standards.

9.12 Measuring Low Resistance Using the Kelvin Double Bridge

Although the Wheatstone bridge has a wide range of utilities, it is not well adapted to the measurement of a very small resistance, that is, resistance in the order of one thousandth of an ohm or even less. This is due to the

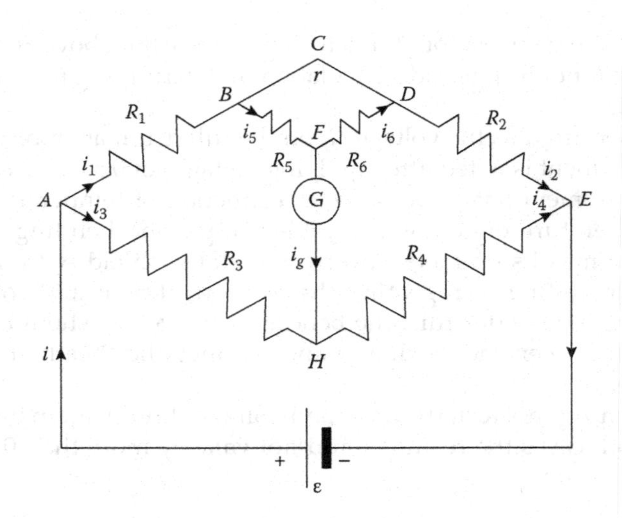

FIGURE 9.6
Principle circuit diagram of a Kelvin double bridge.

fact that the equipment usually available do not have sensitive arrangements because of the difficulty in eliminating contact and junction resistance of the same order of magnitude as the resistance to be measured. The double bridge, as devised by Kelvin, does not suffer from these defects (Figure 9.6).

Here, R_1 is the unknown resistance vis-à-vis the conductor of a 1 m length cable having very low resistance, and R_2 is a continuously variable resistance. The connection between these, which has some resistance as 'r' is unknown, is shunted by a pair of resistances, R_5 and R_6. It is the shunting of the junction resistance that distinguishes the double bridge from the simple Wheatstone bridge.

To find the condition of balance, there will be no current passing through the galvanometer 'G', when junctions F and H are at the same potential.

Therefore

$$R_1 i_1 + R_5 i_5 = R_3 i_3 \quad \text{and} \quad R_6 i_6 + R_2 i_2 = R_4 i_4 \tag{9.1}$$

and

$$i_1 = i_2, \quad i_3 = i_4 \quad \text{and} \quad i_5 = i_6 \tag{9.2}$$

where i_1 is the current flowing through R_1 and i_2 is the current flowing through R_2, then it follows that the current flowing through other resistors will be of similar sequence. Also considering the parallel paths BCD and BFD,

$$i_5 = \frac{r}{r + R_5 + R_6} i_1 \quad \text{and} \quad i_6 = \frac{r}{r + R_5 + R_6} i_2 \tag{9.3}$$

Combining Equations 9.1 and 9.3, we get

$$\left(R_1 + \frac{rR_5}{r + R_5 + R_6}\right) i_1 = R_3 i_3$$

$$\left(R_1 + \frac{rR_5}{r + R_5 + R_6}\right) i_2 = R_4 i_4$$

Dividing one by the other and with the help of Equation 9.2, we get

$$\frac{R_1(r + R_5 + R_6) + rR_5}{R_2(r + R_5 + R_6) + rR_6} = \frac{R_3}{R_4}$$

or

$$(R_1R_4 - R_2R_3)(r + R_5 + R_6) + (R_4R_5 - R_3R_6)r = 0$$

This condition is satisfied when

$$\frac{R_1}{R_2} = \frac{R_3}{R_4} = \frac{R_5}{R_6}$$

This set of conditions represents the only balance independent of the junction resistance 'r', an essential condition.

There are many makes of the Kelvin double bridge available in the market based on the same principle.

9.13 Wheatstone Bridge to Measure Conductor Resistance on Completed Cables

Wheatstone bridge is widely used to measure any electrical resistance accurately. There are two known resistors: a variable resistor and an unknown resistor connected in bridge form, discussed later. By adjusting the variable resistor, the value of the unknown electrical resistance can be easily measured (Figure 9.7).

9.14 Wheatstone Bridge Theory

The general arrangement of the Wheatstone bridge circuit is shown in Figure 9.7. It is a four-arm bridge circuit where the arms, *AB*, *BC*, *CD* and *AD*, consist of electrical resistances, *P*, *Q*, *S* and *R*, respectively. Among these

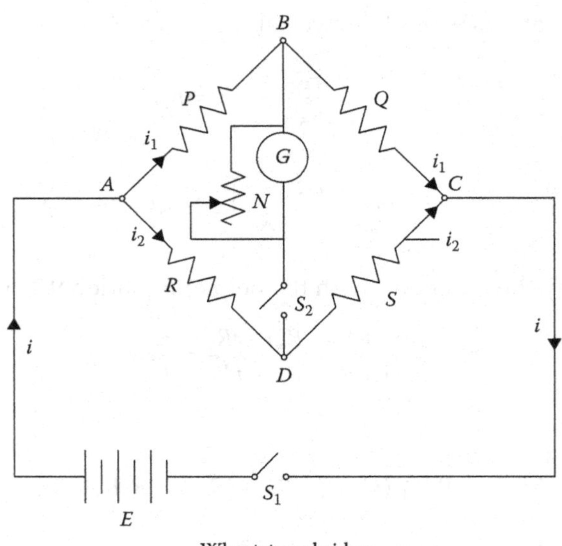

FIGURE 9.7
Wheatstone bridge.

resistances, P and Q are known as fixed electrical resistances and these two arms are referred to as ratio arms. An accurate and sensitive galvanometer is connected between terminals B and D through a switch S_2. The voltage source of this Wheatstone bridge is connected to terminals A and C via a switch S_1 as shown. A variable resistor S is connected between points C and D. The potential at point D can be varied by adjusting the value of the variable resistor. Suppose current I_1 and current I_2 are flowing through paths ABC and ADC, respectively. If the electrical resistance value of arm CD is varied, the value of current I_2 will also be varied as the voltage across A and C is fixed. If the variable resistance is continually adjusted, a situation may come when the voltage drop across resistor S, that is, I_2S, becomes exactly equal to the voltage drop across resistor Q, that is, I_1Q. Thus the potential at point B becomes equal to the potential at point D, and hence the potential difference between these two points is zero, and no current passes through the galvanometer. Then, the deflection in the galvanometer is nil when the switch S_2 is closed.

Now, from the bridge circuit, current $I_1 = V/(P+Q)$ and current $I_2 = V/(R+S)$.

Now the potential of point B in respect of point C is nothing but the voltage drop across resistor Q, which is

$$I_1Q = \frac{VQ}{P+Q} \tag{9.4}$$

Again the potential of point D in respect of point C is nothing but the voltage drop across resistor S, which is

$$I_2S = \frac{VS}{R+S} \tag{9.5}$$

Equating Equations 9.4 and 9.5, we get

$$\frac{VQ}{P+Q} = \frac{VS}{R+S} \quad \text{or} \quad \frac{Q}{P+Q} = \frac{S}{R+S} \text{ by transposing}$$

$$\frac{P+Q}{Q} = \frac{R+S}{S} \quad \text{or} \quad \frac{P}{Q}+1 = \frac{R}{S}+1 \quad \text{or} \quad \frac{P}{Q} = \frac{R}{S} \quad \text{or} \quad R = S \times \frac{P}{Q}$$

In this equation, the values of S and P/Q are known, so the value of R can be easily determined.

The electrical resistances P and Q of the Wheatstone bridge are made of a definite ratio such as 1:1, 10:1 or 100:1, known as ratio arms, and S, the rheostat arm, is made continuously variable from 1 to 1000 Ω or from 1 to 10,000 Ω.

Nowadays, many types of electronic resistance measuring instruments are available in the market, but the classical method of measuring resistance with the Wheatstone bridge is still in vogue and is reliable.

9.15 AC Voltage Test

The AC voltage test should be done on each cable drum length at ambient temperature, using alternating voltage at power frequency (60502-2; IEC:2005-49). For single-core cables, the test voltage shall be applied for 5 min between the conductor and the metallic screen.

Test procedure for three-core cables: For three-core cables with individually screened cores, the test voltage shall be applied for 5 min between each conductor and metallic layer. For three-core cables without individually screened cores, the test voltage shall be applied for 5 min in succession between each insulated conductor and all the other conductors and the collective metallic layers. Three-core cables may be tested in a single operation using a three-phase transformer.

1. *Test voltage*: The power frequency test voltage will be 3.5 U_o. Values of single-phase test voltage for standard rated voltages are given in tables of specification. If, for three-core cables, the voltage test is carried out with a three-phase transformer, the test voltage between the

phases shall be 1.73 times the values given in those tables. In all cases, the test voltage shall be increased gradually to a specified value.

2. *Requirement*: No breakdown of the insulation should occur. The AC voltage test for insulation may be conducted with a partial discharge test.
3. *AC voltage test for sheath*: The sheath can be tested in a spark tester while the inner screen or armouring is earthed and the testing beads are allowed to touch the outer sheath from all sides. In another case, the cable along with the drum can be immersed in a water tank. Voltage is then applied between the water and the screen or armour underneath the sheath. The third option is to coat the sheath with graphite powder. Voltage is then applied between the graphite powder and the screen/metallic sheath or armour. This method is particularly applied for extra high voltage (EHV) cables.

9.16 Partial Discharge Test

Partial discharge tests have to be conducted simultaneously with voltage tests.

It is known that the formation of voids cannot be avoided when any type of polymer is extruded. Various chemicals and fillers during the curing process emit a certain amount of gas, a part of which is entrapped within the material and a larger part is given out into the surroundings. These voids can be of various sizes. Generally, they are kept restricted within a microstructure by monitoring the process parameters carefully. According to international norms, the number of voids and sizes are to be restricted on a per square centimetre basis and are given by calculation as per International Electrotechnical Commission (IEC) or Japanese International Standard (JIS) specifications. Partial discharges within an insulating material are initiated in a void filled with gas (Figure 9.8). The dielectric constant within the void is less than the surrounding solid insulating material. The electric field within the void becomes higher in intensity for the same distance as in a solid dielectric. When voltage is increased across the dielectric above the corona, inception starts within the void.

Partial discharge (PD) action also can start when the void is between the conductor and insulation, or between the metallic screen and insulation, and causes a tangential electric field. An equivalent circuit of partial discharge is shown in Figure 9.8. As the voltage is applied, a spark jumps across the gap within the gas-filled void. A minute current flows in the conductor attenuated by the voltage divider network C_x, C_y, C_z in parallel with capacitance C_b.

When an electric field is applied in the cable, the gaseous materials within the void ionise and tend to polarise in +ve and –ve poles. They start striking

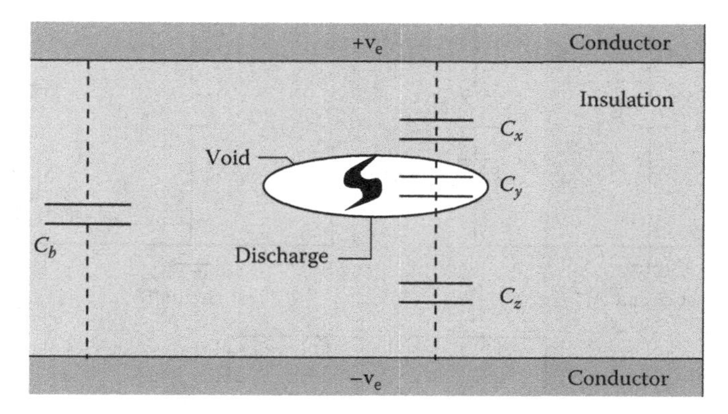

FIGURE 9.8
Discharge pattern within an insulation void.

the wall of the void in random, and if the void is large, they may slowly rupture the insulation creating fault conditions. To detect this, measurement is done by applying a predetermined voltage to know at what value the ionisation starts. At this voltage, due to ionisation, a certain amount of energy is absorbed, which is measured in picocoulomb (pC). When the voltage is lowered, ionisation stops at a certain value. This indicates that ionisation can completely stop at a value, which is reliably safe. If the condition that the absorption of energy does not exceed the given limit (pC) at the higher predetermined voltage is satisfied, then one can assume that the insulation quality of the cable is satisfactory. The maximum peak should not exceed the given limit to ensure soundness of the insulation.

PD is measured by calibrating the voltage of the spikes against the voltage obtained from a calibration unit discharged into the measuring instrument. The calibration unit is simple in operation and comprises merely of a square wave generator in series with a capacitor connected across the sample. Usually these are triggered visually to enable calibration without getting into the HV area. While testing, these are disconnected during the actual measurement of discharge.

The amount of pC to be measured gives the average value of the power absorbed by the voids that are formed within the insulation throughout the length during ionisation.

The standard measuring circuits are given in Figure 9.8, where circuit C_b represents a discharge-free blocking capacitor, C_x the test sample and Z the detecting impedance. The test circuit shown in Figure 9.9a is the popularly accepted circuit and can detect equipment that will not be damaged if a breakdown occurs in the cable during the test. When a discharge occurs within the test cable C_x, a pulse with a quick rise time exists within the circuit of C_x, C_b and Z. The voltage thus developed across the impedance Z is fed into the amplifier of the detector. The bridge, as shown in Figure 9.9c, has two cable samples with identical capacitance values and requires no

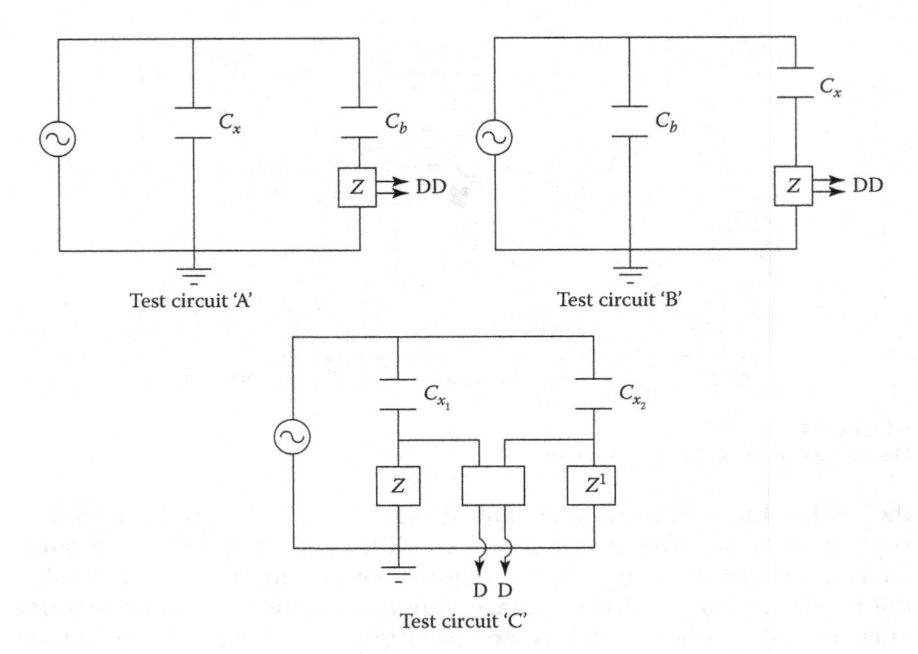

FIGURE 9.9
Test circuit diagrams for partial discharge.

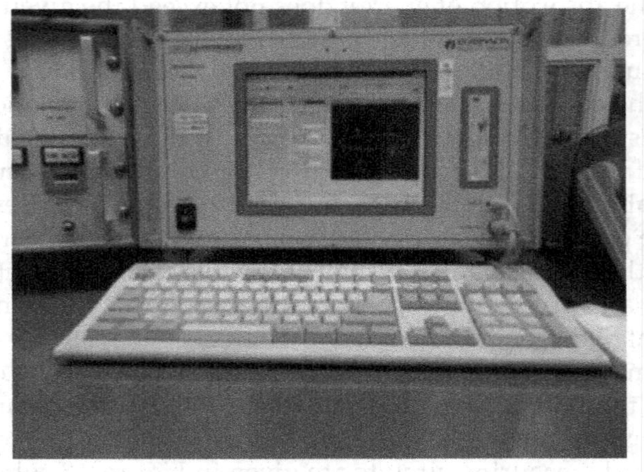

FIGURE 9.10
PD measuring oscillograph.

blocking capacitor, as the circuit remains in a balanced condition rejecting all external noise during operation (Figures 9.10 through 9.12).

During measurement, the voltage is increased slowly until the discharge is shown on the screen as spikes on the rotating ellipse. The intensity of discharge determines the nature of void formation within the insulation and is measured by picocoulomb (pC). By lowering the voltage slowly, the spikes

FIGURE 9.11
Shielded room for PD testing.

FIGURE 9.12
Oil terminations.

will get extinguished. This voltage at which the spike vanishes is called the inception voltage.

The allowable size of the void is calculated using the following formula:

$$2a = \frac{\dfrac{0.71\times 10^4}{1.74\times V}}{\left(r_o \ln \dfrac{R}{R_o}\right) - 37.6}$$

where $2a$ is the diameter of the void (μm) = allowable size of the void

$$V = \frac{\frac{E_0}{\sqrt{3}} \times 1015 \times 1.2}{1.1}$$

E_o is the nominal voltage of the cable (kV)
1.2 is the ratio of inception and extinction voltages of partial discharge
R_o is the radius under the insulation shield
r_o is the radius of the conductor shield

Measuring the smallest possible discharge is recommended to ascertain the smallest size of voids, thereby requiring a highly sensitive detector. Even a small amount of background noise can distort the result giving a false reading. Generally, the equipment and detector are installed in a room completely free of any noise or disturbances.

Thus, testing of PD is done in an atmosphere completely free of electromagnetic and transmitted noise frequency.

The common sources of noise are as follows:

- *Power cables and wire conducting path*: In industrial areas, noise sources are from rotating machine, thyristor-controlled power systems, switching devices and power supply cables.

  In this case a separate supply transformer has to be used and it should not be related to other supplies. The cable should be a shielded one, to be laid in steel pipes with an isolated ground connection.

  A power filter is used and placed near the HV power supply. The capacitor used should be PD-free. A main filter with isolating transformer has to be used for the control desk.
- *Inductive coupling due to induced voltage in a ground loop*: Ground loop should be avoided. Connecting cables for the PD room should not be laid parallel with other power lines and should be away from them.

  A separate earth system should be used for the PD room and should be away from any other factory earthing positions to avoid any ground surface current. The ground resistance should not exceed 2Ω. To avoid inductive coupling, the PD room is placed on PE-/PVC-insulated tray-type basin, keeping the room completely isolated from the adjacent earth layer and the factory floor. The grounding rod should be closed and inside the shielded room. This rod is kept isolated from the soil and about a metre below the ground starting from the floor level.

- Capacitive coupling generating stray capacitance
- Radiated frequency from radiowave and other electromagnetic waveforms

In order to ensure interference-free measurement, the cable and the equipment have to be placed in a specially made isolated shielded room. The room is built on an isolated PVC/PE tray, as stated before. The room is made of an aluminium or a closely knitted copper wire mesh doubled wall, separated by an insulating layer made of either dry wood or polythene foam. The door and windows should also have an interconnecting coupling device. The room has to be solidly grounded as described before.

Partial discharge falls into three classes:

1. Internal discharge
2. Surface discharge
3. Corona discharge (Figure 9.13)

The results obtained after measurement need to be computed and analysed to ascertain the degree of quality in the insulating media. This has to be communicated to the operating personnel to take proper precaution wherever necessary.

The partial discharge test should be carried out in accordance with IEC 60885-3, except for the sensitivity which is defined in IEC 60885-3 as 10 pC or above. For three-core cables, the test shall be carried out on all insulated cores with the voltage being applied between each conductor and screen. The test voltage should be raised gradually to, and held at, $2U_0$ for 10 s and then slowly reduced to $1.73U_0$. There shall be no detectable discharge exceeding the declared sensitivity from the test object at $1.73U_0$.

NOTE: Any partial discharge from the test object may be harmful.

Voids within XLPE insulation can be observed physically by dipping a small piece of insulated conductor within a hot silicon bath. A piece of XLPE-insulated conductor with an inner and outer semicon having a length of 50 mm has to be cut off from an extruded XLPE core. Metal wire strands

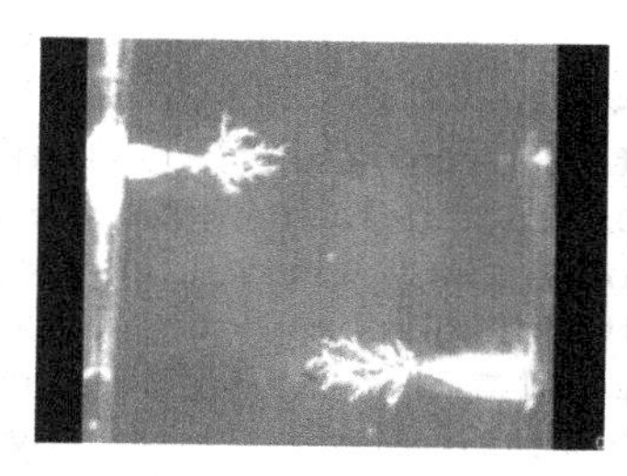
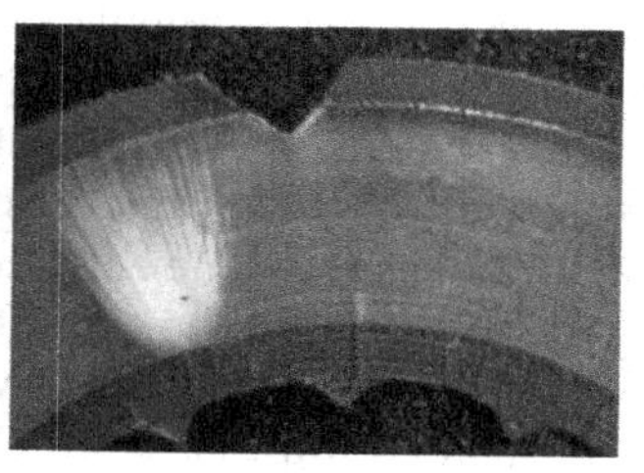

FIGURE 9.13
Corona discharge.

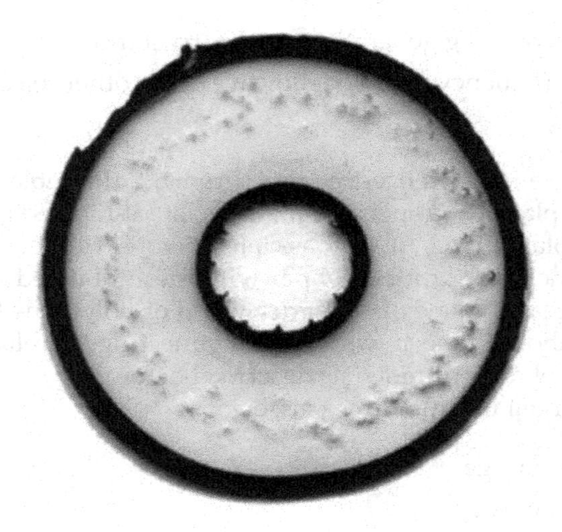

FIGURE 9.14
Void within insulation.

from the conductor are to be removed carefully by pushing each wire out from the insulated part. The inner and outer semiconductor with insulation remaining in position is now dipped in a large glass beaker containing silicon oil. This bath is then heated to 120°C. The XLPE insulation will gradually become transparent revealing all the voids and defects within, including the bonding areas between semiconducting layers and insulation (Figure 9.14).

9.17 Impulse Voltage Test

At the termination point, the cable remains surrounded by the atmospheric environment and may remain connected to a transformer or switchgear enclosed within a console over ground in a substation. During stormy weather and rainy season, occasional lightning may strike in and around the substation. A sudden lightning produces a very high surge voltage within the cable and electrical equipment. This surge voltage increases and induces an abnormal high stress within the insulation of the cable many times more than the normal operating voltage, but for a short while. During this period, the cable should not get damaged or fail due to insulation breakdown. Lightning also strikes from the earth to the clouds when the clouds get negatively charged. To ascertain the stability of insulation under such conditions, artificial lightning, like a momentary pulse, is generated to strike on the insulated cable (called impulse test). The impulse test is followed by a voltage test. This test is performed on the sample in a temperature of 5°C–10°C above the maximum conductor temperature in a normal operation. The impulse

TABLE 9.1

Impulse Test Voltage as per IEC Recommendation

Rated Voltage (U_o/U [kV])	Impulse Test Voltage (kV)
3.8/6.6	60
6.35/11	75
11/11	95
12.7/22	125
19/33	170
38/66	325
64/110	550
76/132	650
127/220	1050
230/400	1425
On this basis, the impulse voltage for 500 kV cable should be	
288/500	1730

voltage is applied according to the procedure given in IEC 60230 and should have a peak value as given in Table 9.1. Each core of the cable should withstand 10 positive and 10 negative voltage impulses as per IEC regulation and Bureau of Indian Standard (BIS) specification. After the impulse test, each core of the cable sample should be subjected, at an ambient temperature, to a power frequency voltage test for 15 min. The test voltage should be as specified. No breakdown of insulation should occur (Figure 9.15).

During the application of voltage, the peak value should rise within 1.2–5 μs and decays to half value should be approximately 50 μs. For voltages of 132 kV and above, the peak value rises in 250 μs and decays to half peak value in approximately 2500 μs. The energy required can be calculated using the capacitance of the cable and the voltage to be applied and is given by the equation ½ CV^2. The voltage is raised in steps through a series of capacitance and inductive free resistance by tuning and oscillating the circuit. The energy of each step is added to raise the final impulse stage and is measured in joules.

9.17.1 Inductance

The flow of current within the conductor generates a magnetic field that produces an opposing current, and this is known as self-inductance. This increases the effective resistance by restricting the flow of current at the outer periphery of the conductor, known as the skin effect. For larger-size conductors and cables with higher voltages, this value must be taken into consideration. Furthermore, the magnetic field generated due to the flow of current influences the adjacent cable, producing an induced voltage and current (longitudinal voltage and eddy current) within the metallic sheath. This creates a loss of power in the form of heat. This phenomenon has been discussed earlier under the skin effect and proximity effect.

FIGURE 9.15
EHV impulse generator set.

9.18 Dimensional Check

Conductor examination and checking of dimensions, measurement of the thickness of insulation and sheath and measurement of the overall diameter should be made on one length from each manufacturing series of the same type and nominal cross section of the cable, but it should be limited to not more than 10% of the number of lengths in any contract.

1. *Electrical and physical tests*: It should be carried out on samples taken from the manufactured cables according to the agreed quality control procedures. In the absence of such an agreement, for contracts, the total length exceeds 2 km for three-core cables, or 4 km for single-core cables.
2. *Repetition of tests*: If any sample fails in any of the tests, two further samples should be taken from the same batch and submitted to the same test or tests in which the original sample failed. If both additional samples pass the tests, all the cables in the batch from which they were taken shall be regarded as complying with the requirements of the standard. If either of the additional samples

fails, the batch from which they were taken shall be regarded as failing to comply.

Conductor examination compliance under the requirements for conductor construction of IEC 60228 should be checked by inspection and measurement when practicable.

The thickness of insulation and of non-metallic sheaths (including extruded separation sheaths, but excluding inner extruded coverings) is measured.

3. *Requirements for insulation*: For each piece of core, the smallest value measured has been specified in the given specification.
4. *Requirements for non-metallic sheaths*: The piece of sheath shall comply with the following:
   a. For unarmoured cables and cables with an oversheath not applied directly over the armour and metallic screen or concentric conductor, the smallest value measured should not fall below 85% of the nominal value by more than 0.1 mm.
   b. For an oversheath applied directly over the armour, metallic screen or concentric conductor and for a separation sheath, the smallest value measured should not fall below 80% of the nominal value by more than 0.2 mm.
   c. The minimum thickness of the lead sheath should be determined by one of the following methods, at the discretion of the manufacturer, and should not fall below 95% of the nominal thickness by more than 0.1 mm.
      i. *Strip method*: The measurement should be made with a micrometre of plane faces of 4–8 mm diameter and an accuracy of ±0.01 mm on a test piece of sheath about 50 mm in length. The piece should be slit longitudinally and flattened carefully. After cleaning the test piece, a sufficient number of measurements should be made along the circumference of the sheath, not less than 10 mm away from the edge of the flattened piece, to ensure that the minimum thickness is measured.
      ii. *Ring method*: The measurements should be made with a micrometre having either one flat nose and one ball nose or one flat nose and a flat rectangular nose 0.8 mm wide and 2.4 mm long. The ball nose or the flat rectangular nose should be applied to the inside of the ring. The accuracy of the micrometre shall be ±0.01 mm. The measurements should be made on a ring of the sheath carefully cut from the sample. The thickness will be determined at a sufficient number of points around the circumference of the ring to ensure that the minimum thickness is measured.

9.19 Measurement of Armour Wires and Tapes

- *Measurement of wires*: The diameter of round wires and the thickness of flat wires should be measured by means of a micrometre having two flat noses to an accuracy of ±0.01 mm. For round wires, two measurements have to be made at right angles to each other at the same position and the average of the two values taken as the diameter.
- *Measurement of tapes*: The measurement should be made with a micrometre having two flat noses of approximately 5 mm in diameter to an accuracy of ±0.01 mm. For tapes up to 40 mm in width, the thickness should be measured at the centre of the width. For wider tapes, the measurements should be made 20 mm from each edge of the tape, with the average of the results taken as the thickness.
- *Measurement of external diameter*: If the measurement of the external diameter of the cable is required as a sample test, it should be carried out in accordance with Clause 8 of IEC 60811-1-1.

9.20 Voltage Test for 4 h

This test is applicable only to cables of rated voltage above 3.6/6 (7.2) kV. *Sampling*: The sample should be a piece of the completed cable at least 5 m in length between the test terminations. A power frequency voltage should be applied for 4 h at an ambient temperature between each conductor and the metallic layer(s). The test voltage should be $4U_o$; it should be increased gradually to the specified value and maintained for 4 h. No breakdown of the insulation should occur.

9.21 Type Tests, Electrical

When type tests have been successfully performed on a type of cable, covered by a standard with a specific conductor cross-sectional area and rated voltage, the type approval should be accepted as valid for cables of the same type with other conductor cross-sectional areas and/or rated voltages, provided the following three conditions are satisfied:

1. The same materials, that is insulation and semiconducting screens, and manufacturing process are used.

2. The conductor cross-sectional area is not larger than that of the tested cable, with the exception that all cross-sectional areas up to and including 630 mm^2 are approved when the cross-sectional area of the previously tested cable is in the range of 95–630 mm^2 inclusive.
3. The rated voltage is not higher than that of the tested cable. The approval should not be based on the conductor material.

9.22 Cables with Conductor and Insulation Screens

A sample of completed cable 10–15 m in length should be subjected to the tests. The normal sequence of tests should be as follows:

a. Bending test, followed by a partial discharge test
b. Tan δ measurement
c. Heating cycle test, followed by a partial discharge test
d. Impulse test, followed by a voltage test
e. Voltage test for 4 h

9.22.1 Special Provisions

Measurement of tan δ may be carried out on sample different from that used for the normal sequence of tests. Measurement of tan δ is not required on cables with a rated voltage below 6/10 (12) kV.

9.23 Bending Test

The sample should be bent around a test cylinder (e.g. the hub of a drum) at an ambient temperature for at least one complete turn. It should then be unwound and the process repeated, except that the bending of the sample should be in the reverse direction without axial rotation. This cycle of operation should be carried out three times. The diameter of the test cylinder should be $-25(d + D) \pm 5\%$ for single-core cables and $-20(d + D) \pm 5\%$ for three-core cables for cables with a lead sheath or with an overlapped metal foil longitudinally applied, and, for other cables, $-20(d + D) \pm 5\%$ for single-core cables and $-15(d + D) \pm 5\%$ for three-core cables, where D is the actual external diameter of the cable sample in millimetres.

Upon completion of this test, the sample should be subjected to a partial discharge test and should comply with the requirements given in the specification.

The sample should then be checked for the following:

1. No crack on the sheath surface should occur.
2. No break of copper or aluminium shield wire should occur.
3. No encroachment against the insulation shield and the outer sheath by screen wires should occur.
4. No cavities, cracks or flaw should be there on and between the conductor shield layer, insulation layer and insulation shield layer.

9.24 Heating Cycle Test

The sample, which has been subjected to previous tests, should be laid out on the floor of the test room and heated by passing a current through the conductor, until it reaches a steady temperature of 5°C–10°C above the maximum conductor temperature in normal operation. For three-core cables, the heating current should be passed through all the conductors. The heating cycle should be of at least 8 h duration. The conductor temperature should be maintained within the stated temperature limits, for at least 2 h of each heating period. This shall be followed by at least 3 h of natural cooling in air to a conductor temperature within 10 K of ambient temperature. This cycle must be carried out 20 times. After the last cycle, the sample should be subjected to a partial discharge test and should comply with the requirements given.

9.25 Resistivity of Semiconducting Screens

The resistivity of the extruded semiconducting screens applied over the conductor and over the insulation must be determined by measuring test pieces taken from the core of a sample of cable as made and a sample of cable which has been subjected to the ageing treatment to test the compatibility of the component materials (Figure 9.16).

The test specimens should be prepared in accordance with the procedure shown in Figure 9.16. The test piece should be of 150 mm length taken from the completed cable. For the conductor screen, the sample of the core must be cut longitudinally in half removing the conductors. All coverings on the insulation have to be removed clean.

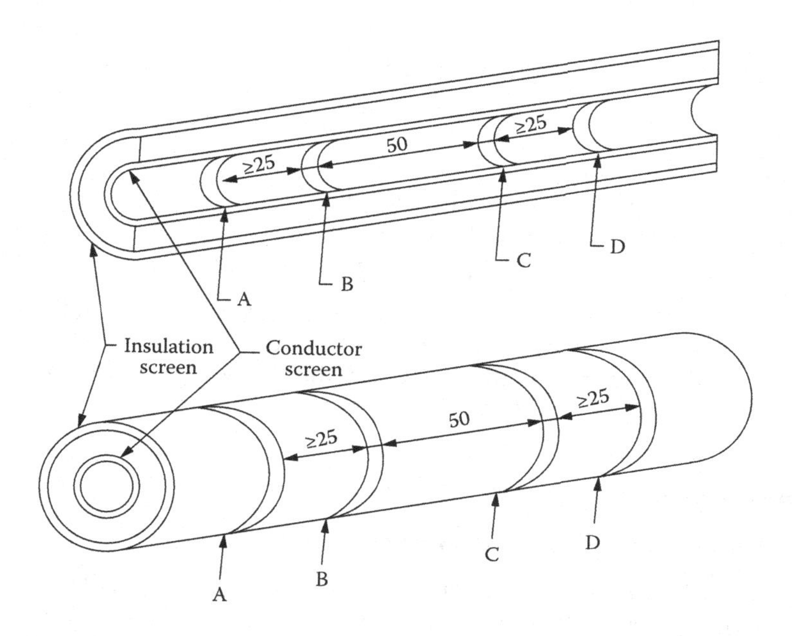

FIGURE 9.16
Sample for measurement of insulation screen and conductor screen. A and D, current electrodes; B and C, potential electrodes; dimensions are in millimetres.

Silver-painted electrodes indicated by A, B, C and D should be marked on the semiconducting surface. Electrodes B and C should be 25 mm apart from the other two electrodes, A and D. Connections have to be made with suitable copper clips. These clips should not touch the insulation screen on the outer surface of the test sample. The assembly should be preheated in an oven at 90°C ± 2°C for 30 min. The resistance between the electrode has to be measured by a suitable high-resistance circuit. After the electrical measurement, the physical dimensions of the thickness of the conductor screen and insulation screen, along with the diameter of insulation and conductor screen, have to be measured by an optical method. The volume resistivity of the conductor and insulation screens is calculated as follows:

$$\text{(a) Conductor screen} = \frac{R \times \pi(D-T) \times T}{2L}$$

where

R is the resistance (Ω)
D is the diameter over the conductor screen (m)
T is the average thickness of the screen (m)
L is the distance between the potentials (m)

$$\text{(b) Insulation screen} = \frac{R \times \pi(D-T) \times T}{L}$$

where
R is the resistance (Ω)
D is the diameter over the insulation (m)
T is the average thickness of the screen (m)
L is the distance between the potential electrodes (m)

The measurements should be made at a temperature within ±2°C of the maximum conductor temperature in normal operation. The resistivity, both before and after ageing, shall not exceed the following: conductor screen 1000 Ωm and insulation screen 500 Ωm.

9.26 Test for Longitudinal Watertightness

Cables are sometimes laid in waterlogged marshy lands or in lowlands where water logging is permanent. The cable may cross rivers or places where heavy rain causes water logging. In such areas, permanently laid cables get damaged, and water ingress becomes a common phenomenon. Water can travel through the outer sheath within the conductor and cause failure. In such cases, the cable is designed and constructed with water-blocking elements. In order to ensure that the performance of cables remains uninterrupted for a period of time, they are tested for internal watertightness. But these cables cannot be compared with submarine cables and are termed' internally watertight cables'. The details of the apparatus and the test method have been elaborated in the Bureau of Indian Standard specification No. 7098 (Part-3) under Appendix D.

9.26.1 Water Absorption Test on Insulation

The sampling and test procedure should be carried out in accordance with 9.1 or 9.2 of IEC 60811-1-3 employing the conditions specified in the corresponding Tables: Requirements. The results of the test should comply with the requirements specified in Tables 18 or 19.
The following tests are also significant:

1. Measurement of the carbon black content in the black PE sheath. This is to protect the outer sheath from ageing due to ultraviolet radiation when exposed to the sun.
2. Shrinkage test for insulation and sheath. This test is done to ensure that the conductor insulation and sheath do not retract too much at low temperatures exposing the ends at the jointed portion and allowing dirt and moisture to contaminate the joint assembly, creating a short-circuit condition.

3. Thermal stability test to ensure that the insulation and sheath do not degrade when exposed to higher temperatures for a long time due to nearby fires and the like.
4. Cold impact test to ensure that the materials used for insulation and sheath do not get brittle or crack when exposed to very cold temperatures when installed at high altitudes. This is more so for strategic reasons so that the power supply to high-altitude inhabitants and army personnel does not get affected.
5. Impact test is carried out to ascertain the mechanical stability of the cable against a predetermined sudden force that may act on the surface of the cable due to falling stones and debris while in operation. This type of force may act when the cable is installed in rocky areas or areas having loose soils. Under such circumstances it is expected that power supply is not going to be affected.
6. Strippability test should be carried out when the manufacturer claims that the extruded semiconducting insulation screen is strippable. The test should be performed three times on both unaged and aged samples by using either three separate pieces of cable or one piece of cable at three positions around the circumference, spaced at approximately 120°. Core lengths of at least 250 mm should be taken from the cable to be tested, before and after being aged. Two cuts should be made in the extruded semiconducting insulation screen of each sample, longitudinally from end to end and radially down to the insulation, the cuts being 10 (± 1) mm apart and parallel to each other. After removing approximately 50 mm length of the 10 mm strip by pulling it in a direction parallel to the core (i.e. a stripping angle of approximately 180°), the core should be mounted vertically in a tensile machine, with one end of the core held in one grip and the 10 mm strip in the other. The force to separate the 10 mm strip from the insulation, removing a length of at least 100 mm, should be measured at a stripping angle of approximately 180° using a pulling speed of 250 (± 50) mm/min. The test should be carried out at a temperature of 20 (± 5)°C. For unaged and aged samples, the stripping force values must be continuously recorded. The force required to remove the extruded semiconducting screen from the insulation should not be less than 4 N and not more than 45 N, before and after ageing. The insulation surface should not be damaged, and no trace of the semiconducting screen must remain on the insulation.
7. Oil resistance test is performed to determine the quality of protective covering of the cable. At times, cable is laid in a diesel engine room or near a refinery or in such places where the surrounding remains contaminated with oil and oil fumes. In these cases, cable sheath must not get affected by oil and fumes. The specimen of the sheath taken from the end of the finished cable should be immersed

in oil as per ASTM No. 2 oil at the specified temperature and for a specified period of time. After the immersion period is over, the tensile and elongation are to be measured and recorded. After the tests, tensile must be retained more than 80% of the original value and elongation shall not fall below 60% of the initial test values.

8. PVC gives out chlorine gas and hydrochloric acid when burnt in fire. It is necessary to determine the amount of acid and toxic gas evolves from a particular PVC compound used. The acid and gas given out by the compound shall not be more than the specified limit when a sudden fire breaks out. To protect the environments and working personnel near the cable, this test is to be carried out as per given specification.

9.27 Flame Propagation Tests under Different Conditions

Nowadays, powerhouses, substations and important installations need to be secured from fire hazards. Naturally, more and more demands are raised for the use of fire-resistant and fire survival cables to be installed within the tunnels of metro railways, metro stations and substations, airports, defence establishments (defence installations, warships, etc.), oil refineries, oil exploration sites, oil wells and other important areas where human life is present and should not be compromised should any fire break out due to any cable fault or short-circuit conditions. Emission of smoke and acid fume must also be kept under specified limits to prevent health hazards and keep the surroundings pollution free.

The insulating compound used has to comply with such tests so as to ensure that smoke and acid fumes are not generated under such conditions, and compounds extinguish fire by suppressing the flames automatically within a given period of time.

In certain cases, the cable after suppressing the fire must remain in a *live* condition for about 3–4 h during which all electrical systems should also remain functional until alternative arrangements are made.

These cables have to be tested as per BIS and IEC specifications as specified under fire-resistant (FR), fire-resistant low-halogen and low-smoke (FRLS) and fire survival (FS) tests, under different categories, such as IEEE-3 Category A, B or C.

The choice of materials and design parameters is set in compliance with the regulations.

Fire survival cables have to be tested for 3 h under intense flame, at a temperature >900°C. In this case, the cable's conductor and insulation are wrapped with mica tape.

BS 8491:2008 specification gives details in the 'Method for assessment of fire integrity of large diameter of power cables, for use as components for smoke and heat control systems and certain other active fire safety systems'.

BS 6387:1994 specifies tests for maintaining the integrity of a circuit after the commencement of fire. This might give time to fire-fighting equipment and firemen to evacuate the premises and keep the systems in operation for a certain period of time until alternative arrangements are made.

Both test specifications allow the manufacturer to select material and design parameters to fulfil the given safety requirements. Test equipments and methods are explained in these specifications in detail.

With the development of new technologies every day, safety norms, choice of material, design parameters, testing of raw materials and finished products are being modified and changed from time to time. Naturally, quality norms and controlling systems have become more sophisticated and stringent to keep the environment unaffected and to reduce loss of life, if an untoward accident occurs at a working place or in an establishment.

Basically, the chemical composition imparts a given characteristic to fulfil the required specific values as established in the given standards.

Though XLPE contains no halogen, it is not flame resistant. In this case, special fire-resistant polymers like ethylene vinyl acetate (EVA) or such additives are blended to make the modified XLPE compound fireproof and devoid of any toxic gas. Tests must be carried out on the input material and on the finished product to ensure that specified parameters are being adhered to.

9.28 LOCA Test for Nuclear Plants

9.28.1 LOCA Qualification

Components in various systems in nuclear power plants may be subjected to harsh environmental conditions, like high humidity, temperature and radiation, during the standard operation and during accident conditions such as loss of coolant accident (LOCA). Hence, it is essential to ensure a reliable operation of these components during such conditions. Towards this objective, a qualification approval and ageing studies on materials to provide a reasonable assurance regarding their survival capability under simulated environments, even at the end of the specified service life, are needed.

Facilities like Panoramic Batch Irradiation Technology (PANBIT) and LOCA simulators have been set up within different nuclear research centres.

Technical services are being regularly provided to upcoming and operating nuclear power stations, which have significantly helped in taking appropriate decisions in areas such as

- Estimation of residual life
- Failure analysis and reliability improvement

These include studies on cables and elastomeric and other cable construction materials, for nuclear power plants and nuclear facilities.

9.29 Test and Measurement Facilities

A LOCA environment simulator is shown in Figure 9.17. A typical cylindrical vessel is made of a 6 mm thick stainless steel sheet. The internal diameter is 100 cm and the straight length is 120 cm. The maximum steam temperature and pressure achievable are 150°C and 3.4 kg(g) (50 psig), respectively. An oil-free air compressor has been connected to the simulator for performance evaluation of pneumatic devices during the LOCA test. Provisions have also been made for recording and scanning of steam temperature, monitoring of pressure, water spray and online measurement of performance parameters. Two safety devices, a pressure relief valve set

(a)

(b)

FIGURE 9.17
(a) LOCA environmental simulator and (b) LOCA test chamber inside view.

at 35 psig and a rupture with 50 psig rupture pressure, are mounted on the simulator. Provision is made for manual release of steam and draining of condensate.

The duration and method of tests have been specified by nuclear authorities to ensure safe performance of cables and accessories.

9.30 Synergism Simulator

In order to study the interaction effects of combined environments prevailing simultaneously in an nuclear power plant (NPP) containment, a synergism simulator has to be used as required.

This facility consists of a temperature humidity chamber and a gamma radiation source along with a provision for applying electrical stresses. The typical internal dimensions of the chamber are 84 × 84 × 90 cm^3. It is possible to vary the magnitudes of these stresses as per the design of the experiment. The temperature can be varied from room ambient to 80°C with relative humidity up to 95% ± 5%. However, it can be varied from room ambient to 150°C when used as a temperature chamber alone. Dose rates can be varied from 2 to 30 krad/h using three lead shields for the attenuation of gamma field. It is also possible to study the dose rate effects. A provision has to be made for an online measurement of the performance parameters. The dose rate outside the synergism simulator (original existing PANBIT facility) can be varied from 1 to 900 krad/h, depending upon the distance of test items with respect to the source.

9.31 Cables for Offshore Wind Firms and Oil Exploration Rigs: Hyperbaric Test

Power, communication and control cables are used in offshore equipment like wind power firms and oil rig exploration platforms.

These cables have to be tested to withstand the water pressure and turbulent forces surrounding these elements constantly. Even for navy cables used for exploring sea levels, enemy ships and mines, the cables are dipped in water at large depths and tested for their life-span and durability (Figure 9.18).

The required tests are performed to assure the leakproof capacity of watertight devices like cables and joints.

With a typical chamber it is possible to perform tests with a maximum pressure of 45 bar or more. It means that a simulated depth of 1000 m or more can be considered. Using this equipment, a set of tests are conducted to

FIGURE 9.18
Hyperbaric test chamber.

guarantee the proper operation of the entire system (cylinders, cabling, and connectors) of the supplied cable project.

One of the main difficulties in this test is that the hyperbaric chamber connects the devices under test (DUT) to the outside of the chamber. The chamber has several penetrators of many sizes to cross the wall using the pressure of watertight cable glands. The problem with these cable glands is that they allow the passage of the cable but not the connector of the cable end. It means that some tests using optical or submarine cables must be carefully planned in advance using accessories that are able to perform the connection.

In order to test the totality of the components of the subsea station, every isolated component should undergo several tests before compatibility tests can be performed with several components.

The components to be tested are

- The main junction box
- The submarine cable termination
- Underwater IP camera system
- Underwater connectors (electrical and electro-optical)
- Underwater electrical and optical cables
- Ethernet to serial special cable
- Some mechanical accessories

The cylinders and watertight boxes should be tested first without cables, connectors or electronics inside.

These tests are to be carried out at different pressures from a few decibars to the maximum allowed depth for the cable and box.

9.32 Checking for Leakages in Each Test

This procedure has been adopted to detect structural defects and assembling faults, without having to destroy any component. With these tests, leakages in the protecting caps of the junction box, and in a welded seam of the termination box, can be detected. Once the integrity of the stainless steel boxes has been asserted, the underwater connectors have to be installed along with the connection to the external devices. The connectors are designed to resist the adverse conditions of the marine environment and allow wet connections of new instruments with the station on the seafloor. After that, new tests have to be done to corroborate the correct assembly of the connectors.

With these tests it has been possible to monitor the heat generated for the power supplies and to redesign the internal distribution to improve dissipation and avoid hot areas.

Finally, it can be asserted that all the components resist more than twice the water pressure of the place where they are being installed.

To produce quality materials, a comprehensive quality system has to be maintained and strictly adhered to, such as an all-round cleanliness and clean atmosphere. The maintenance of machinery and equipments, periodic checks on working personnel for their behavioural patterns, understanding of the subject and adherence to standard practices are to be monitored regularly. A skilled, quality mind is essential to produce a quality product.

In order to become a leader in the cable industry, International Standards Organisation (ISO) 9000 standard series must be followed and adhered to. Periodical internal and external audits have to be undertaken to keep up the spirit of quality maintenance in the system.

10

Special Cables

10.1 Aerial Bunch Cables

Aerial bunch cable is a unique concept in which the cable is installed on overhead poles (Figure 10.1). The cable consists of insulated three-phase conductors and a messenger wire, insulated or bare. Apart from the phase conductors, a neutral conductor and a lighting conductor (both insulated) are also incorporated as required. During installation on poles, walls or any other support, the messenger conductor is hooked onto the supporting structure to carry the load of the hanging cable span. Unlike underground cables, the insulated phase conductors and the neutral and lighting conductors are not covered by any common covering. No armour or outer protective sheath is applied. Although it looks very simple in construction, the cable has its own technical parameters and problems to be considered during designing, manufacturing and installation. Aerial bunch cables are installed in places where cables cannot be laid underground because of hard rocky soil or when excavation is very difficult and costly. These are also useful when cables have to be taken over marshy lands, a river or water basin or through narrow streets and lanes.

The concept of aerial bunch cables was initially developed in Scandinavian countries where hard, rocky earth rendered excavation difficult; in such places power lines had to be taken over fjords and hills, and overhead bare transmission lines were not found suitable as they corroded easily because of the saline atmosphere and had to be used sparingly. Further, when these cables were installed on walls and structures, the conductors carrying load endangered human life and property. Therefore, all the conductors had to be kept isolated from each other. The messenger conductor can be kept bare and act as an earth conductor. If the messenger is used as neutral, it has to be insulated.

In order to protect the phase conductors, they are insulated with polymeric compounds such as PVC for low-voltage cables and polythene or cross-linked polythene (XLPE) for medium-voltage cables. All the insulated single- or

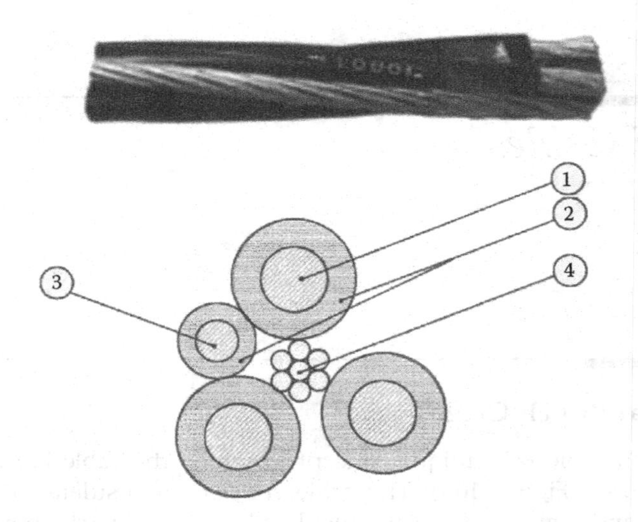

FIGURE 10.1
(1) Aluminium conductor (power), (2) XLPE insulation, (3) aluminium conductor (lighting), (4) bare 'Al-alloy' conductor (messenger).

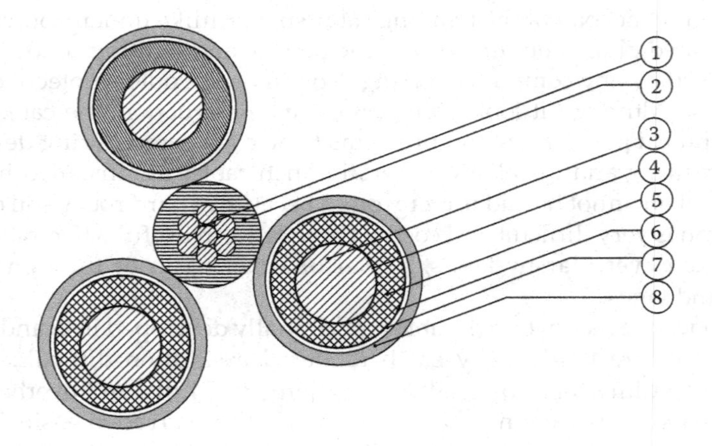

FIGURE 10.2
Medium-voltage aerial bunch cable. (1) Conductor, galvanised steel messenger; (2) insulation, PVC; (3) conductor, aluminium; (4) conductor, screen; (5) insulation, XLPE; (6) insulation screen; (7) copper tape screen; (8) outer sheath, PVC.

three-phase conductors are twisted around a messenger conductor, bare or insulated. This unique feature has the following advantages (Figure 10.2):

- Cable construction is very simple. The inner sheath, armouring and outer sheath are eliminated, making the cables lighter in weight, restricted in diameter and relatively longer in length.

- Manufacturing and testing of these cables are very simple since they require only a few conventional equipment.
- Installation is carried out by stringing on equidistant poles.
- The messenger conductor acts as a supporting element and can be suspended by hooking on poles.
- No insulators are required for the separation of phases.
- The messenger conductor can be utilised as earth or neutral.
- Additional neutral and lighting conductors can be incorporated as required.
- Installation and jointing are simple and easier.
- Supply at different points can be tapped easily from any phase for the easy distribution of power to consumer points.
- All-round saving in costs has made the cable popular.
- In the case of an unexpected failure, cables can be replaced quickly.
- Illegal hooking for pilferage of power is eliminated.
- Due to its simple constructional features, installation can be carried out in difficult terrain, rural areas and narrow urban lanes where excavation is difficult and bare conductors could endanger human life. These cables can be fixed on walls and existing constructions.

Due to these advantages, aerial bunch cables are being extensively used in India and also in developed countries.

10.1.1 Raw Materials

10.1.1.1 Conductor Material

As a conductor material, electrochemical (EC)-grade aluminium is preferred for phase conductors. Alloy aluminium is accepted for messenger conductors. As it is heavy and costly, copper is not considered in this case. Further, copper can enhance the degradation of insulating materials such as polythene by oxidation, which can be rapid at elevated temperatures when the cable is exposed to environmental conditions. Since the cable remains suspended on two points, the weight of the material should be primarily considered. The relation among weight, span of suspension and required tension is given by the following formula (Figure 10.3).

From this equation, it is clear that tension '*T*' is directly proportional to the weight '*W*' of the cable, when sag '*d*' and span length '*L*' remain constant. If '*W*' is increased, '*T*' needs to be applied more or '*L*' has to be smaller or '*d*' has to be increased. Since copper is three times heavier than aluminium, the span length needs to be shortened as sag '*d*' cannot be increased further, which can create problems. Shortening of the span means increasing the number of poles, installation accessories and labour costs.

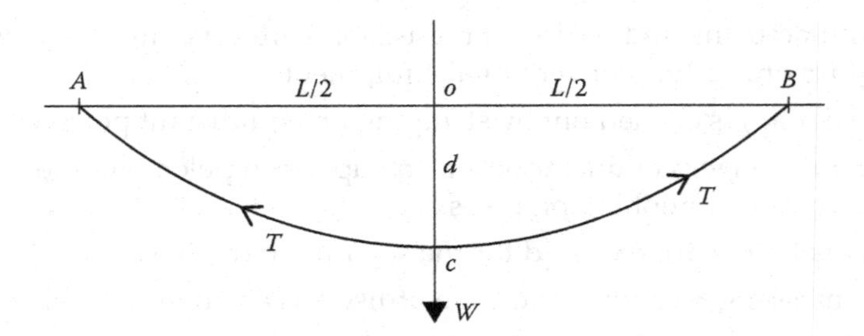

FIGURE 10.3
Suspended cable between two poles *A* and *B*. $T = WL^2/8d$ where *T* is the applied tension on the cable, *L* is the length of span (*A–B*), *W* is the weight of the cable and *d* is the sag of the suspended wire.

Apart from the weight of the cable, wind pressure (along the sea coast and on hilly areas) and the weight of ice (in the case of cold weather as in northern India) also need to be taken into account.

Example 10.1

Let us consider a three-core 95 mm² cable, keeping all other design parameters the same. Only the metal weight is considered here. The specific weight of Cu is 8.89 and that of Al is 2.703.

- For a copper cable, the weight of copper is 2533.00 kg/km, approximately. Keeping the length at 10 m and '*d*' = 1.5 m, we get *T* = 21,108 kg m, approximately. (The catenary constant is 211.08 kg m.)
- For aluminium cable, taking one size higher as three-core 120 mm², the weight of aluminium becomes 973.00 kg/km, approximately. Keeping the length 10 m and '*d*' = 1.5 m, we get *T* = 8,108 kg m, approximately. (The catenary constant is 81.08 kg m.)
- Keeping tension the same as that of copper, the aluminium cable can have a span of 16 m, all other material weight and parameters remaining unaltered; the actual span will be slightly lesser. (Here no other load is considered.)

It shows that even by increasing conductor size, the span length can be increased when aluminium is used as a conductor material. Moreover, aluminium retains a thin layer of oxide on its surface, which acts as an insulation and a protective layer against corrosion. Hence, in the case of a temporary damage to the insulation, the conductor is not immediately affected. For copper, such damages allow moisture to get in and corrode the metal, quickly forming copper oxide and copper sulphate.

Taking the overall cost (including metal price), aluminium has been accepted universally as the conducting material for aerial bunch cables. Aluminium is available in plenty in the form of bauxite ore. Large deposits

TABLE 10.1

Properties of EC Grade Aluminium

Property	Unit	Values
Density at 20°C	kg/m^3	2703
Coefficient of thermal expansion	$\times 10^{-6}$	23
Melting point (approximate)	°C	659
Thermal conductivity	W/cm °C	2.4
Ultimate tensile stress		
Soft temper	MN/m^2	70–90
¾H–H	MN/m^2	125–205
Requirement as per IS 8130	MN/m^2	100–150
Relative conductivity	Taking copper as 100	
Soft (EC grade)		61
½H–H (EC grade)		61
Electrical resistivity at 20°C	Ohm m (10^{-6})	
Soft (EC grade)		2.803
½H–H (EC grade)		2.826
Temperature coefficient of resistance per °C		0.0040

of bauxite ores are found in India. To extract aluminium, bauxite is digested with caustic soda solution to produce alumina (Al_2O_3). Alumina is dissolved in fused cryolite, wherefrom pure metal is obtained by electrolysis. Since aluminium is a trivalent compound, its outer electrons are available for free movement. The metal is softer than copper and can be cold-formed without much change in conductivity.

Typical values of EC-grade aluminium as per international standards are provided in Table 10.1.

10.1.1.2 Messenger Conductor Material

The messenger conductor supports the weight of the cable suspended between the poles. Further, it has to accept wind pressure and ice load if the cable is installed in a cold region. During winter and summer, because of temperature variation, the length of suspended conductors along with the messenger varies. This may create additional stress at the anchoring point of the suspended length of the messenger conductor. Naturally, the mechanical strength of the messenger conductor must be sufficient enough to accept all these stress and strain during its life-span. EC-grade aluminium does not have the strength to bear such load. The mechanical property of the metal needs to be upgraded considerably. Further, while being exposed to adverse environmental conditions, the messenger conductor should not get corroded easily. In order to gain the desired qualities, the metal is alloyed with 0.5%–0.6% magnesium and 0.5% silicon. This alloy can withstand severe environmental conditions and also has the mechanical strength to accept stress and strain as required. The alloy could acquire a tensile strength greater than 300 N/mm^2 on an average. At the initial stage, alloyed wire rods do not give consistent mechanical properties all through. While drawing at lower-size wires to form a stranded messenger conductor having the desired breaking load and elongation, the alloyed metal needs to be treated further to get a uniform dispersion of magnesium and silicon through a heat treatment known as 'the solution treatment' and ageing process.

10.1.1.3 Insulating Materials

As an insulating material, paper dominated the Indian cable market till the 1970s. Thereafter, with the introduction of polymeric compounds, paper-insulated, lead-sheathed and steel-tape-armoured cables started receding. Paper insulated, lead covered, double steel tape armoured (PILCDSTA) cables became nonexistent at the end of them 1980s. Underground laying and installation of heavy weight lead sheathed power and communication cables had to be carried out with much difficulty and precautions. Further, lead is susceptible to cracking if handled carelessly. These cables could not be suspended on poles unless construction and manufacturing processes were changed, which were very rare. An exception was the subsea cable, which was constructed differently. Using aerial cables was out of the question.

With the coming of polymers as insulating materials, these problems were eliminated. In this case, the cable could be laid and installed anywhere. Its manufacturing process also became easier. However, polymers also have problems that need to be studied and considered while designing and manufacturing power cables.

10.1.2 Polymeric Compounds

10.1.2.1 Polyvinyl Chloride

Polyvinyl chloride (PVC) was introduced in India as an insulating material by Siemens (Cable Corporation of India) during the 1970s as 'Protodur' cable. The basic structure of PVC is as follows:

```
–CH2–CH–CH2–CH–
      |       |
      Cl      Cl
```

PVC is produced from vinyl chloride monomer by polymerisation. In earlier stages, vinyl chloride is produced from acetylene and anhydrous hydrogen chloride in the presence of a catalyst. At the beginning, calcium carbide was used to produce acetylene. In the modern process, acetylene is produced from ethylene as feedstock from refineries. Polymerisation of vinyl chloride is done to yield a basic PVC resin. The methods used are (1) suspension, (2) bulk, (3) emulsion and (4) solution. Different polymer resins are categorised by their 'K' values. The higher the 'K' value, the more difficult the processing parameters.

A basic polymer cannot be used as such and needs to be blended and compounded to form different categories of compounds for different applications. Two molecules of chlorine actually make the compound fire resistant. However, in actual case it has been found that chlorine is not very effective with an aluminium conductor. Investigation revealed that during intense fire, emitted chlorine reacts with aluminium to form aluminium chloride, a volatile substance, and evaporates. In such cases, the fire-resisting property does not get activated. To gain flame-retardant low-smoke properties, antimony trioxide or aluminium trihydrate is mixed with the compound. Ingredients used to manufacture PVC compounds are discussed as follows.

10.1.2.1.1 Plasticiser

The resin is blended with a plasticiser to make a homogeneous flexible compound. Different plasticisers are used for different applications. The most popular plasticiser is di-octyl phthalate (DOP). Other types are di-butyl phthalate (DBP), the most volatile grade, di-octyl adipate, tri-cresyl phosphate, tri-octyl phosphate, di-octyl sebacate and di-octyl succinate. To make the compound commercially competitive, some extenders, such as chlorinated paraffin wax or an extender oil, are used to replace a part of the plasticiser. As per applications and given properties, plasticisers are selected. Some of the required properties are as follows:

- Fire-extinguishing property
- Low-smoke and low-halogen compound
- Heat-resisting property
- Lower dielectric loss of compound
- Cold temperature withstanding quality
- UV-resistant property
- High abrasion resistance and flexing properties
- Higher insulation resistant
- Good dielectric strength

10.1.2.1.2 Stabiliser

The compound should be protected from heat, environmental influence, sunlight and ageing. A stabiliser is added to make the compound stable for long performance. Some common stabilisers include lead stearate and barium cadmium stearate. To make the compound restricted practice of hazardous substances (RoHS) compliant, organic stabilisers are now in vogue, such as epoxidised resin and various other organic chemicals. UV stabilisers are used to protect the compound from sunlight. One of the best ingredients for UV protection is carbon black.

10.1.2.1.3 Fillers

To make the compound cheaper and to impart strength and flexibility, different types of fillers are used, such as calcined clay to give higher tensile strength and insulating properties, French chalk to give better flexural and insulating property, coated calcium carbonate (popularly known as Forcal S) and china clay.

10.1.2.1.4 Lubricants

During extrusion, the flow of the compound must be easy and smooth. Also to give a glossy finish, lubricants such as paraffin wax or microcrystalline wax, stearic acid and stearate are used.

10.1.2.1.5 Colourant

For the identification of phase cores or cores of multicore cables, various colours are used. The colours must be compatible with the PVC compound and should not affect its electrical and physical properties. For outer coating, black colour is preferred to protect the cable from UV radiation. For aerial bunch cable for low-voltage insulation, single coating should be black in colour. A minimum of 2%–2.5% carbon black must be used while compounding the material.

10.1.2.1.6 Carbon Black

Of all fillers, carbon black is the most important filler for compounding PVC for aerial bunch cables. The material improves tensile strength and thermal ageing properties, protects against ozone and UV radiation and gives thermal stability. It is an essential filler for PVC outer sheath and for aerial bunch cable low-voltage insulation. But this filler must be used with much consideration. Too much carbon black deteriorates tensile strength and elongation and facilitates absorption of water and moisture beyond the allowable limit. Electrical properties also get affected. The properties of PVC insulating compounds vary as per the formulation to be prepared for a particular application. However, the average data for a few PVC compounds are described as follows: in this case, references have been taken from British Standard (BS), Indian Standard (IS) and International Electrotechnical Commission (IEC) specifications, and for aerial bunch cables, NFC specifications should be referred to in particular (Table 10.2).

10.1.2.2 Polythene

As an insulating material, polythene can match paper. Processing, jointing and installation of polymer-insulated cables have become easier. Polythene is a straight-chain crystalline compound derived from the base compound ethylene directly obtained from refineries. It can be obtained as low-density polythene (LDPE), high-density polythene (HDPE) and linear low-density polymer (LLDPE). All of them differ in density: LDPE has a density of 0.916–0.930 g/cm^3, and HDPE has a density of 0.945–0.960 g/cm^3.

TABLE 10.2

Properties Required as per Different Temperature Conditions as Specified in BS

Material and Type (As per BS-6746)		Tensile Strength (Minutes)	Elongation at Break (Minutes)	Limiting Temperature (°C)[a]	
				Rating	Installation
PVC	T1 1	12.5	125	70	0
PVC	T2 2	18.5	125	70	0
PVC	T1 2	10	150	70	−10
PVC	4	7.5	125–150	85	0
PVC	5	12.5	125	85	0

[a] Temperature condition for current rating and installation.

There are different processes by which this compound can be produced. The popular method is the Ziegler process, which uses an organometal catalyst for the polymerisation of an ethylene monomer. It is a semi-crystalline polymer with a repeat unit of $-CH_2-CH_2-CH_2-$ and is a flexible nonpolar material with excellent electrical properties.

Polythene is used for insulating and sheathing. The material has a lower softening point, so the operating temperature of the cable is restricted to 70°C and the short-circuit rating to 130°C.

Table 10.3 gives the general characteristics of polythene, XLPE and PVC.

Polythene is liable to degrade when exposed to sunlight. It is protected by incorporating 2.5% of fine carbon black (20 μm). The addition of filler is limited as it will make polythene brittle. However, a small amount of butyl or ethylene propylene rubber can be incorporated selectively, which increases its flexing property. Vinyl acetate and alkyl acrylates are widely used as co-monomers. The resulting copolymer, when mixed with channel black, gives a flexible semiconducting material. Copolymerisation also improves the environmental stress-cracking resistance (ESCR) for the outer sheathing material.

For medium- or high-voltage systems, the material must be free from contamination. The extrusion of the material must be done under controlled conditions to avoid any degradation.

TABLE 10.3

Characteristics of Polymeric Insulating Compounds

Characteristics	PE	XLPE	PVC
1. Density (g/cm³)	0.93	0.95	1.38
2. Dielectric constant at 20°C	2.3	2.5	4–8
3. Electrical breakdown (kV/mm) (approximate)	20	25–30	15
4. Volume resistivity (min) (Ωm at 20°C)	1×10^{14}	1×10^{16}	1×10^{11}
5. Tan δ at 50 Hz (20°C)	<0.0003	<0.0001	0.08
6. Operating temperature (°C)	70	90	70
7. Short-circuit rating (°C)	130	250	130

The material becomes brittle when it comes to direct contact with a copper conductor or screen. Brittleness can accelerate with the rise in temperature. This is caused by oxidation in the direct contact of copper. This phenomenon can be inhibited by a suitable choice of antioxidant as per requirement. The choice of aluminium as a conductor has eliminated such problems in aerial bunch cables.

Polythene polymer is highly inflammable. But with a suitable additive or blending with a copolymer, the material can be made flame retardant. Smoke emission, however, is a problem. The material is used for insulation and for sheathing, particularly for low-voltage cables.

10.1.2.3 Cross-Linked Polythene

Polythene is a soft straight-chain polymer. As an insulating material, its operating temperature needs to be restricted to 70°C as it starts to soften at 90°C. The properties of the material are being improved considerably by the cross-linking process. By cross-linking, the material is made 'thermoset' and the operating temperature could thus be raised to 90°C. The short-circuit temperature is also increased to 250°C. The ageing properties of the polymer improve considerably.

The common method of cross-linking is by incorporating a suitable peroxide additive whose melting temperature is high (approximately 170°C), such as di-cumyl peroxide or benzoyl peroxide. At high temperature and pressure, peroxide reacts with the polymer to form a 3D stable compound.

In the early days, the peroxide, catalyst and antioxidant were used to be separately introduced into the extruder by the dosing method along with the polymer during extrusion. At present, a grafted polymer with a peroxide, a catalyst and an antioxidant is supplied by the compound manufacturer, which can be directly fed into the extruder for processing.

In order to obtain grafted XLPE, the polymer is passed through a long, high-pressure stainless tube reactor. The reactor is designed to give a completely contamination-free compound in order to impart higher dielectric and temperature-withstanding properties. Thus, polythene, peroxide, antioxidants and any other ingredients conveyed into the compounding unit must have high purity level. During processing, it is ensured that the materials remain contamination free as the reaction proceeds throughout the passage length. It is also ensured that during movement, the materials do not get coagulated and form any dust particles due to rubbing. Before the incorporation of a cross-linking agent, stabilisers and other chemicals, the materials are passed through extremely fine filters. At every step of the reaction, samples are taken for examination to analyse the quality and contamination level. All the properties along with extrudability are checked before final packing and shipment.

For the peroxide curing system, the cable core is processed in a continuous catenary vulcanising (CCV) line. Here, a fixed length of cable is discarded at the initial and final stages of the operation because of manufacturing constraint. To make the operation economical, longer lengths of the conductor must be kept as feedstock so that the machine can run continuously for a few days.

10.1.2.4 Wet Curing Process (Sioplas-Grafted Compound)

To overcome the limitation of producing smaller lengths as per customer requirements, insulated cores up to 11 kV can be produced by using Sioplas-grafted polythene compound (XLPE) in a batch process. In this case, the insulated cable cores can be cured under hot humid conditions (in a steam bath or in a hot water tank at around 80°C–90°C for a given period of time depending on the thickness of the insulation).

10.1.2.4.1 Monosil System

In this case, ingredients such as silane, peroxide, antioxidant, accelerators and colours are added together and fed into the extruder simultaneously. The reaction starts during the extrusion process where temperature is controlled carefully. In this case, the material cannot be allowed to remain in the extruder for a long period. Mixing of components and feeding must be done under a clean, controlled atmosphere.

10.1.2.4.2 Sioplas Compound

The polymer is blended with a selected silane compound with peroxide. A catalyst is used to initiate the initial cross-linking reaction. Antioxidant, processing aid and colouring agents are also mixed during the extrusion process. The shelf life of this product is very limited and cannot be kept within the extruder for as long as it starts curing within a short period of time. The mixed compound cannot be kept on the shop floor as it will absorb moisture from the atmosphere and starts hardening. The shelf life of this compound is limited.

10.1.2.4.3 Silane Ethylene Vinyl Copolymer

Polythene is grafted with an ethylene vinyl compound when silane is introduced into the polymer chain. The compound comes out ready to use. Catalyst and master batch are added during extrusion. The compound is very stable. By controlling temperature, a limited amount of the compound can be retained within the extruder for a certain period of time. The compound becomes more homogenous and stable in performance. Further, by the use of a base polymer such as LLDPE, the compound is made more resistant to environmental stress cracking.

Nowadays, aerial bunch cables are manufactured up to the 33 kV range. Hence, the quality of the compound must be selected carefully. During extrusion, 2.5% carbon black must be incorporated into the compound.

It is essential that the compound pass the ESCR and UV-resistant tests under moisture, water and damp conditions as the cables will remain exposed to different environmental conditions throughout the year and at different locations. The compound should withstand the impact of ice and severe cold conditions when installed in cold regions such as Kashmir Valley.

The cable is tested for UV and ESCR as per NFC 20540 for ESCR, NFC 33209, ISO 4892-2 (Method A) and ASTM G-155 (Method A), relevant NEMA, and BS and IS specifications. Hence, the compound used should comply with the requirements and pass all the relevant tests.

A typical ESCR compound is tested as per the following methods:

Properties as required	As per specification
• Melt flow index g/10 min ≤ 2.0	ASTM D 1238
• Density g/cm^3 0.920–0.949	ASTM D1505
• Tensile strength MPa ≥ 13.0	ASTM D638
• Elongation at break % ≥ 500	ASTM D638
• Brittleness at low temperature °C ≤ –76	ASTM D746
• ESCR FO H ≥ 96	ASTM D 1693
• Oxidation index ≥ 30 at 200°C min	ASTM D 3895
• Carbon black content 2.6% ± 0.25%	ASTM D1603
• Dielectric strength min ≥ 25 kV/mm	ASTM D149
• Volume resistivity Ωm ≥ 1 × 10^{14}	ASTM D 257
• Dielectric constant ≤ 2.8	ASTM D 1505

By blending the compound with a suitable copolymer such as ethylene vinyl acetate (EVA) and other organic additives, it can be made fire and heat resistant.

10.1.2.4.4 Semiconducting Compound

The conductor and the insulation screen are essential parts of cables with voltage in the range 3.8/6.6 kV and above. It has been found that the wavy contour formed by the spirally twisted wires of a stranded conductor gives rise to high electrical stress on its surface. Further, the distribution of the electrical field within the insulation is never uniform in the radial direction. Distortion of field arises due to variation in dense areas and the formation of numerous non-uniform voids remaining randomly dispersed within the insulation. The thicker the insulation, the more non-radial random electrical stress gets distributed within.

To minimise the effect of localised stress on the stranded conductor and to distribute the electric field radially around the axis of the insulated conductor, a layer of semiconducting material is applied over the conductor and the insulation.

The base material for the semiconducting compound can be polythene or any thermosetting material such as EPR and EVA. High-quality semiconducting compounds are produced with smooth carbon black having an extremely fine particle size. Generally, two types of carbon blacks are used to make semiconducting compounds: 'furnace black', which is produced by burning mineral oil in a furnace, and 'acetylene black', which is produced through a controlled pyrolysis of acetylene. Normally, furnace black contains some ionic molecules, which may initiate the formation of an 'electrochemical tree', but in the case of acetylene black, the material is found to be contamination-free and of extremely fine particle size. Presently, high-quality furnace black has been introduced in the market, which is almost free of contamination.

To produce semiconducting compounds, the base polythene resin used differs from that of the polymer used for manufacturing XLPE compounds. The polymer is mixed with carbon black in high proportion to give an intensively blended smooth conductive compound. The mixing or blending of carbon black with the resin is very tough. Before compounding, all the additives are examined and filtered through a very fine mesh. The compounding machinery is designed to give a proper contamination-free homogenous mix. After mixing, a sample specimen is extruded in the form of a very thin film, which is examined optically to determine the dispersion of carbon black within the molecular matrix of the resin and the contamination level. In this process, the nature of protrusion (called pips) is measured. Their numbers, spread and measures are noted to confirm that these anomalies are within permissible limits. The interface between the insulation and the semiconducting compound is examined optically to determine the quality and smoothness of the semiconducting compound before the material is released for shipment.

The compound must meet the following specifications:

- AEIC CS 6-96
- AEIC CS 8-00
- ICEA S-93-639/NEMA WC 74
- IEC 60502/IEC 60840

Some typical values are as obtained on samples received from suppliers—electrical (for guidance only):

DC Volume	Resistivity	ASTM D991
At 23°C	Ω cm	<20
At 90°C	Ω cm	<100
At 135°C	Ω cm	<500 (Table 10.4)

10.1.2.5 Copper Tape for Screening

In order to allow stress distribution within the insulation to be radial to the axis of the conductor and also to divert a part of the short-circuit current,

TABLE 10.4

Physical Characteristics (as Obtained on Samples after Testing)

Property (for Guidance Only)	Test Method	Unit	Value
1. Density at 23°C	ASTM D	1505 g/cm³	1.12–1.15
2. Ultimate tensile strength	ASTM D	638 kg/cm²	>15
3. Elongation at break	ASTM D	638%	>200
4. Rgeing in air oven at 120°C 168 h	ASTM D		638
Retention of tensile strength		%	>90
Retention of elongation		%	>90
5. Brittleness temperature	ASTM D	746°C	<−50
6. Hot deformation at 121°C × 2 kg	JIS C	3005%	<18
7. Moisture content	Karl Fischer[a]	ppm	<300

[a] As declared by the supplier.

a metallic screen vis-à-vis a copper tape is applied directly covering the semiconducting layer over the insulation.

The copper tape should be uniform in dimension. The surface of the tape should be smooth and bright, devoid of any stain or spot of oxidation and discolouration. Edges should be trimmed and smooth without any sharp points or cut marks. The selection of tapes should be as per the dimensions of the cable. The tape should be EC-grade soft having a resistivity of 17.241 Ω cm/m. Elongation should be 12%–15% minimum. The calculations to determine the tape width in relation to the cable diameter and lapping angle are given in Chapter 8.

10.1.2.6 Constructional Features

10.1.2.6.1 Conductor

1. *Phase conductor*: The conductor material consists of EC-grade aluminium. Aluminium is procured in the form of wire rods in coils having an approximate weight of 2–3 tons. The material should have a purity not less than 99.5%. The resistivity should be less than 0.02803 mm²/m for electrolytic grade. As per IS 8130, the material should have a tensile strength between 100 and 150 N/mm² designated as ½H or H grade.

2. *Wire drawing (for phase conductor)*: To form a conductor, the required size of the wire is drawn from the wire rod to form smaller-diameter wires.

3. *Wire drawing and treatment (for messenger conductor)*: The messenger supports the weight of the cable and also resists environmental pollution and corrosion under adverse conditions such as moist sea weather, thrust of high wind and additional weight of ice

(in cold atmospheres). Being a soft metal, aluminium needs to be strengthened by forming an alloy with a suitable material. In this case, aluminium is made tough by alloying it with a small percentage of magnesium (0.5%) and silicon (0.5%–0.6%). By alloying, material strength is increased considerably (normal alloy designation as per ASTM is 6101-T6).

a. *Solution treatment*: An alloy wire rod is drawn to a diameter of 5.5 mm (approximately). Although alloy materials are spread over the entire region, they are not uniformly distributed throughout the crystalline structure of the metal. Being hard, the metal is drawn better in a bull-block-type wire-drawing machine and collected in continuous coil form. The formed coil is then subjected to solution treatment. The process consists of putting the coils in a basket and placing in a hot enclosed furnace chamber at a uniform temperature of approximately 520°C–550°C (melting temperature of aluminium being around 630°C). At this temperature, the coil becomes red hot. It is heated for about 20–30 min (depending on the original tensile strength of the wire rod) and then dropped in cold water for quenching to make the wire very hard. Solution treatment furnaces can be of two types. One type has a top loading and unloading system. The furnace, in this case, is placed underground, and the top opening is kept above floor level. The furnace is made of a cylindrical steel plate. The inside of the chamber is lined with fire bricks on which heaters are placed spirally all around the periphery of the chamber and covered with perforated stainless steel plates so that heaters do not come in contact with aluminium wires. An air-circulating fan is provided to distribute heat uniformly within the chamber. The chamber is provided with a temperature indicator, a temperature controller and a timer.

   Coils are placed in a basket or a loading casket. The basket or casket is then placed within the chamber by a jib crane. The top lid is closed hermetically. On attaining the required temperature, the timer is set and the coils are heated up. At the end of the set time, the top lid is opened and the coils are quickly lifted out of the chamber and dropped in a basin of cold water, which is kept ready adjacent to the furnace. The time within which the coil should be dropped in water is very important. It must be accomplished within a few seconds. The quicker the dropping time, the better and uniform the result obtained.

   In another type, the furnace is placed above the floor level. In this case, loading is done from the top, but the coil is dropped through the opening of the bottom lid. Here the water chamber is placed directly below the furnace. Naturally, red hot coil can be dropped within 3–4 s. The results obtained here are far better and more uniform than in the former case. In the former case, atmospheric air cools down the top layers faster than bottom layers when the coils are lifted from the top. Hence, the tensile strength is not uniform.

In another type, coils are inserted in a horizontal-type furnace from one end and are ejected out from the other end into the water after the heating cycle is completed. During this process of solution treatment, atoms of alloy metals get disbursed uniformly within the crystal structure of aluminium.

After solution treatment, coils are allowed to cool down to atmospheric temperature. Thereafter, the wire is drawn to the required sizes to be taken on steel bobbins. The wire can be drawn on a tandem-type wire-drawing machine.

b. *Ageing*: Even after wires are drawn, the tensile strength remains very high. Forming of the stranded conductor at this stage is very difficult. These wires are then tempered through the ageing process. In this case, bobbins with wires are placed in a box-type furnace and heated at a given temperature as per requirement depending on the size of the wire, its original tensile strength and final breaking load to be achieved. Normally, temperature varies between 150°C and 180°C for a period of 16–18 h depending on the quality of the product to be obtained (properties must be as per IS 834 part 4). The ultimate breaking load as desired should be 290–295 N/mm^2.

After ageing, samples of wires are tested for breaking load. Once results are found satisfactory, the material is sent for making conductors by the stranding process.

4. *Formation of conductor*: The phase conductor is formed by laying wires concentrically. Each layer should be laid spirally opposite to the other, one above the other. The outer layer should always be in the right-hand direction unless otherwise stated in the specification. The lay length is generally 12–14 times the diameter over the stranded layer as per IS 834-Part-I when the conductor is not compacted. In earlier cases, all the conductors were laid concentrically without compacting. During stranding, a spatial torsional twisting force comes into play with every 360° rotational movement of the carriage when bobbins are kept fixed on the cradle. After a number of turns, these twisting forces become so great that the wire breaks. Further, a snaking effect comes into play on the conductor forming birdcages and the like. To obviate these difficulties, stranding machines are constructed with anti-twist planetary gear to keep the bobbins always on the same plane during the rotational movement of the carriage. Production of

these machines is slow, whereas initial and maintenance costs are high. However, for messenger conductors, these machines are ideal, so were the tubular 7- and 19-wire stranding machines. In tubular-type machines, the threading of wires is done in such a manner that with each rotation of the wire, an anti-twist generates to nullify the initial twist. Naturally, being of higher rotational speed (higher production rate), tubular-stranding machines are ideal for both phase and messenger conductors. However, with the coming of rigid and high-speed machines to form higher sections of conductors, particularly for medium-voltage cables, torsional forces are minimised to a greater extent by compacting the conductor at each stage of stranding. This has the advantage that the wavy contour of the top surface of the conductor is compressed and smooth, giving

a. A lower conductor diameter which reduces the consumption of raw material
b. Lower electrical stress level on the surface of the conductor, where the stress level is given by the formula $E = V_o/\{r \ln(R/r)\}$, where V_o is the operating voltage, R is the radius of the conductor over insulation and r is the radius of the conductor (in this case, r could be taken as the radius of each wire when the conductor is not compacted)
c. Torsional (mechanical) stress level is practically eliminated. No twisting and turning of the conductor is experienced with repeated winding and unwinding during subsequent process and installation of the cable.

In the case of a compacted conductor, the lay length can be 16–18 times the diameter over each layer including the final one. In this case, the air gap within the conductor is reduced, increasing the metal area on the overall cross section of the conductor. The ratio of the metal area to that of the geometrical area of the conductor cross section is called filling factor and is given by $n(d/D)^2 \times 100$ in percentage. The filling factor achieved after compaction could be around 90%.

Here, d is the diameter of the wire, D is the diameter on the stranded conductor and n is the number of wires. The surface of the stranded conductor must be smooth and flawless.

10.1.2.6.2 Insulating the Conductors

The details of insulating a conductor by various extrusion processes are given in Chapter 5. Low-voltage conductors are insulated in an extruder. Normally, black PVC compounds are used to insulate the conductor. These insulated conductors are mostly used in rural areas and in places where the cable is not expected to last long. However, PVC wires can be easily tapped at any point as the jointing is not very difficult. This may lead to easy pilferage of power from any point. Further, the material density being higher than polythene, the span length of the cable is restricted. The material also gets softer when exposed to high solar radiation and becomes brittle at a very low temperature. Current ratings are also restricted to a maximum conductor temperature of 70°C. PVC also cannot be used above 3000/3300 V category cables.

Normal polythene may have a lesser weight ratio compared to PVC, but ageing and operating temperatures are not better than those of PVC. High solar heat or very low temperatures also affect insulating properties and life expectation.

Considering all the factors, XLPE has been the best choice for insulating the conductors in aerial bunch cables. This is done using two methods:

1. Through a normal extrusion process using the Sioplas method for cables ranging from 1,100 to 11,000 V category
2. Using a CCV line to process conductors from 11,000 to 33,000 V category cables

In the Sioplas process, a grafted polymer is used. After extrusion, the insulated cable is conditioned for 24 h on the shop floor, after which the cable is cured by immersing either in hot water at a temperature of 90°C or in a low-pressure steam chamber for a given period of time. After curing, the insulated conductor is kept on the floor for another 24 h before further processing. The curing of the conductor continues for days, though maximum curing is initiated within a few hours of treatment. During conditioning, moisture and unwanted gas exude out of the surface of insulation when shrinkage eliminates/reduces the dimensions of voids within insulation.

In the CCV process, a similar treatment is adopted after the cured material comes out of the cooling process to ensure a good-quality insulated conductor.

10.1.2.6.3 Assembling of Aerial Bunch Cables

Assembling of aerial bunch cables is not done through the normal laying-up technology. In this case, the messenger conductor remains straight in the middle on which the phase and other conductors (such as lighting and earth conductors) are laid in a corkscrew formation. The lay length should be restricted within 35 times the diameter of the phase conductor. For this reason, a 1 + 3 core laying-up machine is found to be ideal. The process can be completed in a drum twister with special arrangement where the central messenger runs straight through the middle of the cable assembly. The ideal machine for this purpose is Skip Strander. Tension control must be regulated properly during the laying-up process. For medium-voltage cables, the screen over the insulation should not crack or get disrupted. Under any circumstances, laying-up should be carried out with 100% back twisting of conductors.

10.1.2.6.4 Testing

The conductor and insulation are tested as per NFC standards. Apart from routine tests, special ESCR and carbon black tests are carried out specifically to ensure that the cable performs in all types of environmental conditions.

10.2 Direct Current Cables

Problems in transferring a large block of power through an AC system have long been recognised, particularly when the distance is long and the power of transmission is raised to the extra-high-voltage range. The larger-size conductor resistance is compounded by the skin and proximity effect. Further, the longer the length, the larger the capacitance values, which in turn increase the charging current. The value of the charging current may become so large that the total transmitted current is neutralised, leaving the line completely useless. To recover the system, costly reactors have to be assembled intermittently.

An alternative is to replace high-voltage AC (HVAC) lines by installing high-voltage DC (HVDC) cables. The economics of transmitting the DC system depends on the cost of the AC/DC terminal convertors. With the introduction of thyristors, high-voltage solid-state terminal equipment have replaced mercury valves and are found to be economical. The design of DC cables involves the following:

- The selection of a transmission system with proper insulation thickness to withstand the operating voltage
- The selection of conductor size as per the required power transmission capability

The HVDC system has many advantages over the HVAC system:

- Absence of a continuous charging current enables the cable to be rated at a much higher voltage range.
- Dielectric loss is negligible. Only a small leakage current is considered.
- Skin and proximity effect being absent, effective resistance is restricted only to conductor resistance. Hence, the induced voltage on the sheath is almost absent. This eliminates the use of cross-bonding-related equipment.
- The risk of void ionisation is negligible. Hence, synthetic materials can be easily regarded as insulating materials.

- DC cables are much lighter and are, therefore, less expensive than AC cables for the same operating voltage.
- This reduces the cost of civil works and installation.
- The cable does not interfere with communication cables laid nearby.

Initially, DC cables were made of paper insulation and are also of oil-filled or gas-filled type. With the coming of XLPE insulation, the problem has been understood in a different way.

In a DC cable, voltage distribution under steady state depends on the electrical conductivity of the dielectric and not on the value of capacitance. The geometry of cable being round, stress distribution remains the same for both AC and DC systems. But in such a case, the electrical conductivity λ_d must remain the same throughout the cross section of insulation of an unloaded DC cable. Considering that I_1 is the charging current for a metre length of cable, when the DC voltage is V between the conductor and sheath, we get

$$I_1 = \frac{2\pi V \lambda_d}{\ln(R/r_0)} \quad \text{A/m} \tag{10.1}$$

where

R is the radius of the conductor under the sheath
r_0 is the radius of the conductor
λ_d is the electrical conductivity (in S/m) of the dielectric

Taking a ring section on the insulation at a radius r within the cable insulation, the voltage across the thickness is

$$dV = \frac{2\pi V \lambda_d}{\ln(R/r_0)} \cdot \frac{dr}{2\pi r \lambda_d} = \frac{V dr}{r \ln(R/r_0)} \tag{10.2}$$

Accordingly, the electrical stress at point r is given by

$$E_r = \frac{V}{r \ln(R/r_0)} \quad \text{V/m} \tag{10.3}$$

On that basis, the maximum stress will be on the surface of the cable conductor.

$$E_{max} = \frac{V}{r \ln(R/r_0)} \quad \text{V/m} \tag{10.4}$$

The minimum stress should be under the sheath on the insulation surface.

$$E_{\min} = \frac{V}{R\ln(R/r_0)} \quad \text{V/m} \tag{10.5}$$

It has been observed that the stress within the dielectric depends on the conductivity of the insulating material. It increases with the increase in temperature. Conductivity is maximum at the surface of the conductor and diminishes towards the outer surface near the screen as the temperature starts dissipating towards the outer surface of the cable. Accordingly, stress is high near the outer surface of the insulation and low at the conductor surface. This phenomenon is called 'stress inversion' due to the accumulation of space charge, which actually determines the criteria for designing a DC cable. It is, therefore, a subject of study as to how space charge behaves in a loaded DC cable. A complete mathematical treatment of the distribution of stress with the rise in temperature within the dielectric of a DC cable is given in King and Halfter (1982).

However, technologists have undertaken experiment to determine the way stress development occurs within the solid dielectric vis-à-vis polythene and XLPE.

The following calculation indicates the way stress development occurs within a solid dielectric. It also shows how the thickness of the insulation could be determined by knowing the allowed temperature rise and thermal resistivity of the material used.

In a loaded DC cable, the electric stress, instead of decreasing from the surface of the conductor towards the sheath, actually increases and the highest value is found at the surface of the sheath. This phenomenon is termed 'stress inversion'. This actually limits the design value of the operating current. The development of stress in a DC cable depends on the conductivity of the insulating material and not on the value of the capacitance. This conductivity increases with the increase in temperature. The operating current is actually determined by the stress that would develop near the outer screen in a loaded cable and is almost equal to the stress near the conductor of an unloaded cable. This phenomenon determines the rise in temperature and is restricted. This rise is a function of dielectric conductance since conductance increases with the rise in temperature. Naturally, the operating current is calculated to keep the temperature rise within the permissible limit. Many experimental evidences have shown that under steady state, the stress distribution in a DC cable is controlled by the conductivity of the dielectric. Any field distortion is also associated with the conductivity of insulation and is determined by various experiments.

The relation between the electrical conductivity of a dielectric, λ, and the temperature at different stress values can be expressed as

$$\lambda_r = \lambda_R e^{a\Delta\theta}\left(\frac{E_r}{E_R}\right)^{\varepsilon} \tag{10.6}$$

where

λ_r is the conductivity of the dielectric at a radial distance r on the insulation from the centre of the cable (Figure 10.4)

λ_R is the conductivity on the insulation on the surface of the dielectric under the cable sheath

E_r is the electric stress on the dielectric at point 'r'

E_R is the electric stress on the surface of the dielectric under the sheath

$\Delta\theta$ is the temperature difference under the sheath of the cable at point 'r' on the dielectric ($\Delta_r - \Delta_R$)

a is the temperature coefficient of the dielectric conductivity of the cable

ε is the dielectric constant of the insulating material

The following equation for a DC cable can be considered to define the electric field:

$$E_r = \frac{V}{\lambda_r \int_{r_0}^{R} \frac{dr}{\lambda_r}}. \tag{10.7}$$

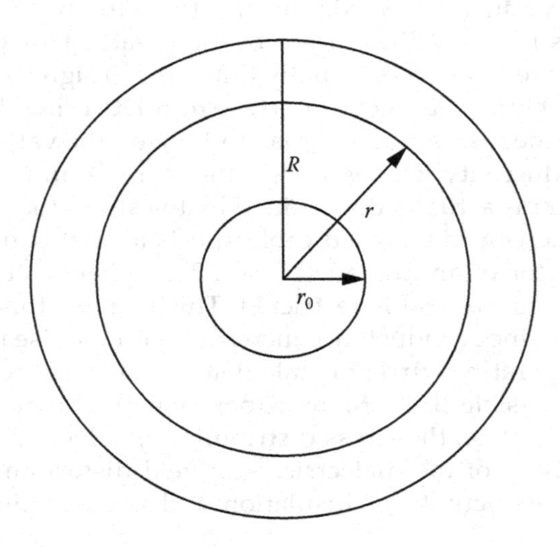

FIGURE 10.4
Cross section of a DC cable.

Now from deduction, it is found that

$$\frac{E_r}{E_R} = \frac{\lambda_R R}{\lambda_r r} \tag{10.8}$$

In a steady-state condition, the temperature difference due to thermal resistivity is g_d °C cm/W, and when the amount of heat generated is t_c, the per unit length in W/m is

$$\Delta\theta = [\theta_r - \theta_R] = \frac{t_c g_d}{2\pi} \ln \frac{R}{r} \tag{10.9}$$

where
 R is the radius of the cable under the sheath and over the insulation
 r is as defined earlier

Substituting Equations 10.8 and 10.9 in Equation 10.6, we get

$$\lambda_r = \lambda_R \left(\frac{\lambda_R R}{\lambda_r r} \right)^{\varepsilon} e^{((a\, t_c g_d / 2\pi) \ln(R/r))} \tag{10.10}$$

Taking $p = at_c g_d / 2\pi$ gives $e^{p \ln(R/r)} = (R/r)^p$

From this, it is deducted that

$$\lambda_r = \lambda_R \left(\frac{\lambda_R R}{\lambda_r r} \right)^{\varepsilon} \left(\frac{R}{r} \right)^{p}$$

or

$$\lambda_r^{\varepsilon+1} = \lambda_R^{\varepsilon+1} \left(\frac{R}{r} \right)^{\varepsilon+p} \tag{10.11}$$

Taking $(\varepsilon + p)/(\varepsilon + 1) = k$, Equation 10.11 can be written as

$$\lambda_r = \lambda_R \left(\frac{R}{r} \right)^{k} \tag{10.12}$$

In the equation, the value of 'p' indicates the thermal resistivity and the temperature coefficient of conductivity and 'ε' is the dielectric constant. In the case of polythene and XLPE, the value of ε can be taken between 2.1 and 2.5.

Taking Equation 10.7 and placing the value of Equation 10.12, it becomes

$$E_r = \lambda_R \left(\frac{R}{r} \right)^{k} \int_{r_0}^{R} \left(\frac{dr}{\lambda_R (R/r)^k} \right) \tag{10.13}$$

Integrating the equation,

$$E_r = \frac{V\left\{\dfrac{2\pi\varepsilon + at_c g_d}{2\pi(\varepsilon+1)}\right\}\times r^{\left\{\frac{at_c g_d - 2\pi}{2\pi(\varepsilon+1)}\right\}}}{R^{\left\{\frac{2\pi\varepsilon + at_c g_d}{2\pi(\varepsilon+1)}\right\}} - r_0^{\left\{\frac{2\pi\varepsilon + at_c g_d}{2\pi(\varepsilon+1)}\right\}}} \tag{10.14}$$

If p =1, the value of k is 1; in such a case, the value of E_r becomes

$$E_r = \frac{V}{R - r_0} \tag{10.15}$$

From this equation, it can be seen that when $k < 1$, the stress level goes down as the thickness increases. Stress increases with the increase in thickness when $k > 1$. Therefore, the best solution is found when $k = 1$, where the conductivity remains inversely proportional to the radius r of the cable. Here λr multiplied by r is a constant.

When the working stress voltage is taken as E_v replacing E_r by E_v, the value of R of the cable can be calculated when the conductor radius r_0 is known. On this basis, the thickness of the insulation can be determined.

When $k < 1$ and replacing r for r_0, Equation 10.14 can be transformed as

$$N = \frac{R}{r_0} = \left(1 + \frac{Vk}{E_{v r_0}}\right)^{1/k} \tag{10.16}$$

And when $k > 1$, the value of N can be calculated from the following equation:

$$N^k\left(1 - \frac{Vk}{E_v r_0 N}\right) = 1 \tag{10.17}$$

In this case, 'R' has been replaced by 'r'.

Finding the value of 'N' and knowing 'r_0', the radius 'R' of the conductor can be calculated such as the thickness of insulation.

The main application of HVDC power cables has been in long-distance power transmission. In recent years, there has been a growing trend towards their application in offshore wind power generation, which is being actively introduced as a renewable natural energy source. After their introduction, the locations of wind power generation facilities shifted from coastal areas to offshore areas due to space constraints and wind velocity. As the power transmission distance increased, HVDC power transmission technology gained more attention. Previously, oil-impregnated insulation cables, such as mass-impregnated (MI) cable and oil-filled (OF) cables, were applied to DC power transmission. In recent years, extruded insulation cables have become desirable since they have no fear of oil leakage or the problem of maintaining pressure. On the other hand, XLPE-insulated cables, which are mainly used for AC power transmission, are known to have a number of problems in insulation when used for DC cables. The main problem is the accumulation of space charge within XLPE-insulating materials. Naturally, studies have to be undertaken to understand the formation and nature of space charge and also how to arrest the space charge movement to restrict reversal of polarity. If these problems are understood and solved, DC cables could be used for very-long-distance transmission of power without significant loss. A new insulating material needs to be developed to suit the purpose.

Keeping this in mind, development has been initiated to have a similar XLPE cable using a newly formulated DC-XLPE material as the insulator. In this context, Japanese scientists and engineers have done extensive work wherein they have been able to offer a workable solution to overcome the problem.

Murata et al. (2013) describe the excellent DC characteristics of the DC-XLPE insulation material developed for DC applications and reports on the implementation status of type tests and pre-qualification (PQ) tests in accordance with the International Council on Large Electric Systems (CIGRE) Technical Brochure on actual cables and accessories.

In 1954, the world's first operation of HVDC power transmission began between the mainland of Sweden and Gotland Island. At that time, MI cables using insulating paper impregnated with high-viscosity insulating oil were used, and for higher-voltage and larger-capacity applications, OF cables were adopted, using insulation paper, which was impregnated with low-viscosity insulating oil and kept in a pressurised condition. Since then, MI cables and OF cables, that is, oil-impregnated paper-insulated cables, have been the mainstream of DC power transmission cables. The oil-impregnated insulation cable technology was developed in response to the demand for higher voltage, larger capacity (larger conductors) and longer power transmission. Extruded insulation cables, in which materials such as XLPE are extruded on the conductor, were first applied in Gotland in an 80 kV DC line in 1999. Here, a voltage source converter (VSC) was used as an AC/DC converter. As the VSC does not require the inversion of voltage polarity (polarity reversal) when reversing the direction of a power flow, the problem of a decrease in insulation performance, such as space charge influences, can be reduced. This has made the use of extruded insulation cables easier for DC applications.

To start with, a project was undertaken to test cables of 200 kV, which was considered the world's highest DC-operating voltage for extruded insulation cables. However, because of the space charge accumulation in the insulation, many problems were encountered in the application of DC insulation.

Based on the results obtained, experimental work started to develop a basic XLPE-insulating material for DC usage. A series of R&D projects have been carried out in collaboration with the Electric Power Development Co. Ltd, Japan. At the beginning of the development, the DC-XLPE cable used a general-purpose inorganic filler material. But to adapt the cable to higher-voltage applications, miniaturisation, higher purification and higher distribution of the inorganic filler material were pursued. As a result, a nanocomposite insulation material was formulated in which inorganic nanoparticles were used and distributed uniformly.

10.2.1 Insulation Material

Cables widely applied in AC power transmission and distribution use XLPE as the insulation material (hereafter denoted as AC XLPE). An AC XLPE insulation material exhibits excellent insulation performance against AC voltages, but it does not exhibit adequate performance against DC voltages because of the accumulation of space charge. By adding nanosized filler materials in the XLPE insulator, some excellent features can be obtained. The XLPE insulation material for DC usage, to which nanoparticles have been added (hereafter referred to as DC XLPE), has the following features in comparison with AC XLPE:

- High volume resistivity
- Low space charge accumulation
- Long DC lifetime
- High DC breakdown strength

10.2.2 Volume Resistivity

The volume resistivities of DC XLPE and AC XLPE were investigated using specimens in the form of a sheet formed by the hot pressing method. The sheet thickness was about 150 μm. The volume resistivity was evaluated with the DC leakage current value measured 10 min after the commencement of the test. Temperatures were set at 30°C, 60°C and 90°C, and electric fields were set at 40, 60 and 80 kV/mm.

It was observed that as per the applied ranges of different temperature and electric field, DC XLPE possesses about 100 times higher volume resistivity than AC XLPE.

10.2.3 Space Charge Characteristics

The space charge distribution in DC XLPE and AC XLPE was evaluated using the pulsed electro-acoustic (PEA) method. Hot-pressed sheet specimens with a thickness of about 200–300 μm were considered. DC voltages with average electric fields of 20 and 50 kV/mm were applied to the specimens, and the change of space charge distribution was measured at 10 s intervals.

It is seen that, as time progresses, space charge gets accumulated and the electric field distribution gets distorted to a large extent. In particular, in the neighbourhood of the anode, negative space charge gets accumulated, leading to a considerable rise in the electric field. In order to express numerically the effect of space charge on the electric field in a concrete manner, the field enhancement factor (FEF) is examined. XLPE shows a clear trend of increase with time. Moreover, the rate of increase in FEF is greater in the case of 50 kV/mm than in the case of 20 kV/mm. To investigate the time-dependent change in the space charge characteristics of DC XLPE, the FEF was evaluated for a longer time. The evaluation was conducted for several days. DC XLPE was found to be stable and remained less than 1.1 for several days from the beginning. These results indicate that, by adding nanosized particles to DC XLPE, the amount of space charge accumulation is kept low in the electric field of several 10 kV/mm as compared with that of AC XLPE.

10.2.4 DC *V–t* Characteristics

The DC voltage–time (*V–t*) characteristics of DC XLPE and AC XLPE were investigated on the pressed sheet specimens with a thickness of about 200 μm. A vacuum-drying treatment was applied to the AC XLPE sheet samples to reduce the by-products of cross-linking as they were known to affect the space charge characteristics. On the other hand, no such treatment was given to the DC XLPE specimens. The sheet specimens were placed between the high-voltage electrode and the ground electrode, and a DC voltage was applied in the silicone oil. The test was conducted at 90°C. As described, DC XLPE has a higher initial DC breakdown strength and longer DC lifetime than AC XLPE. The excellent properties of these cables are as follows:

- In a DC XLPE insulation, the amount of accumulated space charge and the enhancement of the electric field are small.
- In a DC XLPE insulation, the change in the volume resistivity and thermal breakdown resulting from localised joule heating is less likely to occur than in an AC-XLPE insulator.

These two effects actually interact with each other, and the DC breakdown strength is raised to a higher level; further, DC lifetime becomes longer. The positive effects of adding nanosized particles, such as increase in volume resistivity and decrease in the accumulation of space charge, are also observed in low-density polyethylene. Such effects have been reported by many others. Various studies have been conducted on these positive effects, that is, low accumulation of space charge and high volume resistivity.

For example, Ishimoto et al. (2009) measured a thermally stimulated current in a nanocomposite material and suggested that a nanosized filler makes a deep trap site and captures the charge. Maezawa et al. (2007) proposed that an induced potential well is formed around the nanosized particles. The nanosized particles in a high electric field work as a deep trap to capture the charge and prevent the movement of carriers. When the trap site is formed in the insulation material, the internal carriers which play a leading role in electrical conductivity are captured and the hopping conduction is suppressed. As a result, the internal carrier mobility is reduced, and thus the volume resistivity is thought to be raised.

Next, space charge formation by the inner carriers is considered. It is likely that the inner carriers gradually move towards the counter electrode due to the applied electric field and are unevenly distributed in the insulation, which causes the formation of space charge in the dielectric. When nanosized particles are added, carriers moving with electrical charge are trapped by the nanosized particles that are uniformly dispersed in the insulation. This prevents the localisation of inner carriers, and thus prominent differences in the density of space charge are reduced. Therefore, it is thought that space charge accumulation is suppressed. Finally, the injected charge from the electrode is studied. The injected charge is captured by the trap formed by the nanosized particles near the electrode and remains there. This trapped charge causes random dispersion of the electric field near the electrode, which prevents further injection of charge and limits space charge formation. In fact, homo space charge is often formed in front of the electrode in nanocomposite materials.

As described earlier, in the DC XLPE insulation material, uniformly dispersed nanosized particles, which work as traps, increase volume resistivity and suppress space charge accumulation. Due to the synergistic effects of these two, good DC cable characteristics, such as improved DC breakdown strength and long DC lifetime, are exhibited.

Based on experience, XLPE-insulating materials have been developed, which have excellent properties for DC voltage applications. DC XLPE cables using these materials have already completed long-term demonstration tests for high-voltage transmission up to 500 kV. A 250 kV class–type test and a pre-qualification test have also been completed under the test conditions that conform to CIGRE TB 219. Following this success, a 400 kV-type pre-qualification test was conducted under the conditions that conform to CIGRE TB 496. All long-term demonstration tests that were conducted included a polarity reversal test. They were performed at a conductor temperature of 90°C. From these results, it was verified that DC XLPE cables and accessories can be applied in 90°C normal operation and polarity reversal operation in actual HVDC links. DC power transmission technology is not only applied to the conventional long-distance and large-capacity power transmission, but also expected to find wide applications as an environment-friendly technology that enables high-efficiency power transmission in conjunction with renewable energy technologies such as offshore wind power.

10.2.5 Conclusion

A considerable amount of transmission and distribution of electricity is carried out by power cables. At present, most HVDC installations use traditional (oil-impregnated paper insulation and oil-filled types) cables, which pose a risk to environmental pollution in the case of an accident.

Thus, significant advances in and polymeric materials, notably cross-linked polyethylene, have made them the preferred insulation material for power cables due to their economical production, environmental benefits and electrical properties. However, unwanted disadvantages related to performance are featured when operated under DC applications. For instance, the accumulation of comprehensive immobile charges in XLPE superimposes the Laplacian field, resulting in changes in electric stress across the dielectric material. Additionally, in most HVDC transmission systems, both the presence of a temperature gradient across its insulation and bidirectional power flow should be considered. Space charge existence within the insulation is particularly dangerous in the event of polarity reversal, which has been recognised as the root source of breakdown in the early extruded insulation of commercial DC cables. High electric stresses within the insulation may be created, especially in the case when rated voltage is applied on the cable and with the presence of a temperature gradient. Therefore, investigations were carried out both on the space charge dynamics and on the accurate determination of electric field across the insulation of a full-sized cable.

Space charge accumulation within the polymeric material of an XLPE power cable is measured using a modified PEA system with a current transformer attached. The presence of these accumulated space charges, along with the consideration of conductivity, influences the electric field distribution across the insulation material. As it is well known that the conductivity of an insulating material is dependent on both the temperature and the electric field, they make it difficult to determine the electric field distribution in HVDC power cables.

DC cables have to be protected by a metallic sheath. Again this sheath should be protected from the surrounding corrosive elements, particularly from seawater and when buried underground. The return path can be a cable laid parallel to the supply line. In the case of a fault in the return line, earth can become its return path. However, precaution should be taken to prevent ingress of moisture.

10.3 Umbilical Cables

An umbilical cable or an umbilical is a cable which supplies the required consumables to an apparatus. It is named by analogy with the umbilical cord. An umbilical can, for example, supply air and power to a pressure suit or hydraulic power, electrical power and fibre optics to subsea equipment.

Subsea umbilicals are deployed on the seabed (ocean floor) to supply necessary control, energy (electric, hydraulic) and chemicals to subsea oil and gas wells, subsea manifolds and any subsea system requiring remote control, such as a remotely operated vehicle. Subsea intervention umbilicals are also used for offshore drilling or workover activities.

A diver's umbilical is a group of components which supply breathing gas and other services from the surface control point.

For shallow water diving, the diver's umbilical is typically a three-part umbilical comprising a gas hose, a pneumofathometer (pneumo) hose and a diver communications cable, which usually also serves as a lifeline strength member. The 'pneumo' hose is open at the diver's end and connected to a pressure gauge on the surface gas panel, where the supervisor can use it to measure the diver's depth in the water at any time.

More recent umbilicals comprise all the components laid together, such as a twisted rope, so that there is no chance of a kink, no separate lifeline component is required and no tape is required to hold the umbilical together. An additional component, such as a video cable for a diver's camera or a hat light cable, can be added by manually wrapping this additional component into the lay of the existing cabled umbilical (Figure 10.5).

Umbilicals transfer power, chemicals, communications and more to and from subsea developments, and they are literally the lifeline of subsea trees, manifolds, jumpers, sleds and controls. It is a connective medium between surface installations and subsea developments, including electrical, hydraulic and chemical injections and fibre optic connections.

Umbilicals are enclosed within an outer ring specially designed for subsea environments. While these cables must withstand everyday wear and tear, and seabed temperature, umbilicals are also deployed in ultra-deep-water, high-pressure and high-tensile environments.

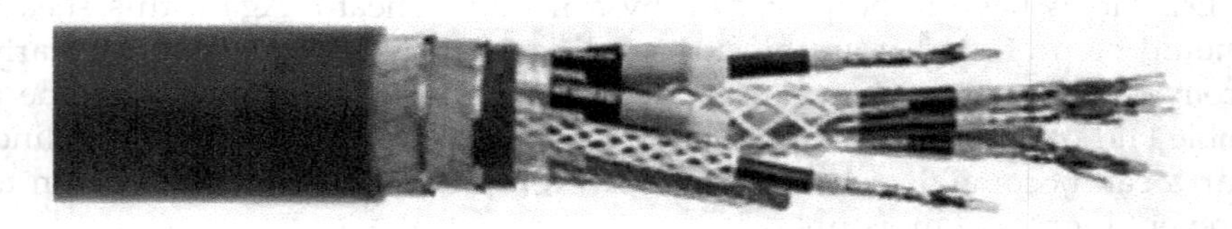

FIGURE 10.5
Umbilical cable system.

FIGURE 10.6
Composite umbilical.

The number of umbilicals used varies as per development because each subsea project is unique. Additionally, there can be a single connection or multiple connections in a single umbilical line. An umbilical might just include chemical injection tubes or communication cables and power cables bundled together and encased in a single line.

Umbilicals that incorporate multiple connections are referred to as integrated umbilicals. Though integrated umbilicals can be used in development and installation programmes, several umbilicals may still be required for further development (Figure 10.6).

There are several umbilicals used for different purposes. Hydraulics are used to bore subsea wells. Some umbilicals pump chemicals to facilitate product extraction. Electrical umbilicals connect to subsea power supply through control panels to different equipment to transmit information about temperature, pressure, etc. Fibre optic cables can relay information instantly to surface monitoring systems.

Recent developments in umbilical technology allowed to integrate the flow line, surrounded by power lines and electrical umbilicals, and the group is then encased in a tubing.

11

Power Cable Laying, Jointing and Installation

For the distribution and transmission of power from generating stations to consumer points, cable lines are laid through different terrains and environments. Considering the diverse surroundings, cables are designed and manufactured. Naturally, laying, installation and environmental conditions should be clearly defined while ordering a cable or transmission line.

Primarily, power is distributed from a power-generating station. Bulk power generated at a nominal voltage is evacuated and transmitted by raising its voltage at times to the extra high-voltage (EHV) range through a high-power transformer. Raising the voltage restricts the conductor size to a rationally accepted level. For a long-distance carrier, overhead transmission lines are used for evacuation. The voltage range can be selected depending on the capacity of the power-generating station. Transmission lines normally selected are of 750, 500, 400, 220 and 132 kV for long-distance, cross-country supplies.

Power is uploaded to a centralised high-power feeder station conveniently located at a distance from large cities and industrial establishments, covering a few square kilometres. To supply power from this main feeder station to different consuming points, substations are installed, which bring down the voltage level to 33/30 or 11/10 kV, allowing each substation to cover a substantial area under its jurisdiction. A main feeder station can control several sub-stations, catering to a few cities and industrial clusters.

If substations are far away from the main distributing station and where there is no danger of carrying power lines through dense localities, the overhead supply line consisting of aluminium conductors steel reinforced (ACSR) or all-aluminium conductors (AAC) is used. In other cases, power is transmitted through cables from the main station to various substations. Internal wiring and supply lines within a powerhouse, main distributing station and substations are maintained through insulated wires and cables.

Power from an overhead line is received by the incoming side of the step-down transformer bushing through wires. The outgoing lower-voltage side is connected to cables or wires for supplying the received power to various consuming points. Supply lines to distant villages or establishments can be of pole-mounted AAC or ACSR for economic reasons. In these areas, safety issues are not so stringent and supply lines are kept at a safe distance from people and locality. However, insulated wires and cables are used to connect the consuming points to avoid direct contact with living beings and equipment.

Cables used between main power points are as follows:

1. From the powerhouse to nearby main distributing stations – EHV and high-voltage (HV) cables for transmitting power as per voltage requirements
2. From main distributing stations to substations – EHV, HV and medium-voltage (MV) transmission cables as per the capacity of substations
3. From substations to distributing transformers within localities, industries, etc. – MV and low-voltage (LV) distribution cables

However, at times, there may be some special requirements which have to be considered based on individual needs.

Whether the route length is short or long, cables should be of a specified length, which are joined together to form a supply line for a longer cable route from the outgoing terminal point to the terminal points at the receiving end; or at times, a shorter length should be terminated from point to point as required. A joint on the straight length is termed 'straight through joint'. Joints at termination points are known as 'end termination'. At times, for providing a service line from the main cable length, a branch line should be taken out, forming a TEE point. This type of joint is termed 'TEE joint'.

Before choosing a cable, the following should be assessed:

1. *Planning the route*: Effort should be made to select the shortest possible route length.
2. *Nature of terrain*: The nature of terrain through which the cable should be laid, such as hard rocky soil, dry land, desert, marshy land, waterlogged region and river crossing, should be studied. It is also required to study whether cables have to pass over bridges, under roads or alongside railway tracks, sea coast, near chemical plants, refineries, etc.
3. *Environmental condition*: Temperature change at a place during summer and winter should be checked along with rainfall. Ice and high wind can prevail at times at high altitudes. Sea sides have high humid conditions. There may be a corrosive atmosphere near a chemical plant or refinery.
4. *Load distribution planning*: Total load requirements and distribution system within a substation and supply network planning to be such that the voltage drop within supply lines are restricted as per norm.
5. *Handling and laying of Cables*: The weight vis-à-vis the length of the cable should be determined so that during handling, proper care could be taken to avoid any mishap or accident.

All factors that influence the installation and operation of cables should be clearly specified while choosing cables. The packing length should be specified, and it length should be such that the weight of the cable with the drum can be easily handled at the site. Cables of longer length should be selected in order to restrict the number of joints and therefore minimise installation cost. Further, every joint is a weak spot and is vulnerable to failure.

While accepting cables from the factory, all lengths should be accepted with test certificate. The certificate should mention the details of the cables. The packing drum should be marked with the following details on the flange:

1. Name and country of the manufacturer
2. Cable code indicating conductor type, number of cores, voltage designation, type of insulation, unarmoured or armoured type, nature of outer sheath
3. Specification reference
4. Drum number
5. Length in metres
6. Gross weight
7. Direction of rotation
8. Year of manufacturing
9. Symbol of standard marking, if any

In the case of a wooden drum, the type of wooden batten should be specified. Nowadays, steel drums with corrugated flange are used. These drums can be made returnable as well. The accepted length of a drum can be calculated as follows (Figure 11.1):

$$\text{Effective diameter of flange} = D - 2s$$

$$\text{Number of turns of cables on the inside traverse } X = t/a$$

$$\text{Mean diameter of the cable layers on the drum } Y = \pi\{d + ((D{-}2s{-}d)/2)\}$$

$$\text{Number of vertical layers of cable } Z = \{(D{-}2s{-}d)/2\}/a$$

The values of X and Z should be rounded off to the whole number, discarding fractional figures.

Now, the cable length which can be accommodated on the drum is 2222$L = X \times (Y/1000) \times Z$ (in metres).

On this basis, a chart can be prepared showing the cable diameter against drum dimensions.

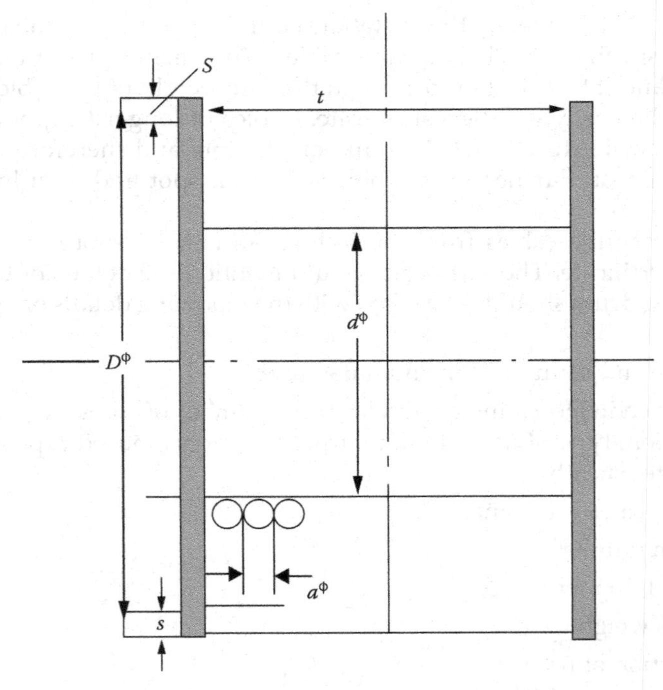

FIGURE 11.1
Cable packing drum. *Notes*: D is the diameter of barrel = d, t is the width of traverse inside, s is clearance kept from the flange edge, and a is the diameter of the cable. All dimensions are in millimetres.

Example 11.1

$D = 1600$ mm
$d = 800$ mm
$t = 1200$ mm
$s = 50$ mm
$a = 55$ mm

Now:
$X = 21.818 \cong 21$ (rounding off to the lowest value to be a whole number)
$Y = (3612/1000) \cong 3.612$
$Z = 6.36 \cong 6$ (rounding off to the lowest value to be a whole number)
The total length to be accommodated in the drum is $21 \times 3.612 \times 6 = 455.112$ m, which is approximately 455 m (or 450 m after rounding off).

After receiving the cable drum at the site, it should be examined for end seals, particularly during monsoon. Seals must remain unbroken at both the ends. Cable drums should be stored in an open or a closed place based on convenience. If possible, the base of the storage place should be made of concrete cement. If drums are placed in a large open space, the floor should be smooth for easy rolling and convenient stacking of drums. The floor and the surroundings must be kept free of weeds and shrubs. During the rainy season, water logging should be avoided. Drums should be protected from moss, fungus, termites and rodents. The area should be secured against pilferage and tampering. Before using, cable ends should be opened and measured for insulation resistance using a 5 kV megger. If the measurement is found to be satisfactory, the cable drum should be transported to the site for laying cables underground or in trenches, ducts or air. During transportation, cranes should be used for lifting the drum in a proper manner. While rolling the drum, the direction indicated by the arrow on the flange should be followed; otherwise, winding will become loose, creating problems. Cable length can also be affected due to loosening and sticking to the floor.

Mainly, public supply distribution cables are laid underground. These cables may run through different areas of a city or industrial locality. Measures should be taken to study the nature of soil and also the way these cables should be laid. Sometimes, excavating soil near the footpath or the drainage system running parallel to a road is necessary. Cables laid should not get damaged due to future excavations for any other utility services. Therefore, cables should be protected by covering them with bricks or tiles. This also blocks the loose soil from subsidence. The thermal characteristic of the soil must be measured for determining the current-carrying capacity of the cables.

The following measures are recommended for handling and laying of transmission and distribution cables:

1. The drum may be transported by a trailer or a specially made vehicle from which the cable can be laid directly into the trench. The entire handling and laying are hydraulically operated.
2. If such arrangements are not available, the cable should be unloaded at the site by a crane. If a crane is not available or the place is not suitable for manoeuvring cranes, the drum should be unloaded by placing wooden ramps. In no case, the drum should be directly dropped from the carrier onto the floor.
3. The drum should be placed on winches, either screw type or hydraulically operated, and lifted to a position for easy rotational movement. An arrangement should be made to apply a proper frictional force on the shaft while pulling the cable so that the drum does not roll by itself under inertia and loosen the layers of winding of the cable and create a messy situation.
4. The cable should be drawn from the top. It should follow the marking on the drum: Roll This Way. If the cable is too heavy, it should be drawn from the bottom sliding over a temporarily created ramp. In this case, rolling will be opposite to the marking.
5. The cable can be pulled by fixing a flexible stocking sliding over the outer sheath. However, precautionary measures should be taken to

see that the sheath does not stretch over the armour. The stocking can be anchored over the armour to avoid the slippage of the outer sheath.

6. In the case of a cable having a lead or aluminium sheath, the pulling eye should be fixed over the metal sheath through soldering. When the cable is pulled by mechanical means, it is advisable to apply pulling force as per recommended practice, particularly at the bends. It is advisable to pull the cable on rollers placed at a regular distance to avoid rubbing of the outer sheath over the rough surface of the ground. At the bends, the cable is likely to get flattened if pressed hard against the roller. Here, the bending radius and the type of guiding rollers and their position should be marked after proper examination. Rollers with skid plates should be used at the bends for smooth running of the cable.

The cable should be drawn continuously. Any intermediate stoppage may increase the frictional load during restarting. The pulling tension is determined by the summation of cable loads up to the point of friction to be multiplied by the friction at the point, which is normally 0.25. When pulled through ducts, it can raise up to 1.0, and graphite lubricants can be used for the smooth sliding of the cable. Where mechanical pulling is not in vogue, a large number of people are used to carry the cable. These people are placed along the trench in a row for carrying the cable forward. In such a case, precautionary measures should be taken to ensure that the cable does not get twisted or bent due to overload and fatigue. At the end, before the termination point, a loop having a larger diameter should be formed. This reserved length helps in making up for any shortfall that may occur at the termination point when a damaged portion of the connected length needs to be discarded. When pulling the cable at longer lengths, clear communication must be ensured at every critical point, particularly at the bends.

If the drum cannot be placed conveniently at the site, the cable should be flaked on the ground before laying. Cables higher than 11 kV range and also of higher diameters should not be flaked. Long lengths of pilot or auxiliary cables are flaked to be laid alongside the main power cables. Pilot cables are supplied in longer length than feeder cables. The bobbin of pilot cables is placed at an advance position alongside the trench. When the pilot cable is laid in the forward position drawing from the top of the drum alongside the first length of the feeder cable, the rest of the length should be laid alongside the next length of the feeder cable. This time, the pilot cable must be pulled from the bottom of the drum. To avoid twisting, flaking is done by placing the cable on the ground in a figure eight formation. The diameter of the loop must be twice the permissible minimum bending radius of the cable. Figure 11.2 shows how the cable should be handled in various figure eight formations.

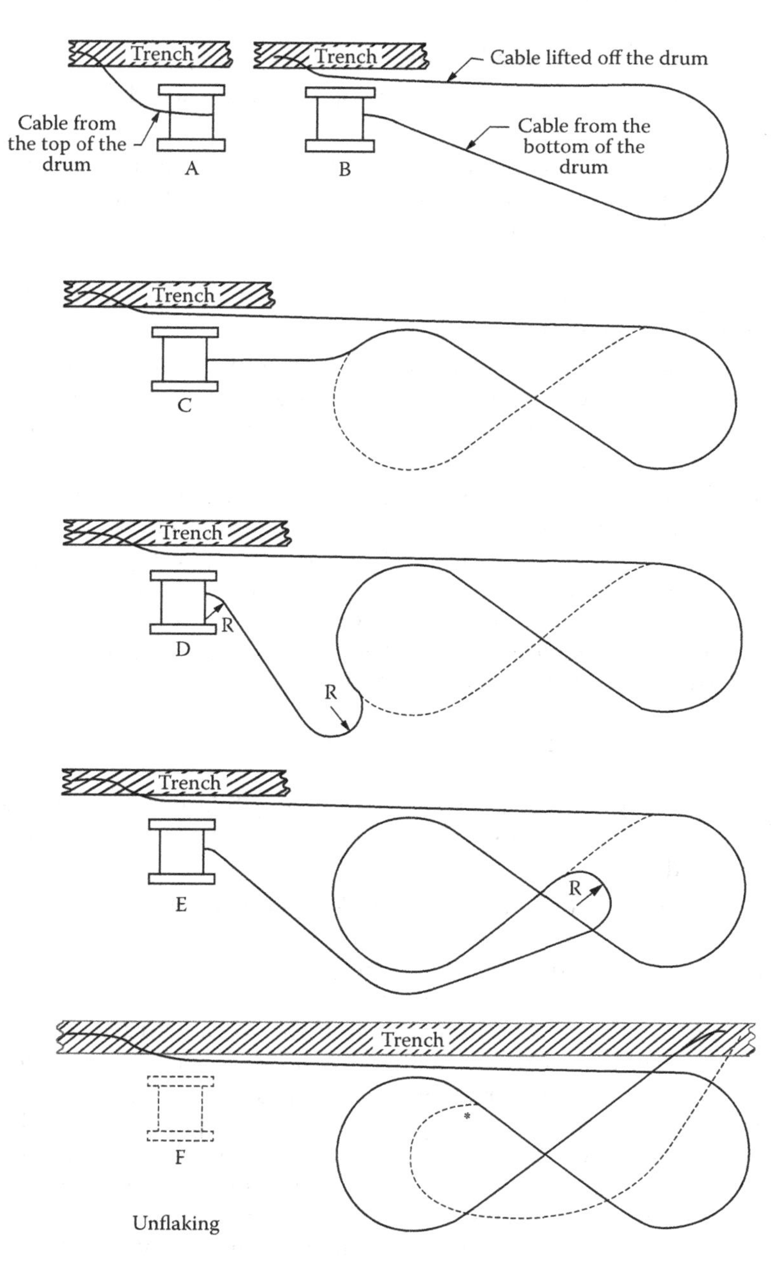

FIGURE 11.2
Asteriks indicates turn away from trench when unflaking.

Before backfilling, the following need to be inspected:

1. The cable is provided with proper bedding.
2. When there is more than one cable, spacing should be correctly arranged.
3. The entry and exit points of the cable duct should be provided with a bellmouth washer fitted properly without damaging the cable. The mouth must be fitted in a manner that no vermin can get in.
4. All laying equipment, including pulling socks, pulling ropes and other implements, are removed.
5. The cable should not be damaged during pulling. Many of the cable faults are caused by installation damages. The soundness of the cable after laying can be examined by running a set of mirrors at the bottom of the cable.
6. The cable should be covered with sand or fine soil to a minimum thickness of 50 mm after compacting. Required earthen wares or concrete cover tiles or a plastic sheet should be placed over the cable in a central position. Then 150 mm backfilling should be done by hand punning. Finally, normal soil should be used to cover the entire trench. Sometimes, a special backfilling material is used to enhance the current rating of cables.

Some special considerations are as follows:

1. The bending radii of the cable must be much larger than the minimum bending radii as specified to prevent undue flattening or twisting on the cable.
2. In cold atmosphere, the cable should be laid when the temperature is above 0°C. The outer sheath made of PVC or PE may crack when the cable is unwounded below the freezing temperature of water and where snow and ice cover the area. At this temperature, no cable-laying work should be undertaken. If the work is urgent, the cable drum should be covered with canvas sheets with a small outlet. The battens over the drum at places, say at 135° and 225° positions, must be removed. At these positions, wax lamps should be placed below the cable, giving a space of 100–150 mm. Thus, a number of wax lamps should be kept burning for a long time until the temperature inside the cable layers rises above 0°C. The measurement should be taken by a mercury thermometer inserted within the cable layer. Care should be taken to see that the cable does not catch fire. Ventilation must be such that a small quantity of hot air can go out, but cold air should not enter the tent, prolonging heating operation.

Once the temperature reaches the desired measure, the laying operation should be completed within a very short time, keeping the cable drum to the nearest possible position of the trench. At the bending, precaution should be taken to see that the cable sheath does not crack. If it happens, the crack should be repaired immediately.

3. Damages in the PVC outer sheath can be easily repaired. It has been found that a thin inner layer of PVC tape can be wetted or made to dissolve in cyclohexanone. If the tape is wetted with the compound solution, it should be wrapped quickly around the PVC sheath. Then, the tape amalgamates with the PVC outer sheath, making a homogeneous ring. A solution of PVC compound in cyclohexanone should be kept ready for this purpose.

Actual process: PVC outer sheath or insulation:

1. Clean the surface of the PVC sheath around the damaged portion extending 30 mm beyond the fractured or cut portion by applying a PVC solution as prepared earlier.
2. Now, apply a second layer of solution and immediately start wrapping clean insulating-grade PVC tape stretching slightly over the damaged part extending 30 mm beyond the damaged area. While applying the tape, it should be ensured that the solution underneath does not get dried up completely.
3. The thickness of the tape should not be more than 0.15–0.2 mm and the width should not be more than 15–20 mm, depending on the diameter of the cable. The tape should be applied with a slight overlap.
4. After every two turns, a layer of thin PVC solution should be applied over the bandage. The operation should be repeated until sufficient thickness is built up.
5. After some time, the wrapped layer of tapes will amalgamate together along with the original outer sheath of the cable, making it a homogeneous body.
6. This repair is completely waterproof and has all the qualities of a homogeneous PVC sheath or an insulating material.

The case of a polythene sheath:

1. Polythene cannot be chemically fused. In earlier cases, attempts were made with hot air welding. But it was not possible to repair a damaged sheath in complex positions through this process. Further, if heat was not applied in a controlled manner, the original sheath could get distorted or remain loose. This practice was abandoned.

2. Another attempt was made using the polyurethane casting method. This was somewhat successful as the mould could be adjusted properly. But cables and fault condition moulds had to be made differently, which was not cost-effective.
3. The solution was found when a self-amalgamating tape was developed. This tape could seal any fault and at any place. However, precaution should be taken to see that repairing is done in a complete pollution-free surface of the cable. The hands and equipment of the repairer must be clean when repairing is done.
4. Nowadays, heat shrink repairing is also done on pollution-free surfaces.

Installation of HV and EHV transmission cables needs special care:

1. Transmission cables are categorised as
   a. Cables used in power stations where bus bar or overhead line connection is not practical
   b. Interconnection of substations in urban areas where an overhead transmission line is not possible or environmentally acceptable
2. These cables are made as per the required length. Delivery lengths are determined based on the total route length.
3. As the rated voltage of the cable increases, the diameter and bending radius also increase. Hence, the minimum hub diameter as well as the diameter of the drum increases. In order to comply with width limitations contained in most sets of highway regulations, the diameter of the drum should be larger. These cables should be loaded on a low-bed trailer to get easy passage below bridges and road tunnels.
4. Once the route length and cable length are fixed, trenches should be excavated in such a manner that cables can be drawn easily and laid side by side. It is important to know the thermal resistivity of the soil where the trench is excavated.
5. The depth of the trench should be as per the International Electrotechnical Commission (IEC) recommendations or as per the local standards. Sufficient space should be given for the snaking of cable due to thermomechanical movements. These trenches are generally deeper than distribution cables, say 2–2.5 m below the surface of the ground level.
6. Jointing bays should be made sufficiently spacious so that air conditioning systems can be installed to keep the temperature of the jointing bay within comfortable limits for working. The atmosphere inside must be kept clean and dust-free.

7. The sides of the trenches should be secured by placing wooden battens along the length of the trench to prevent falling of loose soil and debris within the trench.
8. These cables must be drawn using a mechanical winch which slides the roller. The pulling force applied must conform to the recommended standard. The bend must be sufficiently large to eliminate undue stress.
9. It is important to prepare a sand bed before placing the cable on the floor. The cable lengths should be protected by placing tiles on both sides and a top cover, giving the shape of a rectangular duct. Empty space inside the tiles should be filled with sand if recommended. Thereafter, earth filling should be completed. The cable route should be marked by placing marker plates at certain distances over the ground.

11.1 Backfilling of Trenches and in Pipes

While laying a cable underground, it is normally assumed that the thermal resistivity of the soil is 1.2 km/W. But, in some cases, it has been found by excavating a portion of the soil after a cable failure that the surrounding soil has dried up due to excess heat accumulation at the point. This happened because the thermal resistivity of the soil at the point was much higher than the anticipated value. This led to examining and understanding the effect of thermal resistivity of soil vis-à-vis surroundings on the performance of a cable, particularly HV and EHV cables. It is desirable that before laying and closing trenches, the thermal resistivity of soil at different distances must be obtained and studied. If this practice is too laborious and time consuming, an alternative is to select a backfill material having lower thermal resistivity.

11.2 Selected Materials Such as Sands as Backfill Material

Backfill sands must be selected from sources where the thermal resistivity of the soil does not exceed 2.5–2.7 km/W. These sands should be a mixture of high-grain sands with coarse particles, which can absorb moisture easily. This helps in the dissipation of heat in a better way. If the cost of backfilling is too high, the selection of the cable size should be planned accordingly. Particularly for less than 132 kV cables, the use of backfilling should be considered in a rational manner. However, there are cases where the use of special backfilled material has been found to help increase the cable rating.

11.3 Ducts

When a cable is laid in a duct with a short span, the air space between the duct wall and the cable does not allow efficient heat transfer to the outer space or surroundings. In such a case, the duct should be filled with a mixture of bentonite, cement, sand and water. This mixture can be pumped inside the duct efficiently and closed at both ends. When required, the mixture can be flushed out with a high-power water jet. The mixture being sufficiently flexible, the thermomechanical stress developed is easily absorbed by the filler materials.

11.4 Cable Laying on Steel Trays and Racks (Both for LV Distribution and for MV, HV and EHV Transmission Cables)

In a powerhouse or a factory, cables should be laid inside the building. In these places, cables are generally taken overhead alongside the wall or on roof trusses. These cables should be laid on perforated steel trays/racks attached to walls or laid on the iron angles of roof trusses. Cables, when laid directly touching perforated steel trays/racks, produce an inductive magnetic field, inducing a voltage within the steel trays/racks and creating a hazardous situation particularly when the rating of the cables is very high. This affects the current rating of the cable somewhat. The solution to this problem is to isolate the cable from the metal tray by placing a nonmagnetic spacer below the cable sheath. Synthetic or FRP clamps can be used to keep the cables isolated from the trays running either horizontally or vertically. The spacing of the clamps should be such that the cables do not come in contact with the steel frame of the trays or any steel structure on which the cables run. Nowadays, synthetic trays particularly made of FRP are used extensively, replacing steel trays and racks. This reduces the expenses for clamps as cables can be laid freely on trays. Further, the stress developed at the clamping position during expansion and contraction is also eliminated. Cables should be laid with a small sinusoidal curve to allow expansion and contraction when thermomechanical stress develops within. This can happen during an increase and decrease in the operational load and during summer and winter with the variation of surrounding temperatures. This may also develop a stress on terminating joints.

When cables are laid on stay racks alongside the walls of a tunnel or a building, the spacing of the racks should be fixed at such distances that the cables do not sag heavily. This should be done taking into account the weight of the cable and surrounding temperature conditions. Clamping should be done only to keep the cable in position and should not be tightly fixed. Also, it is better to have stay racks and clamps made of FRP or any other suitable nonmagnetic synthetic material.

After laying the cable, the ends of the cable should be kept sealed properly before jointing starts. They should be secured against the ingress of moisture and polluted material.

11.5 Joints and Terminations

To transmit power from the main station transformer to the consumer transformer or the circuit breaker or any consuming point (say motors, heaters and driers), cable cores need to be connected at both the terminals (end terminals). Further, in the case of a long supply line, several lengths should be laid one after the other. These lengths should be joined end to end by straight through joints. In a distribution system within a city, a long-ring main line is drawn around or across the city. From the main distribution line, secondary branch lines should be taken out to distribute power at various points within the residential areas of the city, such as to feed housing colonies, schools, universities, medical institutions and so on. At these points, branch lines are tapped to feed those consuming points. A TEE joint is made to connect each branch line with the ring main distribution cables.

Every joint made is a weak link and a source of dissipation of energy as heat. Naturally, a joint must be carefully made with a similar material or a material compatible to the cable insulation and other built-up materials. It is expected that a joint must remain in service as long as the cable lasts. The joint is built up in a similar fashion as the cable. It should be ensured that the joint remains water tight.

The design of accessories for joint and termination is based on the design of their associated cables and applications. As it is difficult to conduct a detailed study of all types of cable accessories applicable for all possible situations, only a generalised insight into various accessories, based on the parlance of cable installation practice, has been provided, giving a brief discussion on the desirable design feature and basic jointing technique.

11.6 Desirable Design Features

The philosophy of the design of a joint lies in reproducing the electrical and mechanical parameters of its associated cables as closely as possible to maintain its reliability as a complete system. While this target is more or less met in a fully screened joint, it is not practical to achieve reliability in a joint or termination beyond 80%–85% of the cable itself.

Commensurate with the acceptable limits of design stresses, the quality of materials used in accessories and the available jointing skill, the material requirements and dimensions are kept as low as possible. Taking into consideration the difficult situation in which jointers carry out their work, the design of accessories and adoption of jointing techniques must be kept as simple as possible.

The space within the joint box should be adequate for proper splaying of cores, keeping in mind that the minimum bending radius permitted for the core to be bent does not go below 12 times its diameter.

Hand-applied insulation as a primary layer and filling compound as secondary insulation over the conductor joint must have low dielectric loss. Otherwise, losses in the secondary layer of protection will cause local heating, which will have a multiple effect with that of conductor heating, ultimately resulting in the failure of the joint.

The shapes and sizes of the jointing material should be such that during and after application, there should be no void within the system, which may result in a local ionisation problem. The mechanical design must also be such that the joint can withstand all environmental conditions (absorb the stress developed during operation and short-circuit conditions) and has longer durability.

In particular, a joint and a termination should closely resemble the cable for which they are designed. It is necessary that these accessories should be tested in sequence with the test procedure applicable to the cable itself. It may, at times, be necessary to compromise on certain parameters, but the basic concept of the design and performance should not be sacrificed, though one expects it to be cost-effective and simple to work with.

11.7 Jointing of Cable and Conductor

Jointing of a conductor must satisfy the following requirements:

1. The conductivity of the joint should ideally be maintained at 100%.
2. The tensile strength of the joint should be maintained almost equal to that of the conductor itself but not less than 98%.
3. The surface of the joint must be kept smooth to avoid any undue stress. This is very important for HV and EHV cable conductor joints.

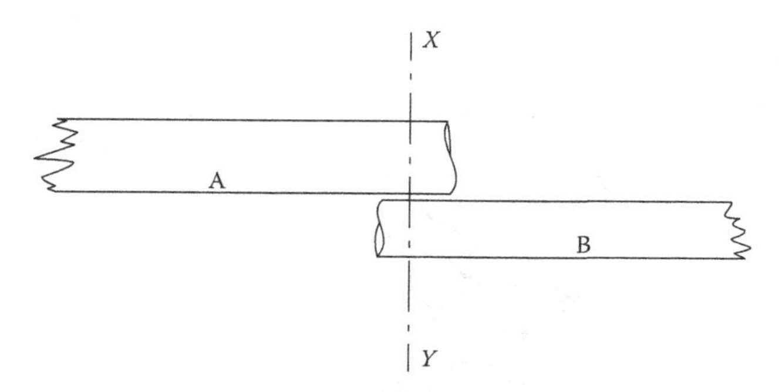

FIGURE 11.3
Positioning of a cable before jointing.

In all the cases, the trench should have sufficient space where the ends of the two cables meet. Figure 11.3 shows the jointing procedure of a distribution cable in general. A sufficient length must be kept for overlapping the cable ends, as shown in Figure 11.3. Cable 'A' and Cable 'B' ends should overlap and must be placed side by side as shown. Once measurements are completed as per the jointing chart, both the cable ends should be trimmed along the *X–Y* line as indicated.

Once the cables are in position, the ends should be spliced as shown in Figure 11.4. The measurement of individual lengths should be as per the relevant jointing instructions to be provided by the supplier of accessories. Referring to Figure 11.4 of a spliced cable end, the details are as follows:
To start with, the conductor of both the cables should be jointed. Jointing of the conductor can be done by

1. Soldering process (classical method)
2. Crimping or compression method

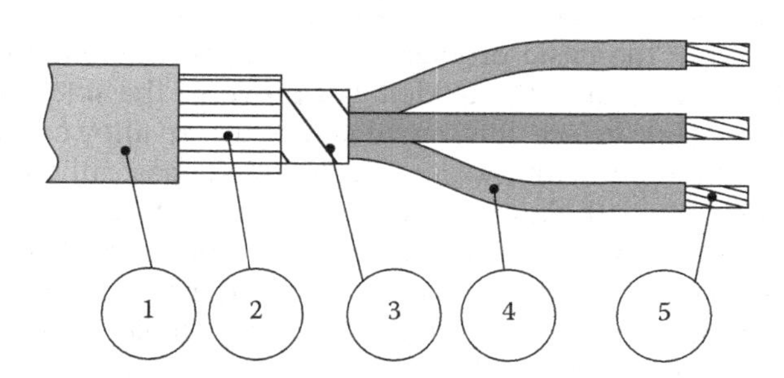

FIGURE 11.4
Spliced end for a straight through joint. (1) Outer sheath of the cable, (2) wire armouring (mechanical and earth protection), (3) taped bedding (it can be of extruded bedding also), (4) insulated conductor and (5) stranded conductor.

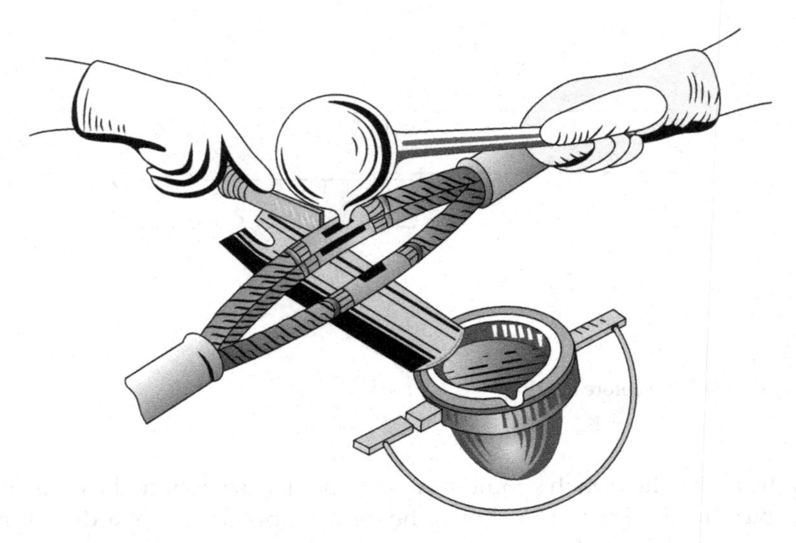

FIGURE 11.5
Conductor joint by soldering process.

Soldering is a classical type of joint. For a solder joint, a flux is used to clean the surface of the wires before the solder is applied. A weak back ferrule or a tubular ferrule with slits is used to cover both halves of the stranded metallic conductor. For cleaning a copper conductor, a flux consisting of rosin and a weak acid is used. The solder consists of tin and lead alloy in a 60:40 ratio. The stranded conductors from both sides of the cables are inserted into the ferrule. The ferrule is then tightened to grip the strands properly. The assembly is heated by a blowlamp. The strands are wetted with a flux. Once the flux penetrates the interstice space of the strand, the molten solder is then poured over, repeating three to four times (Figure 11.5), ensuring that the solder has wetted all the wires of the strand. The assembly is then cooled by wiping the joint with a wet cloth.

An aluminium conductor can also be soldered in a similar manner. In the case of aluminium, the basic problem is to remove the oxide layer on the strands. In the normal case, the oxide layer prevents the solder from firmly adhering to the metal surface. It is customary to use an alloy of lead, zinc and tin solder in conjunction with the organic flux made basically of zinc fluoborate in a solution of triethylamine. At the correct basting temperature, zinc deposits over the aluminium strand while penetrating and dispersing the refractory oxide layer from the surface at the same time. The resulting plated surface having zinc assists in the formation of a stable intermetallic compound. Thus, the flux dissolves the oxide film present on the aluminium surface, and the deposit forms a conducting barrier, preventing the re-oxidation of coated aluminium surface (Figure 11.6).

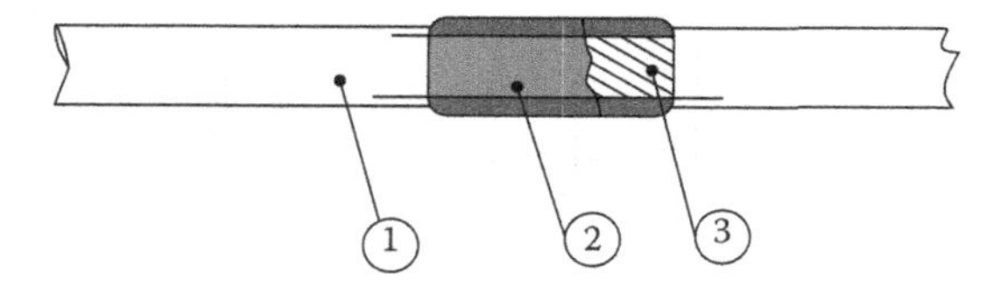

FIGURE 11.6
Conductor joint: (1) insulated conductor, (2) ferrule and soldered joint and (3) section of stranded conductor.

At times, a reaction flux composed of chlorides of heavy metals such as tin, zinc, lead and cadmium in conjunction with an alloy of zinc, tin and cadmium as a solder is also used for aluminium conductor jointing by the solder basting method. However, because of the RoHS regulations, such solders are now restricted for the jointing of conductors.

11.8 Crimping and Compression Joints

Due to the cold flow characteristic of materials, particularly of aluminium, it is possible to join conductors using the compression method. In this case, a relatively soft aluminium tube connector is brought over both the spliced ends of the conductors. A small gap between the conductors is left at the midpoint within the connector tube for accommodating the expansion of lengths during compression. While compressing, it should be ensured that the connector tube grips the conductor, rupturing the oxide layers on the aluminium wires but not breaking them. There should be a gradual metal flow during the amalgamation of the wires together with the ferrule. To smoothly accelerate the process, a suitable coating of flux is applied over the conductors. Some experimental joints have been examined by cutting the joint at an angle of 45° to check the homogeneity of the metal parts. The pressure applied is determined considering the tensile strength, Young's modulus and elongation characteristics of the metal. These joints, if properly carried out, give a good result, retaining 85% of the tensile strength of the original conductor. When a joint is made of a 120 mm^2 stranded conductor, current leakage across the joint as specified should not exceed 0.03 mA for all times when a voltage of 2.5 kV is applied for 15 min. The conductor resistance should be measured across the joint. The temperature rise during the load cycle should be noted. Such joints are being extensively used to connect cable conductors up to the 11 kV range. Care should be taken to see that the extreme edges of the ferrule do not flare up like a bellmouth. Also, the dents and rough edges should be smoothened with extreme care. These are the areas which can create fault conditions. In the case of cables starting from 6/6.6 to 30/33 kV,

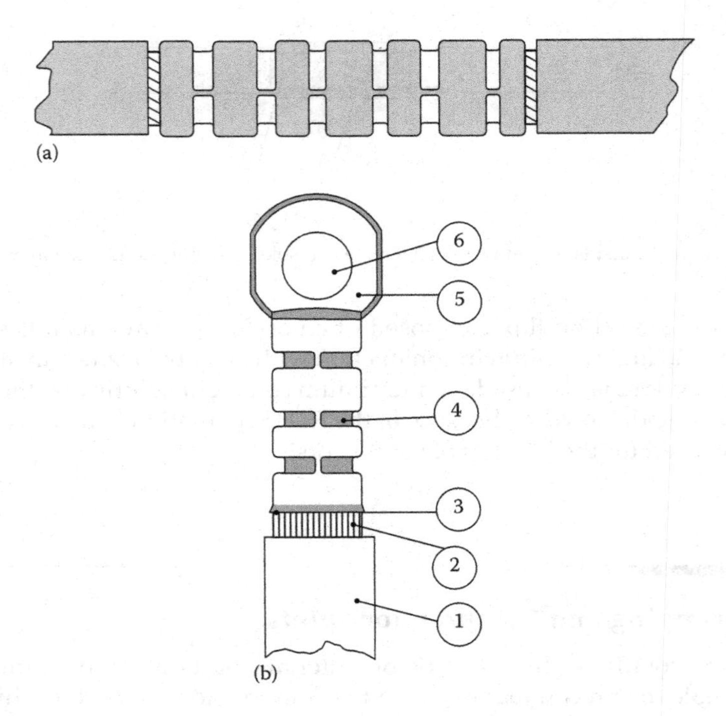

FIGURE 11.7
(a) Crimp joint (straight through)-LV cable conductor; (b) crimp joint (end termination)-LV cable conductor. (1) Insulated conductor; (2) stranded conductor spliced; (3) edge of the lug (ferrule), flaring to be avoided; (4) compressed point (dent); (5) end point of the lug (area more than the conductor area); (6) hole for connecting the lug with bus bar.

a proper semiconducting layer should be applied to minimise the electrical stress over the joint.

Above 30/33 kV, joints are made using the total compression method. In this case, the ferrule is compressed over the total length by hydraulically operated compression tools. These tools can be accurately adjusted for the pressure to be applied up to 100 tons. In this case, the outer surface of the ferrule remains smooth and even. These compression tools can be used for ¾ H aluminium and soft copper conductors for HV and EHV cables (Figure 11.7).

11.9 Welding Methods

Welding is an old Russian technique to join copper and aluminium conductors. A welding transformer is employed to do the job. During welding, a heat sink should be used for protection at both the spliced conductor ends near the insulation. But this method can only be employed where power is available. This method cannot be used at remote, secluded places. Nowadays, the method has become extinct.

Copper thermite can be prepared using either cupric oxide (Cu_2O, red) or cuprous oxide (CuO, black). The burn rate is very fast, and the melting point of copper is relatively low. So the reaction produces a significant amount of molten copper in a very short time. Cuprous oxide thermite reactions can be so fast that copper thermite can be considered a type of flash powder. An explosion can occur and send a spray of copper drops to a considerable distance.

Copper thermite is used in welding thick copper conductors (cadwelding). This kind of welding is being evaluated for cable splicing also. Thermite welding is a very effective way of jointing EHV cable copper conductors. This gives a solid, smooth and long-lasting joint.

11.10 Insulating the Joint

A cable joint should be made as per the pattern of a cable. Moreover, the material used should have similar characteristics of the cable. In this context, one can review the concept of paper cable joints used in the early days. Apparently, it is difficult to match dissimilar materials. However, in some cases, a good contact can be established. Heat shrinkable synthetic polymers, epoxy resin, self-bonding tapes of polychloroprene (PCP), ethylene propylene rubber (EPR) or silicon rubber are some examples. These polymers remain in contact with other polymers undisturbed for a number of years. But, with time, the action of heat, light and ultraviolet rays initiates a slow intermolecular reaction, which gives away the bonding between two dissimilar materials, bringing about a failure in service. In order to overcome this problem, a simple method has been developed by the author.

11.11 Insulation: Low-Voltage Cable from 400/440 to 3000/3300 V PVC Insulated

Cables of various sizes ranging from 400/440 to 3000/3300 V are used in different lengths and quantities for distribution systems. In the earlier days, the jointed part of a conductor was insulated with pre-impregnated paper. To prevent ingress of moisture, an impregnating compound was poured into the sleeve. This compound became an essential part of the joint. Here, the design of the joint should follow the exact pattern of the paper-insulated cable. Even after the arrival of thermoplastic cables, the use of compound pouring (e.g. epoxy or bitumen compound) was retained for some time, which temporarily solved the problem, but materials were not found to be compatible with thermoplastic materials. It needed considerable artwork to build such a joint. With time, recrystallization of compounds created fissures within materials, allowing a capillary passage to moisture. At the crutch point, the material gave way and allowed moisture seepage. The thermal dissipation of these joints was low, limiting overload conditions. During maintenance, the joint should be cut off and replaced with a new one, wasting a portion of the costly cable length.

The idea of using PVC insulation has been developed to eliminate the drawbacks experienced earlier. It is possible to fuse PVC to PVC by applying heat. But the process needs skill and elaborate precaution so that the insulation is not damaged during heating. To avoid such risks, a chemical fusion process has been developed to ensure greater reliability. In this process, the built-up insulation amalgamates with the original insulation having identical parameters. Over the ferrule joint, a specially made self-amalgamating, self-vulcanising elastomeric insulating tape is wrapped tightly. This elastomeric insulation is finally covered with two PVC tapes having a thickness of 0.15 mm and a width of 15–20 mm. A moderately thick solution of PVC in cyclohexanone should be applied intermittently in thin layers during the application of tapes over the joint. All the tapes should be applied with slight stretching and overlapping on both sides of the joint, covering the original insulation of the conductor. The solution should be applied over the conductor insulation before wrapping the tapes. This solution has the characteristic of cleaning all unwanted elements over the joint and insulation of the conductor. After some time, the total assembly gets fused together, forming a continuous homogeneous mass along with the original insulation of the conductor. It is also possible to use coloured tapes to match phase conductors. Such insulated conductor joints of grade 1100 V, as developed by the author, were tested after being immersed in water for 2 weeks by applying a voltage of 20 kV without a rapture. Laboratory tests produced the following results on a completed joint:

1. Insulation resistance.
   a. 60 MΩ (initial).
   b. More than 200 MΩ for all times.
2. Leakage current to be measured by applying a voltage of 2.5 kV to each core and other cores earth. The durations of test are 15 min for initial and final measurements and 1 min for intermediate measurements:
   a. 0.04 mA (initial)
   b. Less than 0.01 mA (all other times)

3. Temperature rise during load cycles: 39°C above the ambient temperature.
4. No deformity externally visible after the completion of tests.

A few joints were installed on the live line of an electric supply and distribution company based on these results. No failure was reported within 10 years of operation, after which the lines had to be re-laid due to a change in the road layout plan.

The requirement of higher skills and the time factor did not make this jointing kit popular, though its reliability and performance were unquestionable. The product cost was slightly higher, but maintenance cost was negligible due to minimal failure records. However, in the meantime, heat shrink joints became popular. The skills required was minimal and jointing time was less, resulting in lower labour cost, and so it became popular in the initial stage. However, the performance of these joints is not so satisfactory even these days. The basic problems of heat shrink joints are as follows:

1. One must be careful about the memory shrinking actuated by the application of heat. Heat must be applied uniformly all over in a consistent manner. Otherwise, kinks and folding on joints will give rise to air pockets. These may get ionised during the operation. Particularly, such voids can create ionisation problems in the 11–33 kV range cable joints and terminations.
2. During heating, an adhesive oozes out and grips the insulation of the conductor when the tube is applied on the joint of the conductor or when the tube is applied to cover the entire joint extending over the outer sheath of the cable. For a few years, these adhesives remain sticking firmly over the insulation or sheath. But, in time, the fluctuation of heat cycles, the action of ultraviolet rays and a slow process of recrystallization weaken the grip of the adhesive. A capillary passage is formed between the sheath and the adhesive of the joint, allowing the ingress of moisture at the ends of the cable joint. For a termination, this is not a problem since cable ends are mostly kept straight in the upward direction and water never travels upwards.

For low-voltage cables, indoor termination type is completed by introducing termination clamps to hold cables in position (Figure 11.8). The clamps grip the outer sheath in position along with the armouring. By fixing the clamps on a metallic structure, the earthing of the armour is ensured. The junction points are protected by rust-proof material to avoid the corrosion of metallic parts. There are various types of clamps available for different applications, including fireproof ones. Outdoor terminations are insulated by applying a heat shrink tube over the insulated conductor. A trifurcated or four-way cover is fitted at the base of the insulated conductors along with the outer sheath of

FIGURE 11.8
Low-voltage cable termination clamps.

the cable. The armour-connecting strip is taken out from the bottom of the heat shrink cap. Open termination lugs are kept within the cover plate of the termination point coated with a waterproof compound after connecting with the bus bar or terminals of equipment such as motor or transformers.

Figure 11.9 shows a completed joint, detailed as follows: The cable (1) is spliced on both sides as per the specified dimensions. The stranded conductors (5) of both the cable ends are inserted in a ferrule (6). The joint is made either by soldering or crimping. For a low voltage, the manner of jointing of conductors does not play a significant role. On the jointed portion, a PVC self-amalgamating and self-cured elastomeric insulation is applied. Over this insulation cover, two insulating-grade PVC tapes with a PVC solution in cyclohexanone (7) are applied, overlapping the conductor insulation on both sides (4). While exposing the conductors, bedding (3) is trimmed as shown in Figure 11.9. The position of the jointed conductors is adjusted by bringing them closer to each other. Over this assembly, an overall bedding of a self-amalgamating, self-cured elastomeric tape (8) is wrapped, extending over the original bedding of the cable (3). Earth continuity is provided by a copper strip (10), which is connected with the armouring (2) of both sides through a set of armour-gripping clamps (11). Earth bonding must be reliable and should have an adequate area to divert a short-circuit current during an earth fault condition. The joint is finally closed by bringing a moulded FRP cover (12) in two halves. The rims

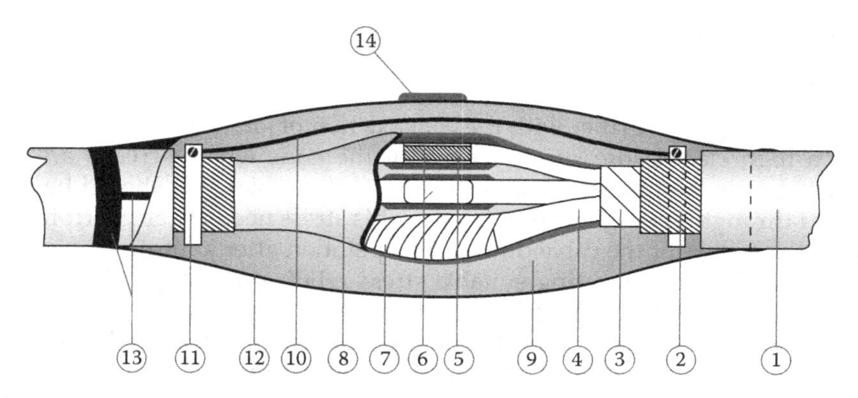

FIGURE 11.9
A complete straight through joint (LV cable). (1) Cable, (2) wire armour, (3) bedding, (4) insulated core, (5) stranded conductor, (6) ferrule on joint, (7) taped insulation of insulating-grade PVC and solution, (8) bedding over the assembly of jointed cores, (9) filling polyurethane compound, (10) armour continuity metal strip, (11) armour-gripping clamp, (12) fibreglass cover, (13) epoxy sealing compound and (14) PU compound pouring hole.

of the joint and the ends are hermetically sealed by applying the epoxy sealing compound (13). A hole is provided in the upper half of the FRP cover (14) through which a PU filling compound (9) is poured inside the joint. The whole is then closed by applying the epoxy sealing compound. This type of joint has been found to give a reliable service for a long time.

In the case of a heat shrink joint, most of the failures were found to be near the crutch point. This is due to the ingress of moisture through the end sealing points, the cause of which has been explained earlier.

11.12 HV and EHV Joints

In the case of HV and EHV cables, a joint needs to be reliable. All the materials used should be free of contamination and void. The insulating materials and compounds used must have high stress-withstanding capacity. HV and EHV cable conductors are joined using the total compression method to obtain a smooth surface over a jointed ferrule. Nowadays, thermite welding is also done to join copper conductors. Whatever be the form of a joint, the surface of the joint must be made smooth and free of any contamination. Any protrusion or contaminated particle on the surface of a joint can create a fault condition. On the jointed portion, a semiconducting tape should be applied, filling up any undulation which may result during jointing. This tape should be properly connected to both sides of the inner semiconducting layer, extending over the jointed ferrule of the conductor to maintain continuity.

11.13 Stress Control

Before proceeding further with the construction of joints, it should be noted that a high electric pressure develops at the ends where screens are terminated over the insulation at both ends of the cables that should be connected through a straight through joint. This stress needs to be controlled to avoid the failure of the cable during the operation after jointing. This stress is restricted by constructing suitable stress relief cones at the cable's both screen termination points and connecting them through a semiconducting layer over the insulation of the joint along with a metallic screen layer.

A cable under operation produces two types of pressure within the cable: (1) radial voltage and (2) longitudinal voltage. A radial voltage is normal to the axis of the cable and can get distorted within the insulation due to localised contamination and void formations. To build a uniform stress level, it is necessary to allow the field to become radial and it is achieved by applying a semiconducting layer of a compound over the insulation. Over the semiconducting layer, a metallic screen is provided, which should be kept on earth potential. Simultaneously, a longitudinal stress also remains active. If the screen breaks at any point on the cable, both longitudinal and radial stresses multiply at the point to match the earth potential, creating a high discharge point. To avoid such a situation, whenever HV or EHV cables are spliced for jointing and termination, a stress relief cone should be provided at the cut point where the screen and semiconducting layers are terminated. The high stress at the point is reduced in steps for smooth operation by lowering the capacitance value at the point of termination of the screen by increasing the thickness of a dielectric in steps until the discharge level comes down below permissible limits. In the case of a polymeric cable, this transition is made smooth by introducing a curved semiconducting layer followed by a metallic screen. Figure 11.10 illustrates the manner in which equipotential surfaces are achieved by introducing a built-up stress cone. The same effect can be achieved by applying a high-permittivity material over the termination point.

The cable joint diameter should be larger than the cable diameter to compensate for the inherent weakness within the joint in the form of a built-up continuity through a compressed or welded ferrule connection on the stranded conductor, the surface of which is not perfectly uniform and polished. Further, a cut is applied to remove a certain length of insulation which should be filled by an artificially built-up insulation on the jointed portion. It is very difficult to produce a homogeneous structure as in the case of an extruded insulating material on the conductor joint. Naturally, the portion remains vulnerable to higher stress development, particularly at the point where the transition is to take place. To overcome this difficulty in building the insulation and the screen at the termination point, a gradual transitional profile is constructed to form equipotential surfaces, reducing the effect of high-stress built-up within the joint. The profile of the stress cone can thus

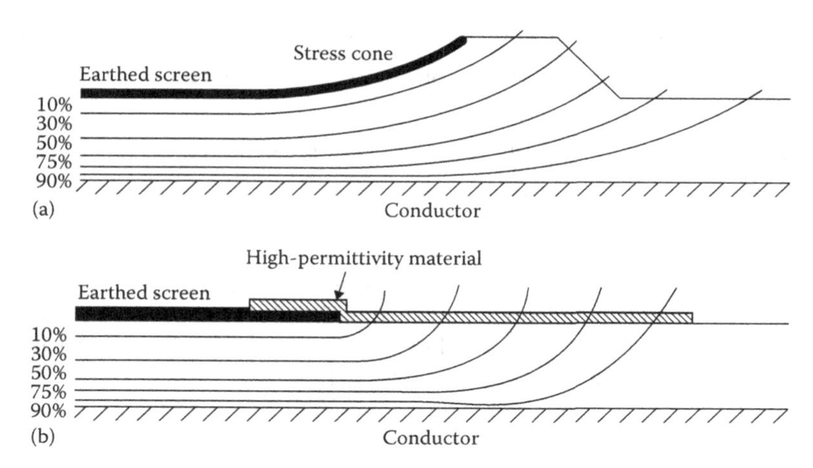

FIGURE 11.10
Stress relief cone and distribution of stress within the stress relief cones: (a) a semiconductor layer built up as per conventional stress cone development method and (b) the type of semiconductor which has been developed to be applied as linear stress control for HV and EHV cables.

be designed considering the constructional features of the cable and the voltage. The characteristic features of an insulation and a semiconducting layer are also important (Figure 11.11).

Assuming the field is radial in all cases considering the sections P_1, P_1 and P_2, P_2 and for any section of the plane to the x-axis of the cable, the field is within the dielectrics of annular cylinders. Assume the following notations:

- r_j is the radius of the insulated part of the joint.
- r_i is the radius of the insulated cable.
- r is the radius of the conductor.

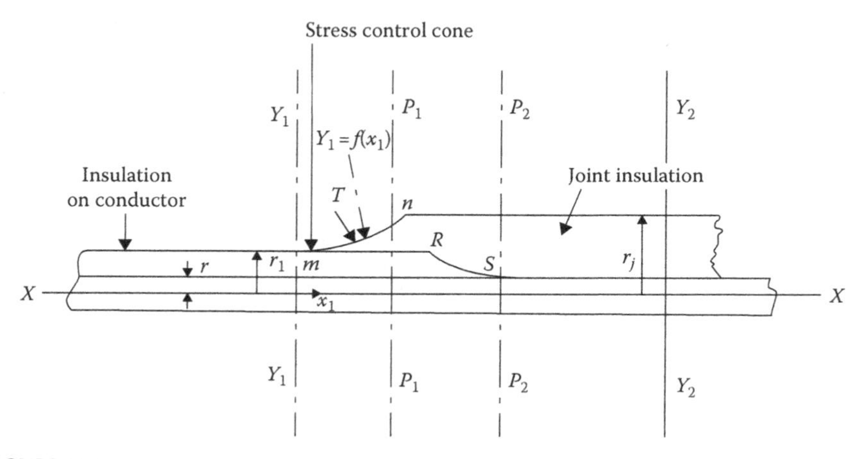

FIGURE 11.11
Stress cone development.

- ε_r is the relative permittivity of the cable insulation and that of the joint.
- V is the voltage between the conductor and the earth.
- E is the stress due to voltage.
- Y_1 is a function of $f(x_1)$, showing the profile of the stress control cone.
- Y_2 determines the transition curve of the cable insulation and is a function of $f(x_2)$.

At a point T on the profile of the stress cone m–n, the field strength can be considered within the dielectric of the joint. The potential drop between the earth and the point T can be calculated as C_1 (the earth stress control cone) and C_2 between the cable conductor and T. Thus,

$$C_1 = \frac{2\pi\varepsilon_0\varepsilon_r}{\ln(Y_1 / y)} \text{ F/m}$$

and

$$C_2 = \frac{2\pi\varepsilon_0\varepsilon_r}{\ln(y / r)} \text{ F/m}$$

It follows that

$$V_1 = \frac{V\left(\frac{1}{C_1}\right)}{\frac{1}{C_1} + \frac{1}{C_2}} = V\frac{\ln\frac{Y_1}{y}}{\ln\frac{Y_1}{r}} \text{ V}$$

If T is very close on the stress control cone profile, the longitudinal stress at the elemental distance dx_1 is found to be

$$E_T = \frac{\partial V_1}{\partial X_1} = \frac{\partial V_1}{\partial Y_1}\cdot\frac{dY_1}{dX_1}$$

$$= V\frac{d}{dY_1}\left(\frac{\ln\frac{Y_1}{y}}{\ln\frac{Y_1}{r}}\right)\frac{dY_1}{dx_1}$$

$$= \frac{V}{Y_1}\left[\frac{\ln\frac{y}{r}}{\left(\ln\frac{Y_1}{r}\right)^2}\right]\frac{dY_1}{dx_1}$$

Now, the point is actually brought over the profile *m–n* when $Y_1 = y$, then

$$E_T = \frac{V}{Y_1} \frac{1}{\ln(Y_1 / r)} \frac{dY_1}{dx_1} \tag{11.1}$$

Since E_T is selected for designing the stress cone and is expressed by the function $Y_1 = f(X_1)$, it should be integrated within the limits of r and Y_1, while Y_1Y_1 is the ordinate axis of measuring X_1 progressively. Thus,

$$E_T X_1 = \int_{r_1}^{Y_1} \left(\frac{V}{\ln \frac{Y_1}{r}} \frac{d(Y_1/r)}{\frac{Y_1}{r}} \right)$$

or

$$V \ln \left(\frac{\ln(Y_1 / r)}{\ln(r_1 / r)} \right) \tag{11.2}$$

or

$$e^{(E_t X_1)/V} = \frac{\ln(Y_1 / r)}{\ln(r_1 / r)} \tag{11.3}$$

Selecting a practical design value for the longitudinal stress E_T, a profile of the stress cone can be worked out. The lower the value of E_T, the larger will be the stress cone. Hence, considering the space limitation and the dielectric strength of the insulation of a joint, a value should be selected to give a workable profile of the stress cone.

Similarly, the profile (*R–S*) of the insulation on the conductor is given by

$$E_T = -\frac{V}{X_2 \ln(r_j / r)} \ln \frac{Y_2}{r} \tag{11.4}$$

The diameter r_j on the jointed ferrule is calculated by assigning a stress value, say E_f, on the surface of the ferrule, when r_f is the radius of the ferrule:

$$E_f = \frac{V}{\ln(r_j / r_f)} \text{ V/m}$$

Knowing the values of r_f, E_f and V, r_j can be calculated as

$$r_j = r_f e^{(V / r_f E_f)} \tag{11.5}$$

TABLE 11.1

Stress Cone Profile Calculation for a 33 kV, 300 mm² Single-Core Cable

$V =$	**19,000**	**Volt**	**19,000**	**19,000**	**19,000**	**19,000**	**19,000**	**19,000**
$E_T =$	1000	V	1000	1000	1000	1000	1000	1000
$X_1 =$	**10**	**mm**	**10**	**12.5**	**15**	**17.5**	**20**	**22.5**
$r =$	11.04	mm	11.04	11.04	11.04	11.04	11.04	11.04
$r_1 =$	20.53	mm	20.53	20.53	20.53	20.53	20.53	20.53
$E_T \times X_1/V =$	0.53		0.53	0.66	0.79	0.92	1.05	1.18
Exp(c14) =	1.69		1.69	1.93	2.20	2.51	2.87	3.27
$\ln(r_1/r) =$	0.62		0.62	0.62	0.62	0.62	0.62	0.62
c15 × c16 =	1.05		1.05	1.20	1.37	1.56	1.78	2.03
exp(c17) =	2.86	y_1/r	2.86	3.31	3.92	4.75	5.91	7.59
$Y_1 =$	**31.54**	**mm**	**31.54**	**36.55**	**43.26**	**52.42**	**65.25**	**83.78**

Note: In the bold values, V_r, is the voltage between conductor and earth, E_r is taken as longitudinal voltage and X_1 is the distances of stress cone along the X-axis. On the basis of these values corresponding values of Y_1 is determined to evaluate the stress cone profile. These are being important data to determine the value of Y-axis.

Example 11.2

By accepting a value of E_f, the diameter r_j can be determined.

Taking $r_f = 13.0$ mm
$E_f = 1000$ V
Hence $r_j = 56.06$ mm (approximately)

Example 11.3

Stress cone profile calculation for a 33 kV, 300 mm² single-core cable (Table 11.1).

Taking the help of Equation 11.2.

Apart from the capacitive method of controlling stress as shown earlier, a second method of controlling stress is the resistive method. The capacitive method requires some skill and an amount of insulation build-up. In the case of the resistive method, non-linear resistive tapes are applied over the stripped cable directly varying with the electrical stress. Application of these tapes is simpler and effective when done with proper care.

11.14 Straight Through Joint on EHV Cable

Figure 11.12a shows a typical EHV straight through joint. It shows that the stranded conductor has been joined by a totally compressed ferrule (1). Over the ferrule, a layer of semiconducting tape has been applied tightly, extending over the extruded inner semiconducting layer on both sides of the cable

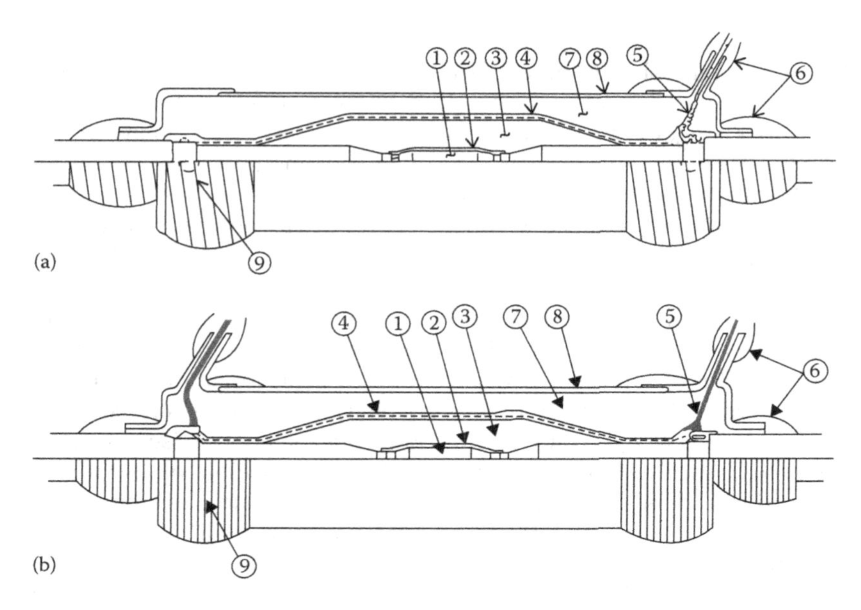

FIGURE 11.12
(a) A straight through joint of a 132 kV cable and (b) an isolating joint with two earthing points. (1) Compressed ferrule joint, (2) semiconducting tape layer, (3) pre-moulded insulating block, (4) braided copper wire screen over outer semiconducting layer, (5) copper earthing conductor connected with the braided screen, (6) self-amalgamated sealing tapes, (7) filled PU compound, (8) FRP-moulded outer protective casing and (9) stress-relieving cone.

conductor (2). The pre-moulded insulating blocks made of the EPR compound, generally made softer and expandable, are brought sliding over the semiconducting layers of the ferrule joint. The inner conductor assembly is tightly gripped by the insulation block, leaving no air gap in between (3). Stress cones (9) are now placed by the sides of the pre-moulded insulating block. It should be noted that the insulating block already has a layer of a semiconducting material embedded on its outer periphery. The end points of the stress cone are connected with the semiconducting layer of the insulating block and with the semiconducting layer on the cable insulation by taping with self-amalgamating semiconducting tapes. A flexible braided tinned copper mesh (4) is tightly wrapped over the total assembly, extending on both sides of the metallic screen/sheath over the cable insulation. The braid is connected with the cable screen by the soldering method. This braid is extended further to connect the earthing conductor (5). An FRP protective cover is placed over the assembly, extending on both sides of the joint (8). The inner space of the cover is filled with a noncorrosive waterproof polyurethane compound (7). All the joints of the casing are completely and hermetically sealed by applying several layers of self-amalgamating and self-curing tapes (6). The materials used are tested for contamination level, voids and for all the required electrical and mechanical parameters in the testing laboratories before use.

HV and EHV cables are made as single-core cables. These cables are protected by a metallic sheath, either of lead or of corrugated aluminium. Magnetic lines of force emanating from the conductor current induce a voltage within the sheath. For a longer cable length, this voltage can rise appreciably, endangering the life of working personnel, and can damage monitoring electrical equipment. As per IEC norms, this voltage must be restricted to 60 V for normal operating conditions and 300 V during short-circuit conditions. If the sheath is earthed at one end, the current starts flowing towards the earthing point, reducing the intensity of the induced voltage. If the cable is short, the rise in voltage will remain restricted. When both the ends are connected to earth, the current flows in both directions. If the length is very long, the voltage at the midpoint will rise abnormally, whereas the current flowing at the ends will help in reducing the intensity of rise in voltage towards the end sections. In such a longer length, if the cable is transposed alternately, the rise in voltage can be restricted. Since such large-diameter cables cannot be laid up as three cores, transposition is done by the cross-bonding of sheath and earthing at regular intervals, say after every three lengths. This may increase a certain amount of circulating current within the transposed sections, but voltage is restricted to the required level. A detailed mathematical treatment is given in Chapter 8.

The solution lies in the fact that cables should be transposed, virtually connecting the sheaths alternately rotating the phasors 120° to each other. In this system, the route is split up into groups of three drum lengths and all joints are fitted with insulated flanges. At each third joint position, sheaths should be connected alternately and earthed solidly, keeping them isolated from the adjacent section. This process of transposing during the jointing of cables and earthing solidly to restrict the sheath voltage is known as cross-bonding.

During cross-bonding, after every three lengths, the joints of all the three cables have two earthing points, isolating them from the sheaths of the next section of running lengths. These types of joints are called isolating joints. Except an additional earthing point, the construction of the joint is exactly the same as stated earlier (Figure 11.12b). Transposition is done through a link box placed near each group of three joints and joining the sheaths alternately as designed with single-point earthing. Figure 11.13b shows a cross-bonding link box with the links to transpose the cable sheath at a minor bonding section. A sheath voltage limiter (SLV) unit is mounted within the box to permit disconnection to safeguard its damage during a 10 kV DC test to ascertain the integrity of the sheath. The SLV unit has three non-linear resistors in star connection, with each resistor connected to one link with the central point earthed so that the insulated flange in each joint shell is bridged by two resistors in series. The main purpose of SLV is to limit surge voltages due to lightning and short-circuit conditions.

The jointing should be carried out in a completely contamination-free atmosphere. Therefore, the jointing bay should be excavated with adequate space with proper side protections. The pit should be covered by a closed tent with the arrangement of air conditioners to keep inside a dust-free atmosphere at

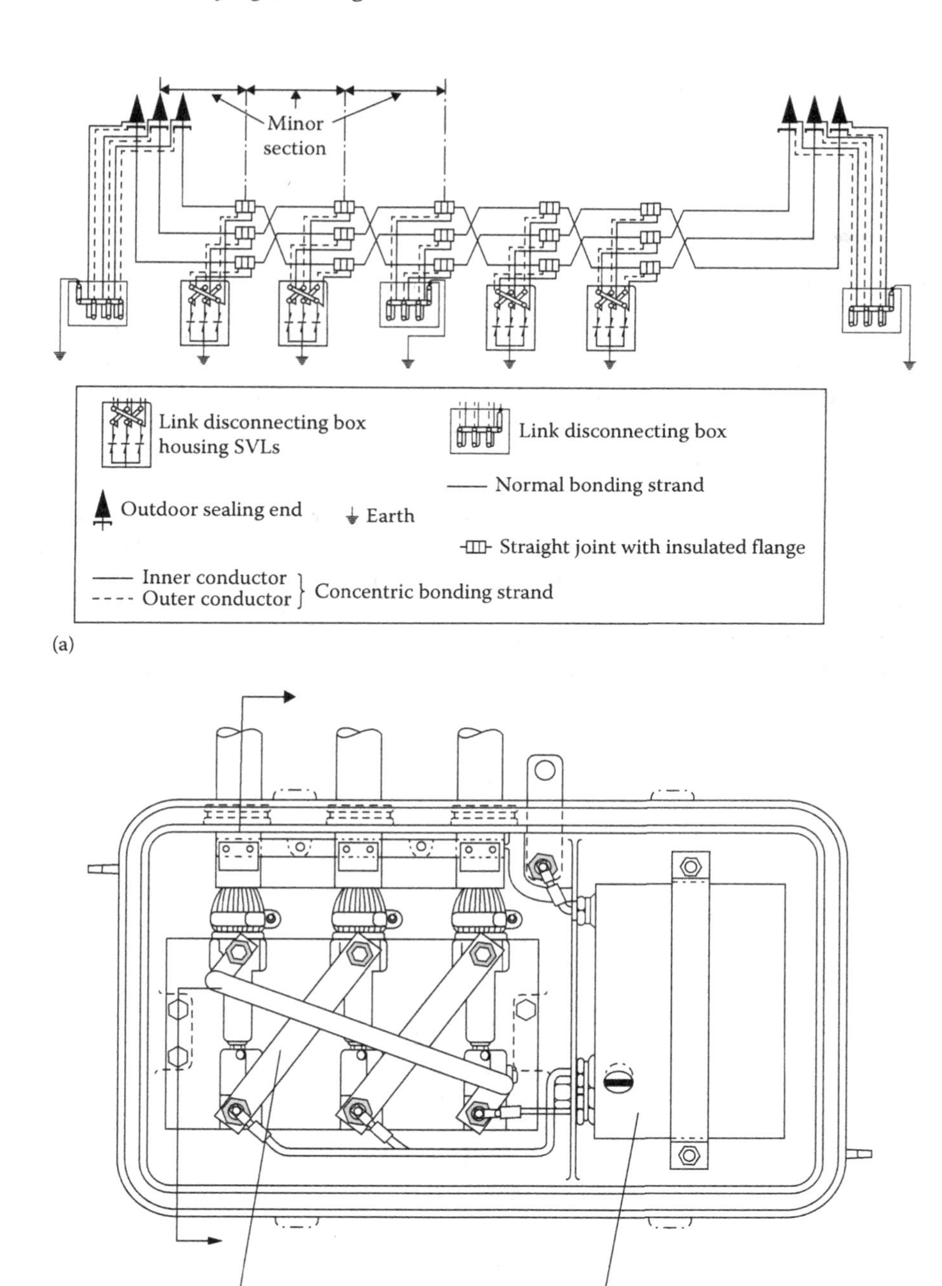

FIGURE 11.13
(a) Cross-bonding of EHV cables and (b) link box.

a constant temperature. Working personnel should clean their hands before wearing working gloves. All the equipment and jointing materials should be kept clean and dry.

11.15 Termination

Nowadays, cable terminations up to 33 kV are of heat shrink type. For LV cables, initially heat shrink tubes are placed over spliced conductors after crimping the lugs on the end point of the conductors. These tubes are then fixed tightly on the conductors through a heat shrinking process. Below the crutch point, an earthing conductor is clamped securely over the armour. A heat shrink cap with three or four fingers (as required) is slipped over the crutch point. This cap is then made to shrink over the conductors and cable sheath completing the joint. The earth conductor is allowed to remain exposed, coming out of the lower part of the cap.

In the case of MV cables up to 33 kV, before slipping the heat shrink tubes with shrouds over the insulated conductors, flexible heat shrink stress cones should be brought over the conductors, covering the semiconducting layer of each core just above the crutch point. The rest of the process is similar to that of LV cable jointing (Figure 11.14).

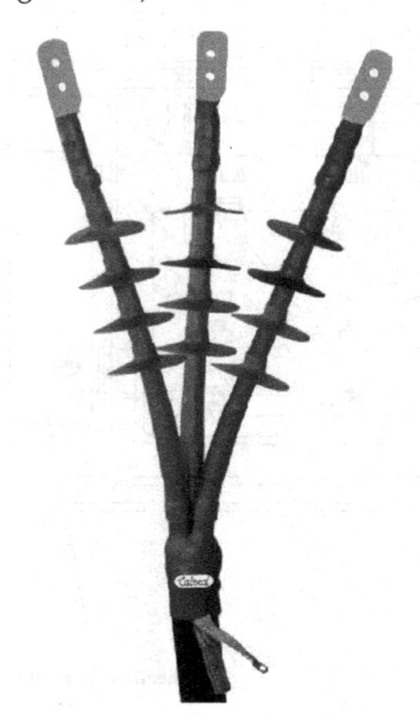

FIGURE 11.14
A 33 kV heat shrink termination joint.

11.16 HV and EHV Cable Terminations

Figure 11.15 shows a sketch of an HV/EHV cable termination joint for an XLPE-insulated cable.

An FRP bushing is fixed, connecting the metal sheath for earth connection. This is sealed with the outer sheath of the cable by means of multiple rubber rings and self-amalgamating waterproof tapes. The total termination assembly is kept fixed on ceramic insulators (6), which in turn are kept fixed on a sturdy metal structure. Thus, the termination remains in an isolated position without any contact with metal frames (5). Over the metal casing, a ceramic insulator bushing is fixed. The stress cone (7) kept enclosed within a copper conical cap (8) is kept fixed over the bottom part of the stress cone. The metal cone is connected with the metallic sheath of the cable at the point where the metal sheath is terminated. This is done by soldering. One or two layers of an insulating tape (11) are wrapped over the stripped insulation (10) of the conductor running upwards, covering the insulation of the stress cone (9). At the top, a ferrule (16) is connected with the conductor using the hydraulically compressed method (15). This is then brought out through a corona shield (17). The connector is kept isolated through a hermetically sealed bushing. The total system remains enclosed within a ceramic bushing (12). The empty space between the conductor and the bushing is filled with contamination-free silicon oil (14). Oil is kept under pressure through a pressure tank. An arching horn (19) is attached to the top with the corona shield for protection against any surge voltage such as lightning. The number of skirts (13) and the length of ceramic bushings are provided, considering the voltage and surrounding atmospheric condition.

The termination should be such that it should remain functional under all weather conditions.

11.17 Thermomechanical Problems

During the operation, a considerable amount of heat is generated within the cable, which adds to temperature fluctuations in the surrounding area, creating thermomechanical problems. The metallic parts of the cable get affected by these types of changes in temperature, subjecting to stress and strain in various metallic components of the cable, including the joints and terminations. Therefore, thermomechanical design must be considered for all types of power cable installations. Special care must be taken for high-voltage power cables, where higher cross-sectional conductors are used.

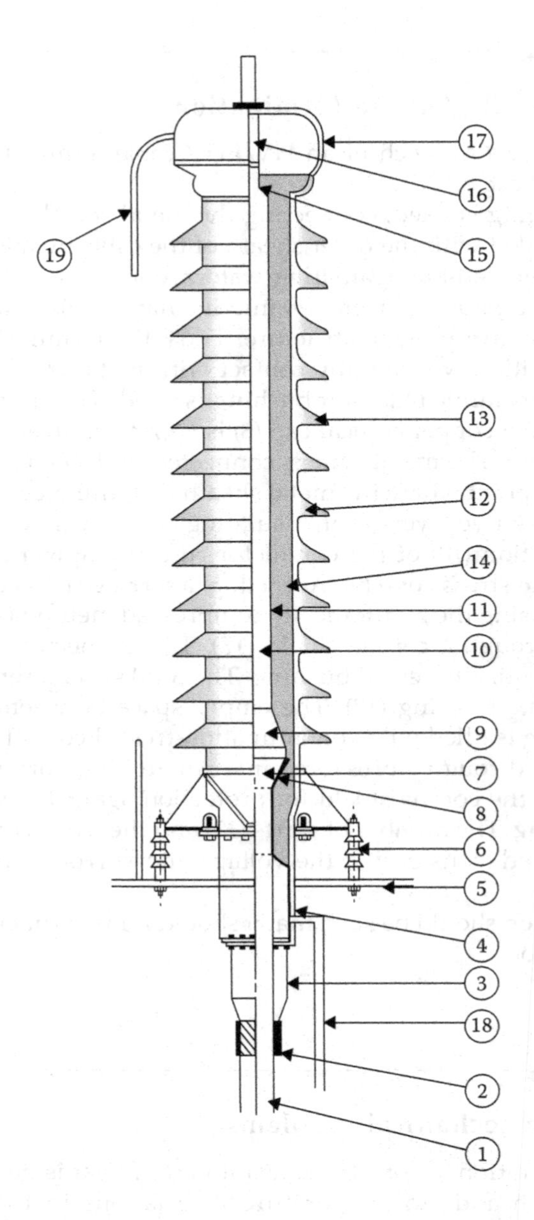

FIGURE 11.15
Schematic diagram of HV and EHV outdoor cable termination. (1) Single-core EHV cable sheath, (2) self-amalgamating tape sealing, (3) FRP cap, (4) metal cover with earthing provision, (5) termination supporting structure, (6) ceramic bushing to keep the termination isolated from metal structure, (7) stress cone, (8) metal screen over stress cone, (9) stress cone insulation, (10) XLPE-insulated conductor, (11) protective tape over insulation, (12) ceramic bushing, (13) skirt of ceramic bushing, (14) silicon oil, (15) termination compressed joint on the conductor, (16) termination ferrule (pin type), (17) corona shield, (18) oil pipe connected to silicon oil pressure reservoir and (19) arching horn.

Two types of practices are adopted for installation procedures:

1. When cables are installed in an unrestricted condition, the cables can have free movement under variable temperature conditions. In this case, compressive or tensile forces generally remain inactive.
2. When cable movements are restricted, thermal expansion and contraction are absorbed fully by the internal compressive or tensile force.

Importantly, different components have different coefficients of thermal expansion and contraction. A free movement may, in such a condition, create internal problems. However, the conductor being the main part of the cable, the forces of stress and strain are generally adopted based on the area of the conductor and the nature of the material. In an installed cable, outer components, such as a metallic sheath, remain more restricted than the conductor, which can move freely. In designing a cable, thermomechanical forces should be considered for the components so that they can withstand thermal expansion and contraction after installation. It is, therefore, necessary to define the method of installation while constructing cables.

In a buried backfilled cable, the movement of conductor and insulation is normally in the longitudinal direction. If the conductor is terminated firmly at the end, virtually no movement can be expected. Then the force developed in the conductor will be

$$F = ES\alpha\theta$$

where

F is the force in the conductor (N/m^2)
S is the effective modulus of the elasticity of the conductor (m^2)
α is the coefficient of the expansion of the conductor (per °C at 20°C)
θ is the temperature rise of the conductor (°C)

According to this equation, the conductor acts as an elastic member. The effective modulus actually depends on the material of the conductor. During the load cycle, the compressive load developed is found to be almost equal in magnitude as per the tensile force within the conductor.

When the cable is laid freely in air, an expansion is not allowed in the rigid clamping system; naturally, the compressive force will be high. To keep the force within the limit, cleats should be spaced in such a manner that the cable does not buckle. Spacing can be calculated using Euler's buckling theory. This theory assumes longitudinal uniformity but is not strictly applicable to corrugated sheathing when the following formula is used:

$$L = \frac{2\pi}{s}\sqrt{\left(\frac{E_{eff}\,I}{1000F_p}\right)}$$

where

L is the maximum support between supports
s is the safety factor
E_{eff} is the effective modulus of the sheathing material (N/m^2)
I is the second movement of area of sheath (mm^4)
F_p is the thermomechanical force developed in the conductor and the sheath (N)

It is common to take a safety factor for an aluminium-corrugated sheath, and E_{eff} should be accepted as 25% of the aluminium.

The second moment of inertia is calculated as

$$I = \frac{\pi(d_0^4 - d_i^4)}{64} \text{ mm}^4 \tag{11.6}$$

Here, d_0 is the diameter of a plain aluminium or lead sheath or the mean crest and trough diameter of a corrugated sheath. Also, $d_i = d_0 - 2t$ (t is the thickness of the metal sheath in millimetres).

At the bend, the spacing of cleats should be reduced by 50% or so depending on the sharpness of the bending. In the case of horizontally laid cables, a sliding roller support may be provided for the smooth movement of the cables.

In the case of vertically installed cables, the spacing of cleats should be such that a small sag can be permitted, as suggested by Holttum. Initially, it was meant for a smooth and corrugated sheath, and the following formula may be considered:

- *For lead-sheathed cable*:

$$L = \sqrt{\left(\frac{Kd_m^2 t}{w \times 10^3}\right)} + 0.2 \text{ m} \tag{11.7}$$

where

L is the spacing (m)
d_m is the mean diameter of the sheath (mm)
t is the sheath thickness (mm)
w is the cable weight (kg/m)
K is the factor as follows:

Material	K (values)
Pure lead	5.98
Alloy E	9.5
Alloy ½ C	9.15
Alloy B	16.19

- *Aluminium-sheathed cable*:

$$L = \sqrt{\frac{0.00244YI}{d_r W}} \text{ m}$$

where
L is the spacing (m)
Y is the effective yield stress of the aluminium sheath (N/m²)
I is the second moment of inertia of the sheath (mm²), to be calculated as per Equation 11.6
d_r is the outside diameter of the smooth sheath or mean square root diameter of the corrugated sheath (mm)
W is the weight of the cable (kg/m)

The effective yield stress of the aluminium-corrugated sheath should be taken as approximately 25% of aluminium. Yield stress is not critical as calculated and can be adjusted as required, gaining practical values.

The thermomechanical problems cannot be taken lightly. It is one of the main considerations to design and install a cable, particularly for HV and EHV power transmission lines. Many cable failures have been attributed to this problem.

11.18 Aerial Bunch Cables and Screen Terminations (11 and 33 kV Cables)

For an aerial bunch cable (in Chapter 8), the screen of an 11 kV × three core 120 mm² conductor has an induced voltage of 30 V per kilometre for a copper tape. In the case of a disruption in the screen, the voltage at the point may rise instantaneously to more than 300 × 500 V, provided the longitudinal voltage is restricted to 500 V. This voltage will certainly disrupt the cable, creating a fault at the point. This is only the case of a single bunch of cable with three cores of 120 mm². At times, several aerial bunch cables are laid on poles running parallel and adjacent to each other. In such cases, the magnetic field of adjacent cables will influence each other, increasing the screen voltage mutually. It is, therefore, necessary that stress cones are provided for 11 and 33 kV cables during straight joints and termination after a proper calculation. If required, earthing should be provided after a certain length to restrict a rise in voltage within the screen.

Nowadays, it is common practice to apply a screen with an aluminium foil to minimise the cost of the cable. This practice needs a rational thinking. The resistance of aluminium is higher than copper; hence, the induced voltage will increase further. A pinhole or any damage to the outer sheath allows ingress of moisture, corroding the screen underneath. The screen gets easily disrupted, creating an instantaneous fault condition. Naturally, the design parameter should consider protecting the aluminium screen by providing a water swellable tape. This may offset the cost and take away the commercial advantage as envisaged.

Earth connections at the required points are also carried out in a simple way. The screen of aluminium or copper is terminated with earthing wires using the twisting method, which is a dangerous practice. Copper screen when twisted can get loose and give away after some time unless the screen is connected with copper wires by soldering. For an aluminium screen, copper wires should not be used. This will create a bi-metallic corrosion. If both copper wires and aluminium strips are required, they must be tinned and soldered to get a perfect connection. Even an aluminium-to-aluminium connection must be done by a special soldering method. Aluminium wires should be avoided for earth connections at all costs. Nowadays, frequent failures of aerial bunch cables are due to this casual approach in which no proper attention is given to earth connections and the method of screen termination. Furthermore, the casual approach of handling cables during winding and unwinding and during installation may disturb the screen and create fault conditions. These cables look very simple in construction, but their handling and installation need careful attention. After all, they carry power to consumers. Any disturbance to power cables will result in heavy commercial losses.

11.19 Tests Conducted after Installation

For distribution cables, two tests should be carried out after installation:

1. Insulation resistance test
2. High-voltage test as per IEC standards or national standards

For transmission cables, the following tests are required:

1. High-voltage test as per IEC standards or national standards as specified
2. Insulation resistance test
3. Voltage test on the outer sheath of the cable

In the design, manufacture, transportation and installation of power cables, every precaution should be taken to satisfy customer service requirements. Modern civilization depends solely on a sound power supply system.

A cable may look simple from the outside, but what transpires within cannot be seen with naked eyes. Its effect can be felt all around us. Mishandling would be disastrous and can ruin our whole system of progress. If we understand and handle it properly, it takes us to the highest point of civilization.

References

ALCAN (UK) Ltd., *Aluminium for Insulated Cables*, Warrington, U.K., 1960.

Andresen, K., F. Dias, N. D. Kenney, *Corrugated Metallic Cable Sheath*. Paper 58-75: Recommended by the AIEE Insulated Conductors Committee and approved by AIEE Winter General Meeting, New York, February 2–7, 1958.

Artbauer, J., Ing., *Kabel Und Leitungen*. VEB Verlagtechnik, Berlin, Germany, 1961.

Chernukhin, A., *Fundamentals of Cable Engineering*. V. Privezentsev, I. Grodnev, S. Khlodny, I. Riazanov. Translated in English by Adolph Chernukhin, Mir Publishers, Moscow, Russia, 1973.

Copper Development Association (CDA), *Copper for Electric Cables* Publication No. 52 and 56, Hemel Hempstead, U.K.

Ganguli, S. K., *Aluminium – Sheathing Material for All Types of Cables*, IEEMA Publication, April 2007.

Ganguli, S. K., *Influence of Thermal Resistivity of Materials and Soil On – Current Carrying Capacity of Underground Power Cables*, IEEMA Publication, August 2006.

Ganguli, S. K., *Vulcanisation of Thermosetting Compound and Design Parameter of a CCV Line*, IEEMA Publication, December 1987.

Hartlein, R., *Long Life XLPE Insulated Power Cables*, National Electric Energy Testing (R&D) Centre, Forest Park, GA, May 2006.

Haefely (Test System), *Trends in Power Cable Technology*, Technical Catalogue Publication, Basel, Switzerland, 1982.

IEC 60287-2-1:1994, *Electric Cables–Calculation of the Current Rating, Part 2: Thermal Resistance, Section 2.1: Calculation of Thermal Resistance*, International Electrotechnical Commission, Geneva, Switzerland, 1994.

IS 7098, *Indian Standard- Cross-Link Polyethylene Insulated Thermoplastic Sheathed Cables-Specification*, Part 2: For Working Voltages from 3.3 kV to and Including 33 kV (Second Revision), Part 3: For Working Voltages from 66 kV to and Including 220 kV.

Ishimoto, K., T. Tanaka, Y. Ohki, Y. Sekiguchi, Y. Murata, Thermally stimulated current in low-density polyethylene/MgO nanocomposite—On the mechanism of its superior dielectric properties, *IEEJ Trans. FM*, 129(2), 97–102, 2009 [in Japanese].

Kwasaki, Y., K. Otani, H. Miyauchi, *Radiant Curing Process (RCP) for Manufacturing HV and EHV XLPE Cables*, Sumitomo Electric Industries Ltd., Osaka, Japan, August 1976.

King, S. Y., N. A. Halfter, *Underground Power Cable*. Longman Group Ltd., London, U.K., 1982.

Mac Allistar, D., *Electric Cables Handbook*, Granada, British Insulated Callender's Cables, London, U.K., 1984.

Maezawa, T., J. Taima, Y. Hayase, Y. Tanaka, T. Takada, Y. Sekiguchi, Y. Murata, Space charge formation in LDPE/MgO nano-composite under high electric field at high temperature, in *2007 Annual Report Conference on Electrical Insulation and Dielectric Phenomena*, Vancouver, British Columbia, Canada, 2007, pp. 271–274.

Murata, Y., M. Sakamaki, K. Abe, Y. Inoue, S. Mashio, S. Kashiyama, O. Matsunaga, T. Igi, M. Watanabe, S. Asai, S. Katakai, Development of high voltage DC-XLPE cable system, *SEI Tech. Rev.*, 76(Special issue), 55–62, April 2013.

Parwal, S. N., *Development of EHV Cables*, Institute of Electrical Engineers, Mumbai Branch, Tutorial on Cables, Mumbai, 1987.

Schenkel, G., Ing., *Plastic Extrusion—Technology and Theory*. Dr. Ing. G. Schenkel; Original German version published by Carl Hanser, Munich, Germany, 1963. English version later published by ILFFE Books Ltd., London, U.K., 1966 and by American Elsevier Publishing Company Inc., New York: Library of Congress Catalogue Card number 66-17240.

Siemens, *Electrical Engineering Handbook*. Siemens, Heyden & Sons Ltd., London, U.K., 1981.

Sumitomo Electric Industries Ltd., *Radiant Curing Process for Manufacturing HV and EHV Cables*, Sumitomo Electric Industries Ltd., Osaka, Japan, ETR No. 3A-3099; August 1976.

Tadmor, Z., I. Klein, *Engineering Principles of Plasticating Extrusion*.

Tanaka, T., A. Greenwood, *Advanced Power Cable Technology*, Volume I, CRC Press, Boca Raton, FL.

Tanaka, T., A. Greenwood, *Advanced Power Cable Technology*, Volume II, CRC Press, Boca Raton, FL.

Vig, B. R., S. K. Ganguli, Processing of copper for insulated cables and wires, in *International Symposium: 'Copper Forum-II'*, December 1987.

Wanser, G., *Kabel mit Geschweissten und Gewellten Metallmateln*, Conference Lecture, Moscow, Russia, October 26, 1965.

Index

Q

X

Z

Printed in the United States
by Baker & Taylor Publisher Services